The Wetland Bird Su
Wildfowl and Wader Counts

Peter Cranswick, Mark Pollitt, Andy Musgrove and Becky Hughes

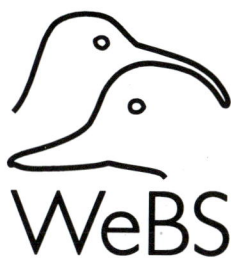

WeBS

Published by

British Trust for Ornithology, The Wildfowl & Wetlands Trust,
Royal Society for the Protection of Birds and Joint Nature Conservation Committee

June 1999

ISBN 0 900806 27 3
ISSN 1353-7792

This publication should be cited as: Cranswick, P.A., Pollitt, M.S., Musgrove, A.J. & Hughes, R.C. 1999. *The Wetland Bird Survey 1997-98: Wildfowl and Wader Counts.* BTO/WWT/RSPB/JNCC, Slimbridge.

Published by: BTO/WWT/RSPB/JNCC.

Cover: Tufted Duck by Martin Ridley

Designed and produced by The Wildfowl & Wetlands Trust, Slimbridge.

Printed by Severnprint Ltd, Gloucester

Printed on Evolve Silk in Cheltenham ITC and Gill Sans.

Available from: WeBS Secretariat, WWT Slimbridge, Glos GL2 7BT, and Natural History Book Service, 2-3 Wills Road, Totnes, Devon TQ9 5XN, UK.

This report is provided free to all WeBS counters, none of whom receive financial reward for their invaluable work. Further feedback is provided to counters through the twice-yearly WeBS Newsletter. For further information please contact the WeBS Secretariat or relevant National Organiser.

ACKNOWLEDGEMENTS

This book represents the nineteenth report of the Wetland Bird Survey. It is entirely dependent on the many thousands of dedicated volunteer ornithologists who supply the data and to whom we are extremely grateful. The Local Organisers who co-ordinate these counts deserve special thanks for their contribution.

We are also grateful to the following people for providing technical assistance, supplementary information and comments on draft texts and additional data:

Ian Andrews, Graham Austin, Kendrew Colhoun, Simon Delany, Tony Fox, Ian Francis, David Gibbons, Susan Guy, Richard Hearn, Steve Holloway, Baz Hughes, Mel Kershaw, Rowena Langston, Margaret McKay, Carl Mitchell, Mark O'Connell, Malcolm Ogilvie, David Paynter, Steve Percival, Eileen Rees, Mark Rehfisch, Jeff Stenning, David Stroud, Paul Thompson, Paul Walkden and Clare Ward. Graham Austin and David Stroud deserve special mention for their extensive contributions. Many amateur observers also provide reports of their studies; these are acknowledged within the text.

The cover painting of a Tufted Duck is by Martin Ridley.

Grateful thanks to all and apologies to anyone who has inadvertently been missed.

WETLAND BIRD SURVEY

Organised and funded by

British Trust for Ornithology
The Nunnery, Nunnery Place,
Thetford, Norfolk IP24 2PU

The Wildfowl & Wetlands Trust
Slimbridge, Gloucester
GL2 7BT

Royal Society for the Protection of Birds
The Lodge, Sandy,
Bedfordshire SG19 2DL

Joint Nature Conservation Committee
Monkstone House, City Road,
Peterborough PE1 1JY

CONTACTS

WeBS Secretariat and Core Counts
  Head of Secretariat: **Peter Cranswick**
  National Organiser (Core Counts): **Mark Pollitt**
  Assistant National Organiser: **Becky Hughes**
The Wildfowl & Wetlands Trust, Slimbridge, Glos
GL2 7BT
Tel: 01453 890333 x 255/280
Fax: 01453 890827
e-mail: firstname.surname@wwt.org.uk

Low Tide Counts
  National Organiser: **Andy Musgrove**
British Trust for Ornithology, The Nunnery,
Thetford, Norfolk IP24 2PU
Tel: 01842 750050
Fax: 01842 750030
e-mail: Andy.Musgrove@bto.org

For general enquiries, please contact the
WeBS Secretariat. More detailed data than
published in this report can be obtained from
Becky Hughes (Core Counts) or Andy
Musgrove (Low Tide Counts).

PREFACE

You will have already noticed at least one change to the report from previous years.

The key aim of the changes herein has been to make the report more useful, both as a reference, but also to show more clearly how the UK meets its obligations to various conservation agreements and legislation. Where possible, analyses and presentation have been standardised across all species, though further work is needed in future reports before this is consistent throughout.

Most importantly, following these changes, I would urge you to read the revised analysis, presentation and interpretation sections at the start of the report. Much remains as before, but the intention has been to provide every last explanation that anyone might want. Inevitably, this section has become longer, but hopefully it is organized logically for the casual reader to find the information they desire. There is more detail, but minutiae have been moved to appendices for the sake of simplicity.

Part of the above aim is that this report provides a 'one-stop-shop' reference for monitoring of non-breeding waterfowl in the UK. Thus, the format has been modified so that, as far as possible, each species account provides all relevant information in a single place, rather than several sections as previously.

Thus, I would be grateful for comments and suggestions for further improvements, but also for instances where information is unclear or even missing, particularly any survey data that has been omitted.

Feedback on the usefulness of these changes would also be gratefully received. Change for the sake of it is not helpful and certainly time-consuming, and it is useful to know how far we should persist with this.

Which leads back to the most obvious change and the one which will probably provoke most comment.

Size, it is well known, is not everything, but the larger format was proving unwieldy and costly, particularly for postage. The report matches that of the Irish Wetland Bird Survey, perhaps the most important point of reference when considering change and standardisation, though it remains the content that is important, and something that both schemes continue to work closely upon together. For those that like to see successive volumes arranged neatly on their bookshelves, I hope this doesn't grate too much.

Peter Cranswick
Head of WeBS Secretariat

ERRATA

The quality of the Low Tide Count distribution maps in the 1996-97 report was well below the desired standard. This was due to differences in software compatability at WWT and BTO and, in the rush to publish the report, lack of time to fully check the output. Our apologies to all concerned.

CONTENTS

SUMMARY

The Wetland Bird Survey and Wildfowl and Wader Counts

The Wetland Bird Survey (WeBS) is a joint scheme of the British Trust for Ornithology (BTO), The Wildfowl & Wetlands Trust (WWT), Royal Society for the Protection of Birds (RSPB) and Joint Nature Conservation Committee (JNCC) to monitor non-breeding waterfowl in the UK. The principal aims of the scheme are to identify population sizes, determine trends in numbers and distribution, and to identify important sites for waterfowl. WeBS Core Counts are made annually at around 2,000 wetland sites of all habitats; estuaries and large still waters predominate. Monthly co-ordinated counts are made mostly by volunteers, principally from September to March, with fewer observations during summer months. Data from other sources, e.g. roost counts of grey geese, are included in this report where relevant.

This report presents total numbers counted for all species in the most recent year in Great Britain and Northern Ireland. Annual indices, calculated using the 'Underhill' method, are provided for the more numerous species. For certain wildfowl species, monthly indices, showing relative abundance during the winter, are also provided.

Species accounts provide yearly maxima for all sites supporting internationally and nationally important numbers. Sites with changed status are highlighted and significant counts at a national or site level are discussed. Counts are placed in an international context where possible, and relevant research is summarised. Waterfowl totals are provided for all sites meeting criteria for international importance and species occurring in internationally important numbers on each are identified. Brief overviews of research initiated by WeBS or using WeBS data, and of conservation issues pertaining to UK waterfowl, are provided.

WeBS Low Tide Counts are made on selected estuaries to determine the distribution of birds during low tide and to identify important feeding areas that may not be recognised during Core Counts which are made mostly at high tide. A summary of results for these estuaries, and distribution maps for selected species, are provided.

Waterfowl totals recorded by the Irish Wetland Bird Survey, a similar scheme operating in the Republic of Ireland, are also included.

Appendices list all UK sites identified under the Ramsar Convention and Special Protection Areas notified under the EC Directive on the Conservation of Wild Birds. Also, waterfowl count totals for the most recent year are provided separately for England, Scotland, Wales, the Isle of Man and the Channel Islands.

The 1997-98 year

This report summarises counts during 1997-98 and previous years (since 1960 for wildfowl, 1969 for waders and the early 1980s or 1990s for other species groups). Coverage remained at the same relatively high levels achieved throughout the 1990s, with around 1,500 sites covered during each winter month.

The winter was generally mild, in contrast to much colder weather in the previous two winters. Consequently, there were marked changes in numbers of several species which occur in the UK chiefly as winter visitors.

Numbers of divers and grebes were broadly similar to recent totals. Although British counted numbers of Great Crested Grebes were the lowest for 10 years, annual indices suggested more stable numbers. Counts of Cormorants in Britain were the lowest since 1991-92 although annual indices in Northern Ireland were the highest since the late 1980s. Numbers of most heron species were around normal.

Mute Swans continued their increase since the ban of lead fishing weights, reaching record levels in 1997-98. Numbers of Bewick's and Whoopers were much lower, a result of poor breeding success and, for the former, birds remaining on the continent during the mild weather. All of the major migratory *Anser* and *Branta* goose populations returned in similar numbers to the previous winter, despite continued poor breeding success in some Brent populations. Numbers of Svalbard Barnacle Geese remained high following the recent large increase. Annual indices for Canada Geese fell 21%.

Wigeon numbers dropped sharply from 1996-97 as a result of the milder weather. Gadwall and Teal numbers continued to rise, reaching their highest levels to date. By contrast, Mallard annual indices fell for the tenth year in succession, with values and counted numbers reaching record low levels. Annual indices for Pochard were low, around half the level of 10 years ago in Northern Ireland where Tufted Duck indices showed a similar picture. Most sea-duck were recorded in average numbers, although Scaup counts were high in Britain but low in Northern Ireland. Goldeneye numbers also fell in Northern Ireland. Smew and Goosander numbers fell with the return to milder weather, though they remained higher than normal. Red-breasted Merganser annual indices in Britain and Northern Ireland reached near record highs. Ruddy Ducks continued to increase.

Rails were recorded in around average numbers, although Coot was somewhat less numerous.

Annual index values and most counts of Oystercatcher, Grey Plover, Know, Sanderling, Dunlin, Bar-tailed and, to a lesser extent, Black-tailed Godwits fell sharply in 1997-98. However, these species had shown elevated values in 1996-97 due to the severe winter and 1997-98 represented a return to average numbers. Counts of several species which favour mild weather also returned to more normal levels, although 1997-98 values represented increases over the previous winter, particularly for Avocet, lapwing, Golden Plover, Curlew and Redshank.

Many primarily non-estuarine coastal species exhibited continuing declines, reinforced from preliminary results from a national survey of the UK's open coast in December and January, particularly Turnstone and Purple Sandpipers.

Counts of species which occur primarily as autumn passage migrants were generally lower than normal.

Counts of Black-headed, Common and Lesser Black-backed Gulls were higher than previous maxima. Other species were recorded in average numbers.

Peak counts of the more numerous terns were lower than usual, with the exception of Arctic Tern, with counts exceeding double the previous record.

The number of escaped waterfowl species and their counts continued the steady increase of recent years.

A summary of WeBS Low Tide data is presented for 18 estuaries counted in 1997-98: Alt, Belfast Lough, Blyth, Chichester Harbour, Cleddau, Dee (England/Wales): North Wirral Shore, Hamford Water, Mersey, Montrose Basin, North Norfolk Coast, Northwest Solent, Pagham Harbour, Portsmouth Harbour, Ribble, Southampton Water, Strangford Lough, Tamar complex and Ythan.

INTRODUCTION

The UK is of outstanding international importance for waterfowl. Lying on some of the major flyways for arctic-nesting species, large numbers of waterfowl are attracted, especially during winter, by the relatively mild climate and extensive areas of wetland, notably estuaries. As such, the UK has an obligation to protect and conserve both these waterfowl and the wetlands upon which they depend.

The UK is signatory to a number of international conservation conventions and, as a member of the EU, is bound by international law. In particular, the 'Ramsar' Convention on Wetlands of International Importance especially as Waterfowl Habitat, the EC Birds Directive and the EU Habitats and Species Directive, between them, require the UK to identify important examples of wetland and other habitats and sites important for birds and designate them for protection. Implicit in this is that regular monitoring is undertaken to identify and monitor such sites. These instruments also lay particular significance on the need to conserve migratory populations, and consequently most of the waterfowl populations in the UK. The Agreement on the Conservation of African-Eurasian Migratory Waterbirds (AEWA) of the 'Bonn' Convention on the Conservation of Migratory Species of Wild Animals, recently ratified by the UK and entereing into force in 1999, represents a specific Agreement requiring nations to take co-ordinated measures to conserve migratory waterbirds given their particular vulnerability due to their migration over long distances and their dependence on networks that are decreasing in extent and becoming degraded through non-sustainable human activities. Article three of the Agreement requires, among other things, that sites and habitats for migratory waterbirds are identified, protected and managed appropriately, that parties initiate or support research into the ecology of these species, and exchange information and results. Explicit in this Agreement is that adequate monitoring programmes are set in place to fulfil these objectives and the Action Plan to the Agreement specifically requires that nations endeavour to monitor waterfowl populations.

Aims and objectives of WeBS

The Wetland Bird Survey (WeBS) aims to monitor all non-breeding waterfowl in the UK to provide the principal data on which the conservation of their populations and wetland habitats is based. To this end, WeBS has three main objectives:

- to assess the size of non-breeding waterfowl populations in the UK;
- to assess trends in their numbers and distribution; and
- to assess the importance of individual sites for waterfowl.

A programme of research, to understand the ecology of waterfowl and investigate the effects of habitat change and anthropogenic impact, underpins and enhances these objectives.

These results also form the basis for informed decision-making by conservation bodies, planners and developers and contribute to the sustainable and wise use and management of wetlands and their dependent waterfowl. The data and the WeBS report also fulfil some of the objectives of the Conventions and Directives listed above. WeBS also provides UK data to Wetlands International to assist their function to co-ordinate and report upon waterfowl monitoring at an international scale.

Structure and organization of WeBS

WeBS is partnership scheme of the British Trust for Ornithology (BTO), The Wildfowl & Wetlands Trust (WWT), Royal Society for the Protection of Birds (RSPB) and the Joint Nature Conservation Committee (JNCC), the last on behalf of English Nature (EN), Scottish Natural Heritage (SNH) and the Countryside Council for Wales (CCW), and the Environment and Heritage Service in Northern Ireland (EHS).

WeBS continues the traditions of two, long-running count schemes which had formed the mainstay of waterfowl monitoring in the UK since 1947 (Cranswick et al. 1997). WeBS Core Counts are made at a wide variety of wetlands throughout the UK. Synchronised counts are conducted once per month, primarily from September to March, to fulfil all three main objectives. In addition, WeBS Low Tide Counts are undertaken on selected estuaries with the aim of identifying key areas used during the low tide period, principally by feeding birds; areas not otherwise noted for their importance by Core Counts which are normally conducted at high tide.

The day-to-day running of the Core and Low Tide Count schemes is the responsibility of the National Organisers, with assistance from a number of other staff.

The success and growth of these count schemes reflects accurately the enthusiasm and dedication of the several thousands of volunteer ornithologists throughout the UK who participate. It is largely due to their efforts that waterfowl monitoring in the UK is held in such international high regard.

Aim of this report

This report presents syntheses of data collected in 1997-98 and previous years in line with the WeBS objectives. Data from other national and local waterfowl monitoring schemes are included where WeBS data alone are insufficient to fulfil this aim, so that the report provides a single, comprehensive source of information on waterfowl status and distribution in the UK. All nationally and internationally important sites for which data exist are listed, as are all sites designated under international law or Conventions (see Appendices 1 & 2).

We recommend that the National Organisers (see *Contacts*) are contacted in the first instance by anyone with queries regarding this report or requiring further information.

PROGRESS AND DEVELOPMENTS

New structure of WeBS

In 1998, the WeBS partnership took the decision to re-organise the day-to-day administration of WeBS, largely for the purposes of efficiency. Consequently, organisation of WeBS Core Counts is undertaken by the WeBS Secretariat, based at WWT, and that of Low Tide Counts by BTO. Contact details are given on page 3. The WeBS scheme, nevertheless, remains a four-way partnership between BTO, WWT, RSPB and JNCC, and the direction of the scheme remains the responsibility of a Steering Group consisting of members from all four partners. We are confident that this new structure will enable a more efficient and effective monitoring scheme, enabling WeBS to deliver greater benefits for the conservation of the UK's waterfowl populations.

Alert limits

Trends calculated using count data reveal fluctuations between years as a result of a number of factors, particularly weather and productivity. Even allowing for these factors in individual years, many populations routinely exhibit general increases or declines over relatively short periods. Consequently, it is difficult to deduce whether these variations are attributable to external factors, e.g. bad spring weather on the breeding grounds, or result from negative influences at important sites.

A system of 'alerts' is currently being developed in order to identify when declines in populations are of sufficient magnitude to give cause for concern, triggering relevant bodies to take appropriate action. Following a workshop in autumn 1998, these methods are being developed by BTO on behalf of a number of organisations for all bird species for which sufficient data exist. It is anticipated that a provisional system of alerts will be put in place in 1999 for a number of bird recording schemes. The large body of WeBS data allows many waterfowl species to be included in this process. It is planned that alert analyses will be undertaken routinely as part of the WeBS programme and, where appropriate, results will be included in future reports.

UK-WeBS/I-WeBS Protocol

A major development in cross-border collaboration was made in May 1998 when partners of the UK and Irish Wetland Bird Surveys signed a protocol formalising the international co-operative relationship between the two schemes.

The new agreement will ensure that collection and dissemination of waterfowl data form the two countries are undertaken in a complementary manner. Common standards are adopted by both schemes and special surveys, where possible, will be co-ordinated throughout Britain and Ireland. Analyses conducted at this level will improve our understanding of the movements and ecology of waterfowl within this biogeographic region.

WEATHER IN 1997-98

UK weather is summarised from the journals *Weather* and *British Wildlife*. Figures in brackets denote the WeBS priority count date in that month. European weather is summarised from *Weather* and arctic breeding conditions in western and central Russia, one of the main breeding grounds for birds wintering in the UK, are summarised from Tomkovich & Zharikov (1998).

United Kingdom

Spring began dry and mild in many areas. The first three weeks of **April** (6th) were almost exclusively dominated by high pressure. Rainfall was well below average in almost all parts of the UK, particularly in the east, with temperatures between 1-2°C above normal. A brief heat wave saw temperatures in early **May** (11th) rise as high as 27°C, though this soon gave way to much colder conditions bringing sharp night frosts to all parts and snow as far south as Derbyshire. An unsettled period mid month then gave way to warm, sunny weather over the final week. Mean temperatures were near normal and rainfall was above average in most areas outside southern and eastern England. The warm weather continued into **June** (22nd), though this gave way to unsettled weather mid month and to prolonged heavy rain in the latter part of the month; England and Wales received double the usual rainfall, making it the wettest June this century. **July** (20th) was warmer than average, temperatures 1-2°C above the long-term norm. Rainfall was very localised with thundery downpours. Heavy rain early in the month caused flooding in Moray and Nairn. Warm conditions continued throughout most of **August** (24th) which was the second hottest on record. Much of Scotland and Northern Ireland was exceptionally dry, though England and Wales recorded slightly above average rainfall.

September (21st) continued the run of warmer than average months and brought little in the way of rainfall to eastern and southeastern Britain. The first week was dominated by changeable westerly weather, then high pressure dominated between the 7th and 10th. Cooler northwesterlies followed backing southwesterly mid month, and high pressure then reasserted itself for the remainder of the month bringing warm settled conditions to most parts.

The long Indian summer continued during the first week of **October** (19th) before the arrival of more disturbed weather with frequent rain and occasional strong winds in the second week. A return to settled anticyclonic conditions brought warm days, much sunshine and cooler nights by mid month. This was followed by widespread, and sometimes severe, night-time frosts, affecting all areas for the remainder of the month.

By contrast, **November** (16th) was warm and wet. Southerly winds dominated throughout, with the only significant widespread frosts occurring mid month either side of the priority count date. Rain was widespread and often heavy with localised flooding in parts of Sussex, Cornwall, Aberdeenshire and parts of Northern Ireland.

December (14th) began cold, with several inches of snow falling in parts of southeast England in the first two days, but mostly dry, with patches of freezing fog, elsewhere. The return of southwesterlies from the 5th brought a period of changeable, windy and very mild weather until mid month, followed by a short easterly incursion during the third week. Mild southerly and southwesterly conditions dominated for the remainder of the month with an intense depression bringing severe gales to Wales and northwest England on Christmas Eve. Average temperatures were 1-2°C above normal.

Stormy weather continued in the first week of **January** (18th). A period of calmer, mild southerly winds between the 9th and 12th saw temperatures reach an exceptional 17°C in Prestatyn on the 10th. Unsettled conditions resumed until the 19th, with England and Wales remaining mild but Scotland and Northern Ireland colder with snow in the Highlands. High pressure dominated the rest of the month, with widespread frost and patchy fog across most of the country. Temperatures were around 2°C above the long-term mean, with most areas receiving above average rainfall.

The mild theme continued into **February** (15th) which equalled the warmest this century (in 1990). England and Wales were exceptionally dry with just 30% of the average precipitation, though parts of northern and western Scotland had two and a half times their normal amount. Sunny days and frosty nights characterised most areas during the first few days. High pressure then developed and remained over or near to southern England for most of the month, while westerly or southwesterly airflows dominated northern Britain giving spells of prolonged heavy rain in parts of western Scotland. On the last two days, cold arctic air plunged southwards across much of the country bringing wintry showers to many areas and blizzards in the Highlands.

Once the cold arctic air of late February passed, the first two weeks of **March** (15th) saw a series of depressions and fronts track eastwards across the country, bringing frequent and often heavy rain to all areas. High pressure during the third week heralded a spell of more settled weather, finally giving way to a series of Atlantic depressions and heavy rain from the 24th. March was yet another warm month, and rainfall in western districts was well above average. In eastern areas of England, however, the long drought continued: since March 1995 there has been a 15% deficiency in rainfall. January 1997 was the last month in which the average temperature was significantly below the respective long term average.

Table i. The proportion of still water count units (lakes, reservoirs and gravel pits) in the UK with any ice and with three-quarters or more of their surface covered by ice during WeBS counts (England divided by a line drawn roughly between the Humber and the Mersey Estuaries).

Region	Ice	S	O	N	D	J	F	M
Northern	>0%	0	0	0	0	0	0	0
Ireland	>74%	0	0	0	0	0	0	0
Scotland	>0%	0	<1	0	2	20	<1	<1
	>74%	0	0	0	<1	12	0	0
N England	>0%	0	0	0	2	<1	0	0
	>74%	0	0	0	0	0	0	0
S England	>0%	0	0	0	<1	<1	0	0
	>74%	0	0	0	0	<1	0	0
Wales	>0%	0	0	0	0	0	0	0
	>74%	0	0	0	0	0	0	0

Northwest Europe

Winter 1997-98 was also mild on the continent. September saw cold and wet conditions in western Russia, though milder in parts of northern Scandinavia. October was the only notably cold month across most of continental Europe, with the Baltic countries, Poland and western Russia recording well below average temperatures. The Netherlands and northern France were notably wet, whilst Iceland was the only country to enjoy warmer than average conditions. November saw mild conditions continue in Iceland though elsewhere, temperatures were closer to normal. With the exception of northern France, several neighbouring countries received only half their usual rainfall. The mild theme continued into December, with only western Russia recording temperatures below the long-term mean. January and February were exceptionally warm throughout the whole of Europe with mean temperatures up to 5-6°C higher than normal in places. The latter month was notably dry in westernmost countries. Winter closed with a mild and wet March.

Arctic breeding conditions in Russian tundras

For the second year in succession, spring in the tundras of European Russian and northernmost Siberia was very late and prolonged, with poor conditions returning to some areas in late May. Summer 1997 was humid and cold in the west, but warm and dry on Taimyr.

Contrary to predictions, large numbers of lemmings were not widespread, and, although locally abundant, densities decreased as summer progressed in many areas. Arctic Foxes were generally numerous and bred successfully, though at no sites were large numbers of lemming-specializing avian predators found, e.g. Snowy Owl and Pomarine Skua.

Consequently, breeding success of waders was evaluated as low (to average for some species) in the west Russian Arctic and average on Taimyr and further east.

WeBS Core Counts

SURVEY METHODS

The main source of data for this report is the WeBS scheme, providing regular monthly counts for most waterfowl species at the majority of the UK's important wetlands. In order to fulfil the WeBS objectives, however, data from a number of additional schemes are included in this report. In particular, a number of species groups necessitate different counting methodologies in order to monitor numbers adequately, notably grey geese and sea-ducks, and the results of other national and local schemes for these species are routinely included. Additional, *ad hoc*, data are also sought for important sites not otherwise covered by regular monitoring, particularly open coast sections in Scotland, whilst the results of periodic, co-ordinated surveys, such as the non-estuarine coastal waterfowl survey, are included where the data collected are compatible with the presentation formats used in this report. The methods for these survey types are outlined below and more detail can be found in Gilbert *et al.* (1998). Although the precise methods for some of the additional count data presented within this report are unknown, it is safe to assume that they will follow closely the general methods presented here.

WeBS Core Counts

WeBS Core Counts are made using so-called "look-see" methodology (Bibby *et al.* 1992), whereby the observer, familiar with the species involved, surveys the whole of a predefined area.

Counts are made at all wetland habitats, including lakes, lochs/loughs, ponds, reservoirs, gravel pits, rivers, freshwater marshes, canals, sections of open coast and estuaries.

Numbers of all waterfowl species, as defined by Wetlands International (Rose & Scott 1997), are recorded. In the UK, this includes divers, grebes, Cormorant, herons, Spoonbill, swans, geese, ducks, rail, cranes, waders and Kingfisher. Counts of gulls and terns are optional. Vagrants, introductions and escapes are included.

Most waterfowl are readily visible. Secretive species, such as snipes, are generally under-recorded. No allowance is made for these habits by the observer and only birds seen or heard are recorded. The species affected by such biases are well known and the problems of interpretation are highlighted individually in the *Species Accounts*.

Most species and many sub-species are readily identifiable during the counts. Categories may be used, e.g. unidentified scoter species, where it is not possible to be confident of identification, e.g. under poor light conditions.

Species present in relatively small numbers or dispersed widely may be counted singly. The number of birds in large flocks is generally estimated by mentally dividing the birds into groups, which may vary from five to 1,000 depending on the size of the flock, and counting the number of groups. Notebooks and tally counters may be used to aid counts.

Counts are made once per month, ideally on predetermined 'priority dates'. This enables counts across the whole country to be synchronised, thus reducing the likelihood of birds being double-counted or missed. Such synchronisation is imperative at large sites which are divided into sectors, each of which can be practically counted by a single person in a reasonable amount of time. Local Organisers ensure co-ordination in these cases due to the high possibility of local movements affecting count totals.

The priority dates are pre-selected with a view to optimising tidal conditions for counters covering coastal sites at high tide on a Sunday (see *Coverage*). The dates used for individual sites may vary due to differences in the tidal regime around the country. Co-ordination within a site takes priority over national synchronisation.

The accuracy of each count is recorded. Counts suspected to be gross underestimates of the true number of non-secretive species present are specifically noted, e.g. a large flock of roosting waders only partially counted before being flushed by a predator, or a distant flock of sea-duck in heavy swell. These counts may then be treated differently when calculating site totals (see *Analysis*).

Observers do not receive official training but most are experienced ornithologists and/or counters. Data are input by a professional data input company. Data are keyed twice by different people and discrepancies identified by computer for correction. Any particularly unusual counts are checked by the National Organisers and are confirmed with the counters if necessary.

Goose roost censuses

Since many 'grey geese' spend daylight hours in agricultural landscapes, most are missed during counts at wetlands by WeBS. These species are usually best censused as they fly to or from their roost sites at dawn or dusk since these are generally discrete wetlands and birds often follow traditional flight lines approaching or leaving the site. Even in half-light, birds can generally be counted with relative ease against the sky, although they may not be specifically identifiable at mixed species roosts.

In order to produce population estimates, counts are synchronised nationally for particular species (see Appendix 3), though normally only one or two such counts are made each year. The priority count dates are determined according to the state of the moon, since large numbers of geese may remain on fields during moonlit nights. Additional counts are made by some observers, particularly during times of high turnover when large numbers may occur for just a few days.

In some areas, where roost sites are poorly known or difficult to access, counts are made during daytime of birds in fields.

As with WeBS Core Counts, the accuracy of the count is noted.

Sea-ducks

The accuracy of counts of waterfowl on the sea is particularly dependent on prevailing weather conditions at the time of or directly preceding the count. Birds are often distant from land, and wind or rain can cause considerable difficulty with identifying and counting birds. Wind not only causes telescope shake, but even moderate swell at all sites except those with high vantage points can hamper counts considerably. Many sites may be best covered using aerial surveys, though these are usually expensive and require experienced, professional counters. In many cases, birds can only be identified to genus, e.g. grebe species or scoter species.

Consequently, the best counts of most divers, grebes and sea-duck at open coast and many estuarine sites are made simply when conditions allow; only rarely will such conditions occur by chance during WeBS counts. Synchronisation between different sites may be difficult or impossible to achieve, and thus co-ordination of most counts to date has occurred at a regional or site level, e.g. within the Moray Firth, within North Cardigan Bay.

Non-estuarine coastal waterfowl survey

Open coast habitats are relatively poorly covered by WeBS and, consequently, the non-estuarine coastal waterfowl survey (NEWS) was undertaken in December 1997 and January 1998. This concentrated on those waders for which a large proportion of the population occur on the coast away from estuaries, notably Ringed Plover, Sanderling, Purple Sandpiper and Turnstone, thus repeating the Winter Shorebird Count of 1984.

Methods were broadly similar to WeBS, except that counts were usually made on the ebbing tide or at low water and observers walked along the intertidal habitat to ensure that birds amongst boulders and weed in particular were not overlooked.

Irish Wetland Bird Survey

The Irish Wetland Bird Survey (I-WeBS) monitors non-breeding waterfowl in the Republic of Ireland (Colhoun 1998). I-WeBS was launched in 1994 as a joint partnership between BirdWatch Ireland, National Parks and Wildlife Service of Dúchas The Heritage Service of the Department of Arts, Heritage, Gaeltacht and the Islands (Ireland), and WWT, supported by the Heritage Council and WWF UK (World Wide Fund for Nature). I-WeBS is complementary to and compatible with the UK scheme. The main methodological difference from UK-WeBS is that counts are made only between September and March, inclusive.

Productivity monitoring

Changes in numbers of waterfowl counted in the UK between years are likely to result from a number of factors, including coverage and weather, particularly for European and Russian-breeding species which may winter further east or west within Europe according to the severity of the winter. However, genuine changes in population size will result from differences in recruitment and mortality between years.

For several species of swans and geese, young of the year can be readily identified in the field and a measure of productivity can be obtained by recording the number of young birds in sampled flocks, expressed as a percentage of the total number of birds aged. Experienced fieldworkers, by observing the behaviour of and relationship between individuals in a flock, can record brood sizes as the number of young birds associating with two adults.

8

ANALYSIS

In fulfilment of the WeBS objectives, results are presented in a number of different sections. An outline of the analyses undertaken for each is given here; further detail is provided in Appendix 3. A number of limitations of the data or these analytical techniques necessitate caution when interpreting the results presented in this report (see *Interpretation of Waterfowl Counts*).

National totals

Population estimates are revised once every three years, in keeping with internationally agreed timetables (Rose & Stroud 1994). UK waterfowl populations will next be revised in 2000, although a number have been revised recently (Appendix 2) for inclusion in the third edition of international *Waterfowl Population Estimates* for presentation to the Seventh Meeting of the Contracting Parties to the Ramsar Convention, in Costa Rica in May 1999.

Total numbers of waterfowl recorded by WeBS and other schemes are presented separately for Great Britain (including the Isle of Man but excluding the Channel Islands) and Northern Ireland in recognition of the different legislation that applies to each. Separate totals for England, Scotland, Wales, the Isle of Man and the Channel Islands are provided in Appendices 4-8. Numbers of waterfowl found on coastal (including estuarine) and inland habitats are provided separately in Appendix 9, particularly for comparison of numbers of waders with those in reports prior to 1994 when waders were not counted at inland sites.

Numbers presented in this report are not rounded. National and site totals calculated as the sum of counts from several sectors or sites may imply a false sense of accuracy if different methods for recording numbers have been used, e.g. 5,000 birds estimated on one sector and a count of seven individuals on another is presented as 5,007. It is safe to assume that any large count includes a proportion of estimated birds. However, reproducing the submitted counts in this way is deemed the most appropriate means of presentation.

The count nearest the priority date or, alternatively, the count co-ordinated with nearby sites if there is considered to be significant interchange, is chosen for use in this report if several accurate counts are available for the same month. A count from any date is used if it is the only one available.

Data from other national surveys are used instead of WeBS counts where the census total provides a better estimate of the total numbers, e.g. the national census of Pink-footed and Greylag Geese in October and November. Totals from different censuses are not combined to produce national totals due to lack of synchronisation (birds counted at roost by one method may be effectively double-counted during the WeBS count at a different site in that month), with the exception of a few goose populations where the risk of double-counting is minimal (see Appendix 2). Consequently, counts from site or regional-based surveys of sea-ducks, for example, are not included in national totals. Data from NEWS are not included in national totals.

For some scarcer species, including many escaped or introduced species, an estimate of the total number recorded by WeBS throughout the country has been provided using summed site maxima, calculated by summing the highest count at each site, irrespective of the month in which it occurred. For some species, this is likely to result in double-counting where birds move between sites.

Annual indices

Because the same WeBS sites are not necessarily covered each year, changes in waterfowl population sizes cannot be determined simply by comparing the total number of birds counted in each year. Consequently, indexing techniques have been developed which allow between-year comparisons of numbers, even if the true population size is unknown.

The 'Underhill index' (Underhill 1989) was specifically developed for waterbird populations and is used in this report for most species. A full explanation of this indexing process is given in Prŷs-Jones *et al.* (1994), Underhill & Prŷs-Jones (1994) and Kirby *et al.* (1995), with additional information on its use in this report in Appendix 3.

In summary, where sites have not been visited, a count for each species is calculated based on counts in other months and years and at other sites. This effectively means that data are available for the same set of sites in each year and counts are thus directly comparable from one year to the next. Changes in the population can be calculated and the relative difference expressed as an index.

Not all species are included in the indexing process. Notably, many of the goose populations are excluded, partly because their reliance on non-wetland sites requires different count methodologies, but also because regular censusing of substantially the whole of the British populations negates the need for an index to be calculated using the Underhill technique. Thus, change indices for Pink-footed, Icelandic Greylag, Greenland White-fronted and Svalbard Barnacle Geese have been derived from the highest total count obtained during censuses of the population in each year (see Appendix 3). Many sea-duck are also excluded from the indexing process because of the extreme counting difficulties involved. Waders excluded from the index include those for which large numbers occur away from wetlands, e.g. Lapwing and Golden Plover, and those that are difficult to count accurately using WeBS methods, e.g. Snipe and Jack Snipe. Waterfowl species which only occur in small numbers in Britain and Ireland have also been excluded.

Index values for wildfowl species have been provided separately for Britain and Northern Ireland. However, values calculated for waders in Northern Ireland were found to be statistically unreliable due to the small number of estuaries contributing to each index value, and consequently indices have been calculated for the UK as a whole for these species.

For all species, the index value has been constrained to equal 100 in the most recent year. In particular, this enables direct comparison of values for wildfowl in Great Britain with Northern Ireland despite the different availability of data as a consequence of the later start of the scheme in the province (see Appendix 3 for availability of data for different species groups and countries).

Monthly indices

The abundance of different wildfowl species varies during the winter due to a number factors, most notably the timing of their movements along the flyway, whilst severe weather, particularly on the continent, may also affect numbers in the UK. However, due to differences in site coverage between months, such patterns cannot be reliably detected using count totals. Consequently, an index is calculated for each month to reflect changes in relative abundance during the season.

The index uses only counts from sites covered in all seven months (September to March). Totals calculated for each month from these sites only can then be compared directly (expressed as a percentage of the maximum numbers), thus revealing patterns of seasonality for the species considered. These are presented as graphs in the species accounts, giving both the value for the 1997-98 winter, and the average value from the five preceding winters, 1992-93 to 1996-97. Non-migratory, scarce and irregularly counted species are omitted and only WeBS Core Counts have been used in the index.

Broad differences in the monthly values between species reflect their status in the UK. Resident species, or those with large UK breeding populations, e.g. some grebes and Mallard, are present in large numbers early in the winter. Declines through the winter result in part from mortality of first year birds, but also birds returning to remote or small breeding sites that are not covered by WeBS. The majority of UK wildfowl either occur solely as winter visitors, or have small breeding populations that are swelled by winter immigrants, with peak abundance generally occurring in mid winter.

The vast majority of the wintering populations of many wader species are found on estuaries, and, since coverage of this habitat is relatively complete and more or less constant throughout winter, meaningful comparisons of total monthly counts can be made for many species. Consequently, monthly indices are not calculated for waders. As counting of gulls and terns is optional, indices are not calculated for these species either.

Site importance

Tables in the *Species Accounts* rank the principal sites for each species according to average seasonal maxima for the last five seasons in line with recommendations of the Ramsar Convention (see Appendix 2 and *Presentation and notation*).

The count nearest the priority date or, alternatively, the count co-ordinated with nearby sites if there is considered to be significant interchange, is chosen for use in this report if several accurate counts are available for the same month. A count from any date is used if it is the only one available.

In accounts for divers, grebes, Cormorant, herons, wildfowl and Kingfisher, annual maxima are derived from any month, with the season running from July to June inclusive. Average maxima for sites listed in the wader accounts are calculated using data from only the winter period, November to March.

Data from other sources, often involving different methods, e.g. goose roost censuses, are used where these provide better, i.e. larger, counts for individual sites. NEWS data have only been presented for selected species (Ringed Plover, Sanderling, Purple Sandpiper, Bar-tailed Godwit and Turnstone) and only for sites previously noted as being of national importance.

In the first instance, average maxima were calculated using only complete counts but, if any incomplete counts exceeded this initial average, they were also incorporated and the averages recalculated. Averages enclosed by brackets are based solely on incomplete counts.

Counts at any site are considered to be incomplete whenever significant under-recording is thought to have occurred, due to part of the site not being counted or adverse counting conditions. When counts from individual count sectors are summed to give an overall species count for complex sites, these counts might have been done under very different conditions, particularly at very large sites and consequently may have quite different qualities assigned to accuracy of the count. Additionally a variable amount of the overall site may have been uncounted.

The importance of the contribution of each count sector to the site total is based on its average contribution to the total at the time of year in question and on recent years (to allow for seasonal and long term trends). Further, consideration is given to the fact that a count sector which normally holds a significant proportion of a site total for species A may hold only a small proportion of the site total for species B. Consequently, if such a count sector is not completely counted, the site total will now be treated as complete for species B but incomplete for species A. These species-by-species qualities are assigned to waders, gulls, terns and herons.

In addition to the assessment of sites in *Species Accounts*, sites are identified for their importance in terms of overall waterfowl numbers in *Principal Sites*. The peak count at each site is calculated by summing the individual species maxima during the season, irrespective of the month in which they occurred. Only WeBS Core Counts and national goose censuses (see Appendix 3) are included in totals. Additional counts made using different methodologies, such as those of sea-ducks on the Moray Firth, are not incorporated.

The locations of all sites named in this report are given in Appendix 10.

PRESENTATION AND NOTATION

Detail is provided here on the format of presentation and the notation used in *Species Accounts* in particular. The information provided in *Analysis* and *Interpretation of Waterfowl Counts* should mean that results presented In other sections are self-explanatory.

The main purpose of the *Species Accounts* is to list important sites for each species, sub-species or populations, as relevant. This is done using certain numerical criteria adopted widely for use in conservation legislation and guidelines for site designation (see Appendix 2), although a number of exceptions have been made in some cases. Where available, the international and national importance thresholds are listed at the start of each account, although, for some numerous species, no population estimates, and therefore no thresholds, are available. Less numerous species, for which thresholds are not likely to be produced, are classified as "scarce" whilst species are classified as a "vagrant" where the UK does not fall within its normal range of distribution. In line with the recommendations of Vinicombe *et al.* (1993), records of all species recorded by WeBS, including escapes, have been published to contribute to the proper assessment of naturalised populations and escaped birds. Following Holmes & Stroud (1995), non-native species which have become established are termed "naturalised". These species are categorised according to the process by which they became established: naturalised feral (domesticated species gone wild); naturalised introduction (introduced by man); naturalised re-establishment (species re-established in a area of former occurrence); or naturalised establishment (a species which occurs, but does not breed naturally, e.g. potentially Barnacle Goose in southern England). With the exception of vagrants, all other non-native species have been classed as "escapes". The native range is given in the species account for naturalised species, escapes and vagrants.

The maximum count in any month of 1997-98, and the month of occurrence, is given for Great Britain and Northern Ireland in each account except for species occurring in very small numbers. Where productivity data have been collected, the proportion of young and mean brood size, where available, are also listed at the start of the account for ease of reference.

Index values, where calculated, are graphed within each account. Annual indices are presented on a log scale, as is the scientific norm for population growth. Where separate British and Northern Ireland values have been calculated (for certain wildfowl species), these are presented on the same graph to allow direct comparison but with different y-axes (vertical axes) for clarity. British indices are denoted using circles and the left-hand axis, and Northern Ireland values using squares and the right hand axis. Where only one index series is presented, circles and the left-hand axis have been used regardless of country.

Monthly indices, where calculated, are graphed within each account. Mean values for the previous five years (1992-93 to 1996-97) are shown using black columns and values for the most recent year using white columns.

Text in each account highlights significant points, e.g. coverage, changes in numbers or indices and at individual sites, and provides an overview of any recently published relevant research or surveys. The terms "recent average" and "previous average" refer to averages based on the winters 1992-93 to 1996-97, i.e. those presented in the previous WeBS report.

Tables provide data for all internationally important sites and all nationally important sites (either in a Great Britain context or, for sites in Northern Ireland, in an all-Ireland context) monitored by WeBS or other appropriate surveys. For each site, the maximum count in each of the five most recent years, the month of occurrence of the 1997-98 peak and the mean of the maxima is given. Incomplete counts are bracketed and missing counts are denoted using a dash "-".

Sites are selected for presentation using a strict interpretation of the 1% threshold (for convenience, sites in the Channel Islands and Isle of Man are identified using 1% thresholds for Great Britain and included under the Great Britain section of the tables). For some species with very small national populations, and consequently very low 1% thresholds, an arbitrary, higher level has been chosen for the inclusion of sites and is highlighted in the text. Where no thresholds are given, e.g. for introduced species, and where no or very few sites in the UK reach the relevant national qualifying levels, an arbitrary threshold has been chosen to select a list of sites for this report. These thresholds are highlighted in the text,

whilst a blank line has been inserted in the table to separate sites that qualify as nationally important from those selected for the purposes of this report using lower thresholds, including 1% thresholds of less than 50 birds.

Where the importance of a site has changed as a result of the 1997-98 count, i.e. it has become nationally or internationally important but was not following the previous year, or it has changed from international to national importance or vice versa, this is indicated in the table. Sites with elevated status have a black triangle pointing up (▲) to the right of the average, whilst those with lowered status are indicated using a triangle pointing down (▼). Sites for which the average fell below the threshold for national importance following 1997-98 are listed under the heading "Sites no longer meeting table qualifying levels".

A few sites that have not been counted in recent years, in most cases due to their isolated location, but were of national or international importance for one or more species when last counted (and thus retain that status in the absence of data to the contrary), are listed in the accounts under the section "Internationally or nationally important sites not counted in last five years". This also serves to highlight the need for counting to be resumed.

All sites which, in 1997-98, held numbers exceeding the relevant national threshold (or adopted qualifying level), but with five year means below this value are listed under "Other sites surpassing table qualifying levels in 1997-98". This serves to highlight important sites worthy of continued close attention. For waders, this includes counts from any month of the year.

It should be noted that a site may appear to have been flagged erroneously as having elevated status if the most recent count was below the relevant threshold. However, a particularly low count six years previously will have depressed the mean in the previous report. The converse may be true for sites with lowered status and thus, in exceptional circumstances, a site may be listed in the relevant sections of the table as both no longer being of national importance and with a peak count in the most recent year exceeding the national threshold.

For a number of wader species, different thresholds exist for passage periods. The list of "sites surpassing passage thresholds in 1997-98" includes all those with counts above the relevant number, even if already listed in the main part of the table by virtue of the winter mean surpassing the national threshold.

Key to symbols commonly used in the species accounts

As footnotes to thresholds (see Appendix 2)
?   population size not accurately known
+   population too small for meaningful threshold
*   where 1% of the national population is less than 50 birds, 50 is normally used as a minimum threshold for national importance
**   a site regularly holding more than 20,000 waterfowl (excluding non-native species) qualifies as internationally important by virtue of absolute numbers
†   denotes that a qualifying level different to the national threshold has been used for the purposes of presenting sites in this report

In tables of important sites:
-    no data available
( )   incomplete count
†   same meaning as when used for thresholds
▲   site was of a lower importance status in the previous year
▼   site was of a higher importance status in the previous year
R   count obtained using roost survey methodology
1, 2 count obtained using different survey methodology (cited at bottom of table)

INTERPRETATION OF WATERFOWL COUNTS

Caution is always necessary in the interpretation and application of waterfowl counts given the limitations of these data. This is especially true of the summary form which, by necessity, is used in this report. A primary aim here remains the rapid feedback of key results to the many participants in the WeBS scheme. More detailed information on how to make use of the data for research or site assessment purposes can be obtained from the appropriate National Organisers.

Information collated by WeBS and other surveys can be held or used in a variety of ways. Data may also be summarised and analysed differently depending on the requirements of the user. Consequently, calculations used to interpret data and their presentation may vary between this and other publications, and indeed between organisations or individual users. The terminology used by different organisations may not always highlight these differences. This particularly applies to summary data. Such variations do not detract from the value of each different method, but offer greater choice to users according to the different questions being addressed. This should always be borne in mind when using data presented here.

For ease of reference, the caveats provided below are broadly categorised according to the presentation of results for each of the key objectives of WeBS. Several points, however, are general in nature and apply to a broad range of uses of the data.

National totals

The majority of count data are collected between September and March, when most species of waterfowl are present in the UK in highest numbers. Data are collected during other months and have been presented where relevant. However, caution is urged regarding their interpretation both due to the relative sparsity of counts from this period and the different count effort for different sites.

A number of systematic biases of WeBS or other count methodology must be borne in mind when considering the data. Coverage of estuarine habitats and large, standing waters by WeBS is good or excellent. Consequently, counted totals of those species which occur wholly or primarily on this habitat during winter will approximate the true number. However, those species dispersed widely over rivers, non-estuarine coast or small inland waters are likely to be considerably under-represented, as will secretive or cryptic species, such as snipes, or those which occur on non-wetlands, e.g. grassland plovers. Species which occur in large numbers during passage are also likely to be under-represented, not only because of poorer coverage at this time, but due to the high turnover of birds in a short period. Further, since counts of gulls and terns are optional, national totals are likely to be considerable underestimates of the number using the WeBS network of sites. Only for a handful of species, primarily geese, do count totals approach the true number in the UK.

One instance of possible over-estimation is the use of summed site maxima to determine the total number of scarcer species. For species with mobile flocks in an area well covered by WeBS, e.g. Snow Goose in south-east England, it is likely that a degree of double-counting will occur, particularly if birds move between sites at different times of the year. These cases are highlighted in the *Species Accounts.*

The publication of records of vagrants in this report does not imply acceptance by the *British Birds* Rarities Committee (e.g. Rogers and the Rarities Committee 1998).

Annual indices

For all species, the long-term trends in index values can be used with confidence to assess changes in overall wintering populations. Because short-term fluctuations provide a less rigorous indication of population changes, care should be taken in their interpretation.

Caution should be used in interpreting figures for species which only occur in small numbers. Thus, numbers tend to fluctuate more widely for many species in Northern Ireland, largely as a result of the smaller numbers of birds involved but also, being at the westernmost limit of their range, due to variable use being made of Ireland by wintering wildfowl.

It should be borne in mind that the missing values used in the Underhill index are calculated anew each year. Because the index formula uses data from all years, each new year's counts will slightly alter the site, month and year factors. In turn, the missing counts may differ slightly and, as a result, the index values produced each year are likely to differ from those published in the previous *Wildfowl and Wader Counts.* The indices published here represent an improvement on previous figures as the additional year's data allow calculation of the site, month and year factors with greater confidence.

The use of a log scale to present indices means that the graph describes the rate of change, irrespective of the population size. Thus, a line showing a change from 10 to 100 has the same slope as from 100 to 1,000 since both represent a 10-fold increase in numbers. This has the effect of reducing the apparent magnitude of changes in numbers at the top end of the scale since a straight line increase over time represents a logarithmic, rather than linear, growth in numbers. Index values are given in Appendix 3.

Monthly indices

As for annual indices, the reduced numbers of both sites and birds in Northern Ireland result in a greater degree of fluctuation in numbers used in the analyses of data from the province.

Site importance

Criteria for assessing the international importance of wetlands have been agreed by the Contracting Parties to the Ramsar Convention on Wetlands of International Importance (Ramsar Convention Bureau 1998). Under criterion 3c, a wetland is considered internationally important if it regularly holds at least 1% of the individuals in a population of one species or subspecies of waterfowl, whilst any site regularly holding a total of 20,000 or more waterfowl also qualifies under criterion 3a. Similar criteria have been adopted for identification of SPAs under the EC Birds Directive in the UK legislation. A wetland in Britain is considered nationally important if it regularly holds 1% or more of the estimated British population of one species or subspecies of waterfowl, and in Northern Ireland, important in an all-Ireland context if it holds 1% or more of the estimated all-Ireland population. The relevant 1% thresholds are given in Appendix 2.

Sites are selected for presentation in this report using a strict interpretation of the 1% threshold. However, it should be noted that, where 1% of the national population is less than 50 birds, 50 is normally used as a minimum qualifying threshold for the designation of sites of national importance. It should also be noted that the 'qualifying levels' used for introduced species are used purely as a guide for presentation of sites in this report and do not infer any conservation importance for the species or the sites concerned since protected sites would not be identified for these non-native birds.

It is necessary to bear in mind the distinction between sites that *regularly* hold wintering populations of national or international importance and those which may happen to exceed the appropriate qualifying levels only in occasional winters. This follows the Ramsar Convention, which states that key sites must be identified on the basis of demonstrated regular use (calculated as the mean winter maxima from the last five seasons for most species in this report), otherwise a large number of sites might qualify as a consequence of irregular visitation by one-off large numbers of waterfowl. However, the Convention also indicates that provisional assessments may be made on the basis of a minimum of three years' data. These rules of thumb are applied to SPAs and national assessments also. Sites with just one or two years' data are also included in the tables if the mean exceeds the relevant threshold for completeness but this does not, as such, imply qualification.

Nevertheless, sites which irregularly support nationally or internationally important numbers may be extremely important at certain times, e.g.

when the UK population is high, during the main migratory periods, or during cold weather, when they may act as refuges for birds away from traditionally used sites. For this reason also, the ranking of sites according to the total numbers of birds they support (particularly in *Principal Sites*) should not be taken as a rank order of the conservation importance of these sites, since certain sites, perhaps low down in terms of their total 'average' numbers, may nevertheless be of critical importance to certain species or populations at particular times.

Peak counts derived from a number of visits to a particular site in a given season will reflect more accurately the relative importance of the site for the species than do single visits. It is important to bear this in mind since, despite considerable improvements in coverage, data for a few sites presented in this report derive from single counts in some years. Similarly, in assessing the importance of a site, peak counts from several winters should ideally be used, as the peak count made in any one year may be unreliable due to gaps in coverage and disturbance- or weather-induced effects. The short-term movement of birds between closely adjacent sites may lead to altered assessments of a site's apparent importance for a particular species. More frequent counts than the once-monthly WeBS visits are necessary to assess more accurately the rapid turnover of waterfowl populations that occurs during migration or cold weather movements.

This list of potential sources of error in counting wetland birds, though not exhaustive, suggests that the net effect tends towards under- rather than over-estimation of numbers and provides justification for the use of maximum counts for the assessment of site importance or the size of a populations. Factors causing under-estimation are normally constant at a given site in a given month, so that while under-estimates may occur, comparisons between sites and years remain valid.

It should be recognised that, in presenting only sites of national importance, this report provides just one means of identifying important sites and does not provide a definitive statement on the conservation value of individual sites for waterfowl, let alone other conservation interests. The national thresholds have been chosen to provide a reasonable amount of information in the context of this report only. Thus, for example, many sites of regional importance or those of importance because of the assemblage of species present are not included here. European Directives and conservation Conventions stress the need for a holistic approach to effect successful conservation, and lay great importance on maintaining the distribution and range of species, in addition to the conservation of networks of individual key sites.

For the above reasons of poor coverage, geographically or temporally, outlined above, it should be recognised that lists of internationally and nationally important sites are limited by the availability of WeBS and other survey data. Whilst the counter network is likely to cover the vast majority of important sites, others may be missed and therefore will not be listed in the tables due to lack of appropriate data.

Some counts in this report differ from those presented previously. This results from the submission of late data and corrections, and in some cases, the use of different count seasons or changes to site structures. Additionally, some sites may have been omitted from tables previously due to oversight. It is likely that small changes will continue as part of the current site mapping project and as the database, developed initially for waders, is brought on line for wildfowl. Most changes are minor, but comment is made in the text where they are significant. Where a site has apparently changed status as a result of recalculations or omissions, comment is made in the text but it is not flagged in the tables in the *Species Accounts*.

Note that sites listed under "Sites no longer of national/all-Ireland importance" represent those that were listed in the 1996-97 report as of national importance but which, following the 1997-98 counts, no longer meet the relevant threshold. It is not an exhaustive list of sites which, at any time in the past, have been of national or all-Ireland importance.

Counts made using non-WeBS methodologies, such as those of sea-ducks on the Moray Firth, are not incorporated into the site totals presented in *Principal Sites*, with the exception of goose roost counts. Thus, it should be borne in mind that other sites that are important for certain waterfowl species are not included in the table, whilst the sites listed may be of 'greater importance' for the species listed if additional data were included.

Lastly, owing to possible boundary differences, totals given for WeBS sites in this report are not necessarily the same as totals for designated statutory sites (ASSIs/SSSIs, SPAs or Ramsar Sites) having the same or similar names.

COVERAGE

WeBS Core Counts

Co-ordinated, synchronous counts are advocated to prevent double-counting or birds being missed. Consequently, priority dates are recommended nationally. Due to differences in tidal regimes around the country, counts at a few estuaries were made on other dates to match the most suitable conditions. Weather and counter availability also result in some counts being made on altèrnative dates.

Table ii. WeBS Core Count priority count dates in 1997-98

6 April	19 October
11 May	16 November
22 June	14 December
20 July	18 January
24 August	15 February
21 September	15 March

Counts were received from 1,983 sites of all habitats for the period April 1997 to March 1998, comprising 3,238 count units (the sub-divisions of large sites for which separate counts are provided). The number of sites remains at the high level of recent years, whilst the increased number of count units continues the trend of greater detail being provided. Of the key waterfowl sites, 1997-98 counts for Chew Valley Lake were not received in time for inclusion in this report.

WeBS and I-WeBS coverage in 1997-98 is shown by 10 km squares in Figure 1. The location of each count unit is shown using only its central grid reference. Thus, for example, the 19 count sectors of the North Norfolk Coast fall in four 10 km squares, broadly indicating the extent of the whole site. In all, WeBS count units were visited in 1,091 different 10 km squares during 1997-98, around average for recent years. As ever, areas with few wetlands or small human populations are apparent on the map as areas with little coverage. The location of many of the key sites mentioned in the report and all estuaries is shown in Appendix 10. The county and grid reference of all sites mentioned by name in this report are given in Appendix 10.

Goose roost censuses

In 1997-98, as in previous years, national surveys of Pink-footed and Icelandic Greylag Geese were undertaken in October and November (Hearn 1998), involving counts of birds arriving at or leaving roosts. Censuses of the native Scottish Greylag population on the Uists were made in August and February (R. MacDonald *in litt.*), and a national census of all key areas was undertaken in late summer 1997 (Mitchell *et al.* in press). Full censuses of Greenland White-fronted Geese, including birds in Ireland, were undertaken in autumn 1997 and spring 1998 by the Greenland White-fronted Goose Study and Irish National Parks and Wildlife Service (Fox & Francis 1998). Greenland Barnacle Geese were counted regularly by SNH and others on Islay and main islands in Argyll (M. McKay *in litt.*) and the Svalbard population was counted frequently on the Solway Firth by WWT staff (WWT unpubl. data). Dark-bellied Brent Geese were censused in January and February by the WeBS network, with counters at key sites making special effort to locate birds using adjacent areas, particularly fields, which would ordinarily be missed during normal Core Counts.

Sea-duck surveys

Data were received from the following regional or site-based surveys for counts of sea-duck, divers and grebes at coastal sites: counts in the Moray Firth between November and January (Stenning 1998); at least once monthly aerial and/or land-based counts of Common Scoter in Carmarthen Bay Between April and March (Cranswick *et al.* 1998); and regular counts of grebes and Red-breasted Merganser off Lafan Sands (M. Howe *in litt.*). However, no data were received for Cardigan Bay or parts of SE Scotland where dedicated counts have been made in recent years.

NEWS

All UK counties with coastal stretches were at least partly covered by the non-estuarine coastal waterfowl survey, primarily in December 1997 or January 1998. In total, 4,824 km (36%) of the non-estuarine coast was covered, divided as follows: 2,646 km in Scotland, 1,504 km in England, 518 km in Wales, 137 km in Northern Ireland, and 19 km in the Isle of Man.

Figure 1. Coverage by 10-km grid squares for WeBS Core Counts in the UK, Isle of Man and the Channel Islands and for I-WeBS in the Republic of Ireland in 1997-98. Small dots represent 1-2 count units per 10-km square, medium dots represent 3-4 units and large dots represent five or more units.

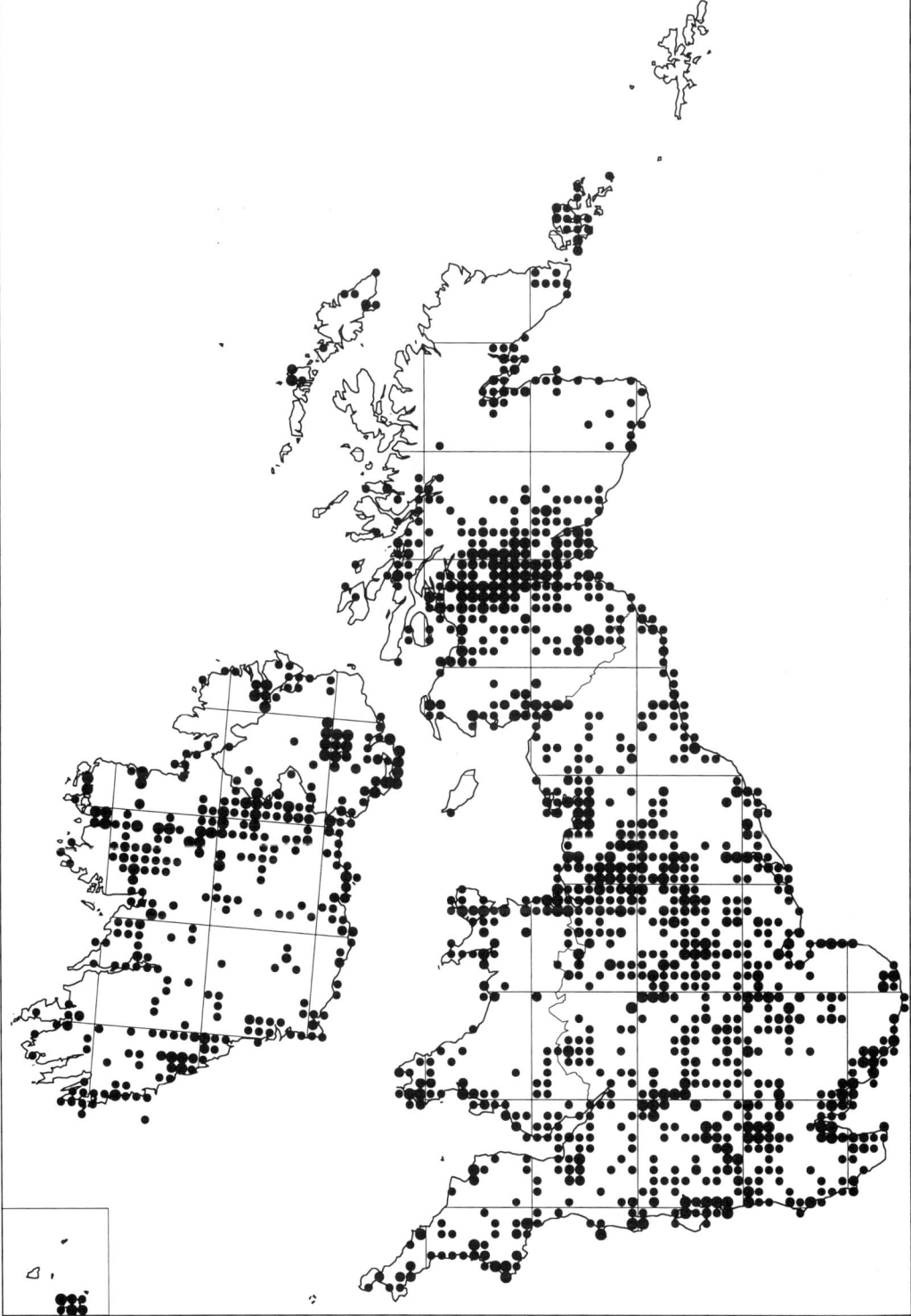

TOTAL NUMBERS

The total numbers of waterfowl recorded by WeBS in 1997-98 are given in Tables 1 & 2 for Great Britain (including the Isle of Man, but excluding the Channel Islands) and Northern Ireland, respectively. Brief comment on these figures are provided below. In addition, counts of waterfowl in the Republic of Ireland by I-WeBS are provided in Table 3.

Site coverage for gulls and terns is given separately since counts of these species were optional.

With a change to generally mild weather in 1997-98, following cold winters in 1995-96 and 1996-97, many of the migrant species which occur in Britain and Ireland chiefly as winter visitors, having bred in Europe and Russia, were able to remain on the continent in larger numbers. Consequently, counts and index values for many of these species returned to more normal values in 1997-98 following high or even record levels.

Throughout the text below, differences in annual index values between 1996-97 and 1997-98 are given in brackets for all species where the change was 20% or more. Values for wildfowl and their allies are for Great Britain unless otherwise specified; those for waders are for the UK.

Divers, grebes, herons and Cormorant

Numbers of divers were similar to those recorded previously by WeBS. Although the peak of Great Northern Divers in Britain exceeded all WeBS counts to date, it was only one greater than the previous highest. Numbers of both Little and Great Crested Grebes were rather lower than normal. The former is perhaps a result of the two preceding cold winters, whilst the peak count of the latter was the lowest since 1988-89 following a steady decline over the last five years, although annual indices suggest more stable numbers; indeed, values for Northern Ireland show a marked increase (+47%). Amongst the rarer grebes, the British peak of Slavonian Grebes in 1997-98 represented the highest count by WeBS to date.

The peak count of Cormorants in Great Britain was the lowest since 1991-92, with annual indices showing a similar picture. In Northern Ireland, however, annual indices (+49%) reached their highest level since the late 1980s. Numbers of the more common herons were around normal, though Little Egret numbers were

slightly lower than the record highs of the previous two winters.

Wildfowl

Mute Swans numbers and annual indices reached record highs in 1997-98, continuing their steady growth since the ban of lead fishing weights. By contrast, Bewick's (-46% in Great Britain; -70% in Northern Ireland) and Whooper Swans (-34%) were much less numerous than in recent years. Poor breeding success on arctic breeding areas in both species, and a mild winter enabling Bewick's Swans to remain on the continent, will have been responsible.

All of the major *Anser* goose populations returned in numbers very similar to the previous winter, following moderate breeding success in 1997. Numbers of naturalised geese and most migratory *Branta* populations were also similar to those of previous years, despite continued poor breeding success in most Brent populations, although index values fell (-21%) for Canada Geese. Numbers of Svalbard Barnacle Geese remained high, following the dramatic rise, whilst Svalbard Light-bellied Brent Geese returned to more normal levels after the influx in 1996-97.

Notable changes in numbers and annual indices were observed for Wigeon, dropping sharply in Great Britain (-23%), although rising in Northern Ireland. Gadwall remained the fastest growing wildfowl population in the UK, passing 13,000 for the first time, although indices declined in Northern Ireland (-30%). Teal numbers and annual indices in Britain both reached new highs, though only marginally higher than for recent years. By contrast, the situation for Mallard worsened further, with the lowest numbers and index values on record, though to what extent the consistent decline over the last 10 years reflects the picture in the wild populations is unclear, given the large number released for hunting. There is a clear need for further investigation. Shoveler annual indices in Northern Ireland (-21%) fell to their lowest ever level.

Annual indices for Pochard were the second lowest for over 30 years in Britain and were around half the level of just 10 years ago in Northern Ireland. A similar picture was seen for Tufted Duck in the province, though British numbers were normal. Scaup numbers were high in Britain but low in Northern Ireland, whilst most other sea-duck were recorded in average numbers for recent years. Numbers of

Goldeneye fell in both Northern Ireland and Great Britain (-31%), to their lowest value on record in the former, though simply returning to normal levels from last year's high in the latter. Numbers of Smew and Goosander (-38%), as predicted, were lower than in the previous two years as a result of the mild weather, though they remained higher than expected. Numbers of Red-breasted Merganser were around normal, though indices increased in both Britain and Northern Ireland (+22%) to near record levels. Available data suggest that numbers of Ruddy Duck continued to increase.

Numbers of rails were around average, though Coot were somewhat less numerous than in recent years.

Waders

Annual index values and most counts of Oystercatcher, Grey Plover (-22%), Knot, Sanderling, Dunlin (-21%), Bar-tailed Godwit (-39%) and, to a lesser extent, Black-tailed Godwit, all dropped sharply in 1997-98, equal and opposite, however, to the large increases seen during the cold winter of 1996-97. Consequently, most were present in around average numbers for recent years.

The reverse was true for a number species favouring milder weather, with high counts of Avocet (+64%), Lapwing and Golden Plover. Curlew (+25%) and Redshank also fared well, perhaps similarly influenced by the weather. Numbers of primarily non-estuarine species, such as Turnstone (-36%) and Purple Sandpipers, showed continuing declines, a picture re-reinforced by preliminary analyses of the 1997-98 NEWS data which showed declines of 27% in Ringed Plover and of 44% in Sanderling since the mid 1980s.

Numbers of species which occur primarily during autumn passage, e.g. stints and Wood Sandpiper, were generally low, following two years of high numbers. This is likely to result simply from the chances of appropriate weather conditions during autumn.

Gulls and terns

The peak counts of both Black-headed and Common Gulls were more than 20% higher than previous maxima. The peak count of Lesser Black-backed Gulls was almost 50% higher than previously, but almost wholly as a result of a large late summer count at just one site in which there is a large breeding colony. Otherwise, counts of

this species and the other common large gulls were around average.

Peak counts of all of the more numerous terns were lower than usual, with the exception of that for Arctic Tern which was more than double the previous record.

Introduced and escaped waterfowl

Many species of waterfowl occur in the UK as a result of introductions, particularly through escape from collections. Several have become established, such as Canada Goose and Ruddy Duck. The British Ornithologists' Union Records Committee recently established a category 'E' for "Species that have been recorded as introductions, transportees or escapees from captivity, and whose breeding populations (if any) are not thought to be self-sustaining" (BOURC 1999).

WeBS records of these species are included in this report both for the sake of completeness and in order to assess their status and monitor any changes in numbers, a key requirement given the need, under the African-Eurasian Waterbird Agreement of the Bonn Convention ". . . to prevent the unintentional release of such species . . ." and, once introduced, the need ". . . to prevent these species becoming a potential threat to indigenous species" (Holmes *et al.* 1998).

The number of species recorded by WeBS in 1997-98 reached a new peak of 22, having grown more or less steadily from 16 five years previously (Figure 2), though this excludes species which occur in both category A and E, e.g. Pink-footed Geese, since separation of escaped from wild birds is not readily possible using WeBS methods. A total of 197 sites held at least one species from category E, whilst summed site maxima reached a new high of 746 birds.

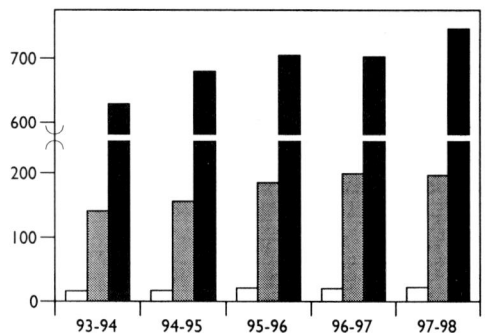

Figure 2. Number of species (white bars), number of sites at which birds were recorded (grey bars) and summed site maxima (black bars) for escaped waterfowl species.

Table 1. Total numbers of waterfowl counted by WeBS in Great Britain, 1997-98[†].

		Apr	May	Jun	Jul	Aug
	Number of sites visited	*889*	*820*	*747*	*781*	*792*
	Numbers of count units visited	*1,.454*	*1,237*	*1,116*	*1,154*	*1,251*
RH	Red-throated Diver	323	63	4	140	27
BV	Black-throated Diver	16	14	0	0	3
ND	Great Northern Diver	34	20	2	2	1
PJ	Pied-billed Grebe	0	0	1	0	0
LG	Little Grebe	950	569	616	874	1,711
GG	Great Crested Grebe	3,937	2,939	3,018	3,536	5,237
RX	Red-necked Grebe	12	3	2	21	70
SZ	Slavonian Grebe	55	4	1	1	3
BN	Black-necked Grebe	8	13	10	8	18
CA	Cormorant	5,976	4,248	3,428	4,333	7,296
BI	Bittern	1	0	0	1	0
LL	Little Bittern	0	0	1	0	0
ET	Little Egret	149	23	50	194	414
HW	Great White Egret	1	0	0	0	0
H.	Grey Heron	1,606	1,188	1,520	2,038	2,681
OR	White Stork	1	1	1	1	1
NB	Spoonbill	0	2	2	4	2
FM	Chilean Flamingo	1	0	1	0	0
FK	Lesser Flamingo	0	0	0	0	1
FL	Greater Flamingo	1	1	1	0	1
YV	Fulvous Whistling Duck	0	0	0	0	0
MS	Mute Swan	8,316	6,443	8,105	9,662	11,218
AS	Black Swan	11	6	13	10	14
TJ	Trumpeter Swan	0	0	0	0	0
BS	Bewick's Swan	20	0	2	3	0
WS	Whooper Swan	405	13	10	9	5
HN	Swan Goose	0	0	11	13	14
BE	Bean Goose	4	1	1	1	0
PG	Pink-footed Goose	19,318	644	15	19	260
WG	White-fronted Goose[1]	0	0	1	0	0
EW	European White-fronted Goose	233	0	0	1	2
NW	Greenland White-fronted Goose	189	0	0	0	0
LC	Lesser White-fronted Goose	0	0	1	1	1
GJ1	Greylag Goose (Iceland)	5,471	316	938	1078	732
GJ2	Greylag Goose (NW Scotland)	52	4	7	9	9,793
GJ3	Greylag Goose (naturalised)	4,662	3,342	6,395	6,949	12,406
HD	Bar-headed Goose	5	4	5	7	10
SJ	Snow Goose	27	10	24	25	4
RJ	Ross's Goose	2	1	1	0	6
EM	Emperor Goose	0	0	1	1	2
CG	Canada Goose	10,985	9,128	19,383	19,518	27,945
BY1	Barnacle Goose (Greenland)	0	0	0	0	3
BY2	Barnacle Goose (Svalbard)	3,599	7,513	1	1	2
BY3	Barnacle Goose (naturalised)	83	43	31	98	147
BG	Brent Goose[1]	0	0	0	0	0
DB	Dark-bellied Brent Goose	16,673	2,012	41	26	41
BB	Black Brant	0	0	0	0	0
PB1	Light-bellied Brent Goose (Svalbard)	6	0	0	0	0
PB2	Light-bellied Brent Goose (Canada)	4	0	0	0	0
EB	Red-breasted Goose	0	0	0	0	0
EG	Egyptian Goose	27	17	29	93	373
ZL	Feral/hybrid Goose	33	42	54	63	44
UO	Unidentified Goose	79	0	0	0	0

20

Table 1. continued

	Sep	Oct	Nov	Dec	Jan	Feb	Mar
Sites	*1,286*	*1,589*	*1,633*	*1,329*	*1,671*	*1,656*	*1,620*
Count units	*1,968*	*2,442*	*2,475*	*2,516*	*2,596*	*2,584*	*2,493*
RH	240	238	395	571	431	501	362
BV	2	28	39	35	15	21	24
ND	16	20	24	73	43	35	44
PJ	0	0	0	0	0	0	0
LG	3,321	3,273	3,023	3,003	2,447	2,353	2,369
GG	7,254	7,991	7,978	7,701	6,899	7,300	7,419
RX	35	47	31	41	36	38	29
SZ	36	129	146	315	242	192	156
BN	25	24	32	47	48	14	28
CA	11,359	13,658	12,830	11,932	12,177	11,140	10,011
BI	0	8	7	14	12	10	7
LL	0	0	0	0	0	0	0
ET	389	430	277	289	278	272	324
HW	0	0	0	0	2	0	1
H.	3,559	3,684	3,123	3,294	2,956	2,929	2,827
OR	1	3	3	3	3	2	2
NB	2	5	5	11	8	12	6
FM	0	0	0	0	0	0	0
FK	1	1	1	1	1	0	1
FL	1	1	2	1	1	0	0
YV	2	0	0	0	0	0	1
MS	13,842	16,683	18,170	17,364	16,763	15,413	14,368
AS	20	21	26	31	16	15	26
TJ	3	3	3	3	2	3	2
BS	0	127	373	5,441	4,110	4,900	104
WS	47	1,619	2,355	3,171	2,450	3,781	3,065
HN	16	33	34	26	38	30	25
BE	1	36	164	49	45	55	3
PG	63,989	235,559	182,182	120,375	67,979	72,401	54,171
WG	1	0	0	0	0	0	0
EW	3	52	474	2,155	4,095	5,665	664
NW	8	545	20,654	600	592	203	19,756
LC	0	1	1	1	0	2	0
GJ1	3,139	39,399	79,477	28,235	16,325	13,847	13,012
GJ2	242	228	183	129	229	2,535	293
GJ3	15,977	17,297	15,422	15,733	12,950	13,062	10,068
HD	15	9	14	14	6	11	16
SJ	16	70	68	35	49	64	69
RJ	0	0	2	2	1	1	1
EM	2	3	4	6	5	3	3
CG	40,395	41,786	43,225	39,642	37,021	30,464	22,980
BY1	37	85	35,123	34	48	28	33,841
BY2	631	19,882	22,244	15,819	18,138	21,777	23,863
BY3	336	619	584	304	388	394	435
BG	2	2	1	3	11	0	1
DB	262	52,193	84,403	99,045	92,569	79,028	52,082
BB	0	0	1	0	1	1	0
PB1	1,660	2,385	2,583	2,507	753	431	30
PB2	0	13	15	23	22	35	26
EB	0	0	0	0	0	0	1
EG	219	232	167	163	152	163	101
ZL	167	279	263	339	439	310	260
UO	0	0	9	0	0	0	0

Table 1. Great Britain, continued

		Apr	May	Jun	Jul	Aug
UD	Ruddy Shelduck	3	0	2	7	3
UE	Cape Shelduck	0	2	2	4	0
UB	Paradise Shelduck	0	0	1	0	0
SU	Shelduck	30,058	15,074	15,742	15,874	31,012
MY	Muscovy Duck	25	29	28	27	13
DC	Wood Duck	3	0	0	1	1
MN	Mandarin	77	54	122	72	49
WN	Wigeon	11,286	434	211	1,105	1,018
AW	American Wigeon	0	0	0	0	0
HL	Chiloe Wigeon	0	0	0	1	0
FT	Falcated Duck	0	0	0	0	0
GA	Gadwall	2,253	1,177	1,245	1,056	3,834
T.	Teal	14,977	374	466	1,740	13,465
KQ	Speckled Teal	0	1	0	1	1
MA	Mallard	25,568	20,003	30,421	37,820	69,925
QB	Chestnut Teal	0	0	0	0	0
PT	Pintail	733	16	13	67	162
PN	Bahama Pintail	0	0	0	0	0
QC	Cape Teal	0	0	0	0	0
GY	Garganey	12	26	6	19	37
SV	Shoveler	2,794	344	291	614	3,242
IE	Ringed Teal	0	0	0	0	1
RQ	Red-crested Pochard	0	2	3	1	8
PO	Pochard	1,211	514	885	2,035	6,636
NG	Ring-necked Duck	0	0	0	0	1
FD	Ferruginous Duck	1	1	0	0	1
TU	Tufted Duck	19,238	6,804	7,755	17,157	29,145
SP	Scaup	1,737	25	5	10	37
AY	Lesser Scaup	1	0	0	0	0
E.	Eider	20,057	14,221	15,484	16,227	21,577
KE	King Eider	0	0	0	0	0
LN	Long-tailed Duck	1,186	22	0	1	0
CX	Common Scoter	4,056	1,268	536	501	889
FS	Surf Scoter	3	0	0	0	0
VS	Velvet Scoter	576	172	101	18	84
UX	Unidentified scoter sp.	0	0	0	0	0
GN	Goldeneye	4,116	166	82	132	101
HO	Hooded Merganser	0	0	0	0	1
SY	Smew	6	2	0	0	0
RM	Red-breasted Merganser	2,340	704	871	701	1,032
GD	Goosander	519	455	431	986	1,031
RY	Ruddy Duck	1,326	628	552	645	1,332
ZF	Feral/hybrid Mallard type	91	135	138	116	144
ZD	Hybrid Aythya	1	0	0	0	0
UM	Unidentified duck	0	0	0	0	0
WA	Water Rail	97	47	44	48	72
AK	Spotted Crake	0	0	2	0	1
MH	Moorhen	4,753	3,119	2,630	4,140	5,924
CO	Coot	20,188	11,820	17,827	25,732	46,076
AN	Crane	0	0	0	0	0
	TOTAL WILDFOWL[2]	250,807	107,539	138,047	173,360	314,243

Table 1. continued.

	Sep	Oct	Nov	Dec	Jan	Feb	Mar
UD	10	8	7	9	9	5	7
UE	0	1	0	0	0	0	0
UB	0	1	0	0	1	0	0
SU	33,766	57,960	63,139	74,352	66,534	65,819	51,716
MY	32	101	116	127	99	93	36
DC	0	4	8	8	2	5	9
MN	156	180	315	266	235	173	181
WN	37,196	208,099	258,030	314,821	327,099	268,666	183,100
AW	0	0	1	1	0	3	4
HL	0	3	4	1	3	4	4
FT	0	0	0	0	0	0	2
GA	7,013	8,524	13,073	12,739	10,956	10,856	6,896
T.	51,584	86,662	122,957	137,754	116,949	80,165	49,071
KQ	1	0	0	0	0	1	2
MA	101,036	123,070	140,213	134,675	122,955	88,168	54,683
QB	0	0	1	0	0	0	0
PT	7,221	14,497	17,472	24,517	14,834	15,762	7,996
PN	1	0	0	0	0	0	0
QC	0	1	1	1	0	0	0
GY	17	4	4	0	0	1	5
SV	7,770	9,268	8,630	7,858	7,410	7,740	7,257
IE	0	0	0	0	0	0	0
RQ	16	46	94	71	85	41	46
PO	8,102	15,515	24,076	29,732	42,091	29,878	11,315
NG	2	0	1	1	1	2	3
FD	1	2	1	2	1	2	1
TU	37,757	38,789	47,954	52,004	50,537	45,558	40,115
SP	215	1,056	1,409	7,529	4,397	4,173	1,314
AY	0	0	0	0	0	0	0
E.	22,965	24,579	19,137	19,349	15,381	14,569	15,981
KE	1	1	0	0	0	0	0
LN	1	376	670	1,793	1,514	732	826
CX	3,433	3,871	5,240	8,565	5,488	3,417	2,271
FS	0	1	3	6	3	2	3
VS	61	311	744	792	390	454	528
UX	0	0	1	0	0	0	0
GN	241	3,455	8,235	16,355	14,518	15,062	13,141
HO	1	0	0	0	0	0	0
SY	0	3	15	149	300	236	72
RM	1,964	2,763	3,034	4,168	4,016	3,897	4,270
GD	817	955	2,079	3,245	3,513	3,628	2,166
RY	2,187	2,503	2,787	3,585	2,992	3,105	2,656
ZF	168	170	165	158	181	132	131
ZD	1	0	1	1	0	1	2
UM	2	3	0	0	0	0	0
WA	124	200	270	345	256	267	239
AK	2	0	0	0	0	0	0
MH	8,870	10,710	11,801	11,114	10,819	11,843	10,888
CO	74,109	87,192	102,507	95,392	78,892	66,005	45,388
AN	0	0	6	4	6	6	0
WILDFOWL	570,163	1,155,453	1,396,917	1,336,452	1,199,996	1,026,681	782,034

Table 1. Great Britain, continued

		Apr	May	Jun	Jul	Aug
OC	Oystercatcher	80,588	41,199	34,719	62,780	171,902
IT	Black-winged Stilt	0	0	0	0	1
AV	Avocet	1,035	227	237	453	819
KW	Black-winged Pratincole	0	0	0	0	1
LP	Little Ringed Plover	194	288	249	140	52
RP	Ringed Plover	4,336	9,349	1,191	1,602	23,163
KP	Kentish Plover	1	0	0	0	0
DO	Dotterel	1	4	0	0	0
GP	Golden Plover	8,493	395	166	2,360	27,802
GV	Grey Plover	33,955	16,808	1,034	4,856	27,375
U.	Unidentified wader	0	0	0	0	0
L.	Lapwing	9,698	3,930	13,112	54,629	68,769
KN	Knot	71,325	8,292	3,106	12,753	70,555
SS	Sanderling	7,749	11,329	137	9,218	7,609
ER	Western Sandpiper	0	0	0	0	1
LX	Little Stint	1	12	5	8	28
PP	Pectoral Sandpiper	0	0	0	0	0
CV	Curlew Sandpiper	0	7	2	16	50
PS	Purple Sandpiper	707	22	0	12	40
DN	Dunlin	97,576	92,251	1,714	44,799	80,110
RU	Ruff	282	94	4	139	480
JS	Jack Snipe	13	1	0	0	0
SN	Snipe	934	138	48	140	948
DS	Great Snipe	0	0	0	0	1
LD	Long-billed Dowitcher	0	0	0	1	0
WK	Woodcock	2	2	1	0	1
BW	Black-tailed Godwit	10,516	1,708	1,602	3,479	16,944
BA	Bar-tailed Godwit	4,271	1,306	1,312	8,984	17,227
WM	Whimbrel	117	1,735	223	712	600
CU	Curlew	32,515	5,124	11,623	52,145	73,421
DR	Spotted Redshank	72	23	14	166	221
RK	Redshank	40,774	3,585	4,081	18,830	51,756
GK	Greenshank	105	99	26	712	1,830
LY	Lesser Yellowlegs	0	0	0	0	0
GE	Green Sandpiper	55	6	37	216	452
OD	Wood Sandpiper	3	3	1	12	46
CS	Common Sandpiper	47	461	324	912	1,336
TT	Turnstone	8,044	2,443	353	1,267	5,483
PL	Grey Phalarope	0	0	0	1	1
	TOTAL WADERS	413,409	200,841	75,321	281,342	649,024
	TOTAL WATERFOWL[3]	665,976	309,595	214,944	456,940	966,367

24

Table 1. continued

	Sep	Oct	Nov	Dec	Jan	Feb	Mar
OC	226,099	236,395	251,410	239,580	213,278	227,196	148,904
IT	0	0	1	1	1	1	1
AV	1,550	2,554	2,409	3,859	3,464	3,232	2,469
KW	0	0	0	0	0	0	0
LP	18	5	1	0	0	0	9
RP	12,580	9,367	9,658	8,549	7,371	7,340	4,781
KP	0	0	0	0	0	0	0
DO	0	0	0	1	0	0	0
GP	36,613	70,650	164,677	175,445	138,161	146,928	44,235
GV	35,519	33,212	45,084	34,543	45,696	46,776	42,659
U.	0	0	0	0	0	1	0
L.	100,215	172,794	398,736	464,466	435,488	347,278	37,291
KN	136,101	184,547	294,025	257,397	179,217	225,010	131,131
SS	8,431	9,023	6,482	6,713	5,916	5,273	7,288
ER	0	0	0	0	0	0	0
LX	47	15	5	2	1	1	2
PP	3	0	0	0	0	0	0
CV	116	42	1	0	0	1	0
PS	55	228	545	1,003	1,061	658	622
DN	87,837	232,464	376,588	462,582	448,820	405,090	169,197
RU	587	215	183	263	384	424	399
JS	3	65	55	78	60	103	74
SN	1,529	4,827	6,026	7,404	4,621	5,038	3,541
DS	0	0	0	0	0	0	0
LD	0	2	1	0	0	0	0
WK	1	10	19	40	29	25	9
BW	16,712	15,048	13,948	13,199	12,247	12,522	15,461
BA	28,998	21,046	40,681	36,406	48,313	42,761	16,425
WM	296	44	12	3	3	5	9
CU	86,358	85,621	61,960	81,567	85,692	91,637	68,173
DR	205	147	179	98	55	142	60
RK	71,450	84,659	73,611	75,751	72,800	80,623	72,412
GK	1,785	713	262	192	527	158	180
LY	0	0	0	0	0	2	0
GE	245	157	161	132	78	100	88
OD	17	1	0	0	0	0	0
CS	302	43	29	19	20	44	41
TT	9,712	14,099	12,926	12,747	10,754	10,263	11,350
PL	1	3	0	0	2	0	0
WADERS	863,385	1,177,996	1,759,675	1,882,040	1,714,059	1,658,632	776,811
WATERFOWL	1,437,501	2,337,581	3,150,016	3,222,109	2,917,322	2,688,544	1,562,013

Table I. Great Britain, continued

		Apr	May	Jun	Jul	Aug
	Number of sites where gulls were counted[4]	*312*	*315*	*300*	*219*	*222*
MU	Mediterranean Gull	17	11	13	30	30
LU	Little Gull	47	30	5	39	28
AB	Sabine's Gull	0	0	0	0	1
BH	Black-headed Gull	53,474	32,252	35,133	82,246	120,772
IN	Ring-billed Gull	1	0	1	0	0
CM	Common Gull	8,699	3,495	3,207	5,114	13,580
LB	Lesser Black-backed Gull	35,859	38,821	49,111	59,085	23,411
HG	Herring Gull	35,165	32,287	36,401	39,933	32,770
IG	Iceland Gull	2	2	0	0	0
GZ	Glaucous Gull	4	0	0	0	0
GB	Great Black-backed Gull	1,711	1,655	1,806	2,804	5,235
KI	Kittiwake	1,297	1,077	403	1,106	4,151
UU	Unidentified gull	19	310	53	305	3,742
	TOTAL GULLS	136,295	109,940	126,133	190,662	203,720
	Number of sites where terns were counted[4]	*671*	*555*	*485*	*489*	*522*
TE	Sandwich Tern	386	2,221	876	3,857	5,718
RS	Roseate Tern	0	0	0	0	2
CN	Common Tern	224	2,092	2,494	4,023	3,696
AE	Arctic Tern	7	249	385	1,007	1,337
AF	Little Tern	3	457	432	236	138
BJ	Black Tern	0	23	2	0	24
WJ	White-winged Black Tern	0	0	0	0	2
UT	Unidentified tern	0	18	16	445	64
	TOTAL TERNS	620	5,060	4,205	9,567	10,981
KF	Kingfisher	63	66	81	123	153

†	*See Appendix 3 for calculation of national totals for goose populations*
1	*Indicates White-fronted and Brent Geese not identified to race*
2	*Total wildfowl and allies represents numbers of all divers, grebes, Cormorant, swans, geese, ducks and rails*
3	*Total waterfowl represents numbers of all species except gulls and terns*
4	*Counting gulls and terns was optional, thus totals are incomplete at a national level*

Table 1. continued

	Sep	Oct	Nov	Dec	Jan	Feb	Mar
Sites	*406*	*467*	*471*	*423*	*506*	*511*	*466*
MU	19	50	35	15	73	65	48
LU	9	18	16	0	9	0	1
AB	3	0	0	0	0	0	0
BH	140,922	138,029	251,574	253,921	232,467	190,033	154,322
IN	1	2	1	3	2	5	4
CM	23,744	33,111	71,677	72,950	70,090	86,528	40,207
LB	14,901	12,470	21,928	8,501	6,231	20,569	36,966
HG	49,164	46,166	67,430	51,394	63,245	61,790	49,834
IG	0	0	2	0	3	5	3
GZ	2	0	1	1	8	8	4
GB	6,499	6,342	13,850	9,048	9,506	4,456	2,805
KI	894	894	237	198	183	16	337
UU	1,616	1,387	930	6,495	3,242	5,760	600
GULLS	237,774	238,469	427,681	402,526	385,059	369,235	285,131
Sites	*839*	*1,116*	*1,142*	*1,130*	*1,174*	*1,129*	*1,085*
TE	2,675	87	3	0	1	0	6
RS	0	0	0	0	0	0	0
CN	904	40	1	0	0	0	1
AE	78	8	2	0	0	0	0
AF	23	1	0	0	0	0	0
BJ	6	1	0	0	0	0	0
WJ	0	0	0	0	0	0	0
UT	20	0	0	0	0	0	0
TERNS	3,706	137	6	0	1	0	7
KF	260	280	210	210	139	178	188

Table 2. Total numbers of waterfowl counted by WeBS in Northern Ireland, 1997-98[†]

		Apr	May	Jun	Jul	Aug
	Number of sites visited	4	3	3	3	5
	Number of count units visited	16	15	15	15	102
RH	Red-throated Diver	0	0	0	0	0
ND	Great Northern Diver	0	1	0	0	0
LG	Little Grebe	1	2	0	0	212
GG	Great Crested Grebe	2	5	0	3	903
SZ	Slavonian Grebe	0	0	0	0	0
BN	Black-necked Grebe	0	0	0	0	0
CA	Cormorant	80	104	59	123	1,117
H.	Grey Heron	16	35	24	48	260
MS	Mute Swan	166	116	74	56	1,547
BS	Bewick's Swan	3	0	0	0	0
WS	Whooper Swan	154	3	0	0	1
BE	Bean Goose	0	0	0	0	0
PG	Pink-footed Goose	0	0	0	0	0
NW	Greenland White-fronted Goose	36	0	0	0	0
GJ	Greylag Goose (Iceland)	88	0	0	0	0
CG	Canada Goose	0	0	0	0	0
BY	Barnacle Goose (naturalised)	2	0	0	0	0
DB	Dark-bellied Brent Goose	0	0	0	0	0
PB	Light-bellied Brent Goose (Canada)	198	0	0	0	0
SU	Shelduck	264	148	80	69	34
MN	Mandarin	3	2	5	0	0
WN	Wigeon	82	3	0	1	2
GA	Gadwall	0	0	0	0	61
T.	Teal	44	11	0	0	76
MA	Mallard	75	96	184	126	6,273
PT	Pintail	0	0	0	0	0
SV	Shoveler	0	0	0	0	14
PO	Pochard	0	0	0	0	359
TU	Tufted Duck	0	2	0	0	2,911
SP	Scaup	0	0	0	0	0
E.	Eider	19	58	23	125	162
LN	Long-tailed Duck	0	0	0	0	0
CX	Common Scoter	0	0	0	0	0
VS	Velvet Scoter	0	0	0	0	2
GN	Goldeneye	25	0	0	0	21
SY	Smew	0	0	0	0	0
RM	Red-breasted Merganser	12	6	15	11	319
GD	Goosander	0	0	0	0	0
RY	Ruddy Duck	0	0	0	0	23
WA	Water Rail	0	0	0	0	0
MH	Moorhen	1	3	2	1	133
CO	Coot	0	0	0	0	3,023
	TOTAL WILDFOWL[1]	1,255	560	442	515	17,193

Table 2. continued

	Sep	Oct	Nov	Dec	Jan	Feb	Mar
Sites	*13*	*24*	*25*	*30*	*32*	*24*	*28*
Count units	*132*	*143*	*174*	*191*	*195*	*187*	*150*
RH	2	4	16	19	21	0	42
ND	0	0	2	2	6	1	4
LG	495	377	529	512	315	273	169
GG	2,428	2,190	2,695	2,457	1,579	671	2,994
SZ	2	5	2	3	8	1	0
BN	0	1	0	0	0	0	0
CA	1,726	1,977	1,716	1,722	1,133	1,201	943
H.	334	283	179	285	272	188	147
MS	1,873	1,966	2,133	2,050	1,749	1,948	1,841
BS	0	0	47	46	133	75	46
WS	34	864	1,142	676	1,570	2,127	1,817
BE	0	1	0	0	0	0	0
PG	30	2	0	0	0	0	0
NW	0	111	8	0	19	91	37
GJ	191	117	191	597	317	399	1,159
CG	41	73	6	57	23	456	118
BY	148	134	123	130	131	126	117
DB	0	0	0	0	6	66	0
PB	12,805	14,910	9,303	5,675	3,323	3,727	2,580
SU	145	809	2,564	3,445	4,685	3,568	2,501
MN	0	0	2	0	0	0	0
WN	2,159	11,278	9,973	10,060	3,565	4,739	3,738
GA	129	80	116	142	107	126	154
T.	765	2,070	2,417	4,823	3,797	3,718	2,343
MA	8,623	7,171	6,781	6,856	4,303	3,883	2,172
PT	10	29	26	318	119	358	106
SV	36	112	98	207	107	81	80
PO	294	1,702	9,256	19,309	18,921	8,296	2,014
TU	5,838	7,195	19,021	16,801	16,395	13,868	8,906
SP	1,427	12	1,330	882	3,816	3,748	2,250
E.	421	797	981	1,091	521	413	293
LN	0	0	10	20	8	11	18
CX	0	0	0	0	1	0	1
VS	0	0	4	1	0	0	1
GN	69	556	6,107	4,779	4,888	4,792	5,694
SY	0	0	0	0	1	1	1
RM	484	606	501	609	425	259	577
GD	0	0	0	1	1	0	1
RY	24	8	7	23	0	28	14
WA	0	0	1	1	0	2	0
MH	233	212	266	189	169	258	264
CO	5,500	5,811	6,645	5,345	3,140	3,295	2,935
WILDFOWL	45,932	61,180	84,019	88,848	75,302	62,606	45,903

Table 2. Northern Ireland, continued.

		Apr	May	Jun	Jul	Aug
OC	Oystercatcher	1,192	925	583	2,220	3,887
RP	Ringed Plover	63	7	3	5	8
GP	Golden Plover	1,614	0	0	0	1
GV	Grey Plover	2	0	0	0	2
L.	Lapwing	40	95	63	275	1,410
KN	Knot	0	0	0	0	2
SS	Sanderling	43	37	0	0	0
CV	Curlew Sandpiper	0	0	0	0	0
PS	Purple Sandpiper	0	0	0	0	0
DN	Dunlin	45	55	16	50	115
RU	Ruff	0	0	0	1	0
JS	Jack Snipe	0	0	0	0	0
SN	Snipe	5	0	0	0	8
BW	Black-tailed Godwit	30	0	2	6	21
BA	Bar-tailed Godwit	3	4	11	66	25
WM	Whimbrel	0	331	1	14	2
CU	Curlew	611	159	649	2,044	2,004
DR	Spotted Redshank	2	0	0	0	0
RK	Redshank	940	76	50	386	1,076
GK	Greenshank	4	0	5	27	35
GE	Green Sandpiper	0	0	0	0	0
CS	Common Sandpiper	0	2	1	3	1
TT	Turnstone	85	1	0	0	74
	TOTAL WADERS	4,679	1,692	1,384	5,097	8,671
	TOTAL WATERFOWL[2]	5,950	2,287	1,850	5,660	26,124
	Number of sites where gulls were counted	2	*1*	*1*	*0*	*1*
LU	Little Gull	0	0	0	0	0
BH	Black-headed Gull	149	110	375	1,518	4,923
IN	Ring-billed Gull	0	0	0	0	0
CM	Common Gull	81	119	109	469	1,570
LB	Lesser Black-backed Gull	44	4	1	30	717
HG	Herring Gull	96	230	210	287	463
IG	Iceland Gull	0	0	0	0	0
GZ	Glaucous Gull	0	0	0	0	0
GB	Great Black-backed Gull	110	482	114	126	211
KI	Kittiwake	0	0	0	0	76
	TOTAL GULLS	480	945	809	2,430	7,960
	Number of sites where terns were counted	2	*1*	*1*	*1*	*1*
TE	Sandwich Tern	26	69	130	296	606
CN	Common Tern	0	10	0	0	0
BJ	Black Tern	0	0	0	0	1
UT	Unidentified tern	0	0	0	43	7
	TOTAL TERNS	26	79	130	339	614
KF	Kingfisher	0	0	0	0	0

†   *See Table 1 for footnotes*

Table 2. continued.

	Sep	Oct	Nov	Dec	Jan	Feb	Mar
OC	12,975	15,319	15,617	17,628	17,799	11,256	8,476
RP	233	232	396	659	525	290	41
GP	820	4,125	11,242	14,380	14,093	11,995	7,684
GV	82	67	153	280	352	285	181
L.	2,569	5,878	13,476	28,936	28,263	15,191	601
KN	166	137	3,772	8,184	9,655	4,426	512
SS	55	0	15	41	1	46	0
CV	6	3	0	0	0	0	0
PS	0	10	44	76	70	20	29
DN	845	1,061	11,829	16,803	13,696	14,313	1,334
RU	2	1	0	0	0	0	0
JS	0	0	2	5	0	0	0
SN	12	53	210	189	135	173	179
BW	386	374	219	293	404	236	396
BA	300	284	514	3,011	3,353	857	453
WM	5	0	0	0	0	0	0
CU	5,162	3,961	3,751	5,972	7,629	6,645	4,096
DR	5	2	2	2	2	0	1
RK	6,430	7,109	7,159	7,129	6,094	5,791	5,974
GK	91	93	57	93	65	57	62
GE	9	0	0	0	0	0	0
CS	1	0	0	0	0	0	0
TT	754	1,070	1,515	1,384	1,573	932	825
WADERS	30,908	39,779	69,973	105,065	103,709	72,513	30,844
WATERFOWL	77,147	101,242	154,171	194,198	179,283	135,307	76,921
Sites	*1*	*4*	*4*	*5*	*6*	*5*	*6*
LU	0	0	0	0	1	0	0
BH	10,329	8,441	7,221	8,656	11,837	8,532	7,780
IN	0	1	0	0	1	0	0
CM	3,092	2,458	1,717	1,490	2,220	3,467	1,179
LB	1,024	599	275	37	175	80	202
HG	3,861	3,159	3,646	2,482	3,750	1,673	3,141
IG	0	0	0	0	4	0	2
GZ	0	0	0	0	6	0	1
GB	596	410	370	252	537	150	308
KI	0	0	1	0	3	0	0
GULLS	18,902	16,068	13,230	12,917	18,534	13,902	12,613
Sites	*1*	*5*	*6*	*8*	*9*	*8*	*6*
TE	537	13	0	0	0	0	0
CN	0	0	0	0	0	0	0
BJ	2	1	0	0	0	0	0
UT	0	0	0	0	0	0	0
TERNS	539	14	0	0	0	0	0
KF	0	1	1	1	1	0	0

Table 3. Total numbers of waterfowl counted by I-WeBS in the Republic of Ireland, 1997-98.

	Sep	Oct	Nov	Dec	Jan	Feb	Mar
Number of sites visited	*130*	*132*	*173*	*156*	*258*	*176*	*169*
Number of count units visited	*231*	*213*	*242*	*238*	*558*	*254*	*261*
Red-throated Diver	41	62	31	128	193	35	167
Black-throated Diver	0	0	8	1	8	0	13
Great Northern Diver	6	21	58	117	251	80	159
Little Grebe	616	461	189	423	556	320	287
Great Crested Grebe	555	379	381	367	984	489	431
Red-necked Grebe	0	0	0	0	1	1	1
Slavonian Grebe	0	0	17	1	12	1	3
Black-necked Grebe	1	2	0	0	2	0	0
Cormorant	2,778	1,734	405	1,410	3,144	1,356	1,641
Grey Heron	596	382	174	357	633	268	339
Little Egret	41	21	0	22	30	33	31
Mute Swan	2,420	2,080	884	1,781	3,120	2,028	1,858
Bewick's Swan	0	0	6	214	520	154	9
Whooper Swan	21	320	1,208	1,508	4,208	1,707	1,757
Pink-footed Goose	0	2	0	7	36	6	12
European White-fronted Goose	0	0	0	0	7	7	2
Greenland White-fr. Goose	38	2,044	928	8,195	10,938	9,728	10,175
Greylag Goose	586	291	1,350	2,476	3,511	1,581	2,038
Canada Goose	70	5	27	75	181	84	15
Barnacle Goose	4	1	303	4	1,920	601	259
Dark-bellied Brent Goose	0	1	1	0	0	0	0
Light-bellied Brent Goose	61	1,266	1,422	5,366	8,545	6,748	6,521
Feral/hybrid Goose	47	50	0	51	73	88	40
Shelduck	123	1,382	328	3,297	8,121	4,149	4,607
Wigeon	3,580	19,898	12,699	17,974	41,547	22,302	8,700
American Wigeon	0	0	1	3	0	1	0
Gadwall	67	237	142	242	326	253	74
Teal	5,839	7,248	5,243	14,814	25,410	12,843	7,891
Mallard	13,488	10,940	4,651	7,620	11,951	5,104	3,412
Pintail	61	364	65	224	626	234	147
Shoveler	258	730	694	1,408	1,709	776	448
Red-crested Pochard	0	0	0	0	0	2	1
Pochard	2,314	12,156	538	1,867	7,857	4,959	256
Ring-necked Duck	0	0	0	0	1	0	0
Ferruginous Duck	0	0	0	1	1	1	1
Tufted Duck	5,054	11,412	602	2,020	7,597	4,045	1,859
Scaup	1	355	1,178	54	1,374	214	45
Eider	0	1	0	4	32	0	14
Long-tailed Duck	0	3	1	4	11	13	6
Common Scoter	811	55	42	1,899	4,035	2,658	4,045
Surf Scoter	0	0	0	0	1	1	0
Velvet Scoter	0	0	0	4	2	0	10
Goldeneye	2	59	188	562	1,976	1,015	584
Smew	0	0	0	1	10	3	11
Red-breasted Merganser	383	470	459	534	1,167	287	910
Goosander	0	0	0	15	15	1	6
Ruddy Duck	1	0	0	6	7	5	14
Hybrid/Feral Mallard type	5	4	0	1	3	8	4
Water Rail	22	15	14	14	37	19	33
Moorhen	357	375	62	385	562	540	482
Coot	9,443	14,929	1,328	2,734	4,184	2,483	1,421
Total wildfowl and allies	49,690	89,755	35,626	78,190	157,435	87,231	60,739

Table 3. continued.

	Sep	Oct	Nov	Dec	Jan	Feb	Mar
Oystercatcher	25,361	20,363	3,216	9,558	22,699	11,676	13,863
Ringed Plover	1,814	2,555	773	1,670	3,136	836	424
Golden Plover	1,252	46,374	21,419	50,045	79,206	53,838	23,414
Grey Plover	1,773	1,103	30	2,201	4,387	2,420	1,249
Lapwing	3,049	17,325	20,009	59,001	126,854	33,451	3,552
Knot	1,092	2,149	57	4,859	19,408	2,905	2,147
Sanderling	1,219	1,351	150	466	1,071	472	547
Little Stint	6	0	1	0	0	0	0
Curlew Sandpiper	19	7	1	0	0	1	0
Purple Sandpiper	52	0	0	80	157	101	279
Dunlin	6,399	8,323	9,907	27,105	69,793	22,786	11,297
Ruff	23	2	0	1	1	4	0
Jack Snipe	0	7	2	21	19	24	20
Snipe	137	544	267	862	1,315	777	394
Woodcock	0	0	1	3	1	1	0
Black-tailed Godwit	9,064	5,470	1,267	3,900	5,426	2,612	3,832
Bar-tailed Godwit	4,549	3,174	1,032	2,923	8,917	3,201	1,583
Whimbrel	5	2	0	2	1	1	2
Curlew	15,009	11,676	3,420	13,900	29,478	13,362	8,372
Spotted Redshank	8	16	0	16	10	12	10
Redshank	10,764	9,200	3,405	6,574	12,965	6,588	6,886
Greenshank	367	428	95	237	455	259	230
Green Sandpiper	4	6	0	1	3	1	4
Wood Sandpiper	1	0	0	0	0	0	0
Common Sandpiper	5	0	0	3	5	1	4
Turnstone	1,376	1,254	340	1,212	1,977	1,320	1,301
Grey Phalarope	1	0	0	0	0	0	0
Total waders	83,349	131,329	65,392	184,640	387,284	156,649	79,410
Mediterranean Gull	1	1	0	1	6	9	8
Little Gull	0	0	0	0	25	4	21
Black-headed Gull	18,608	12,814	5,591	14,830	38,192	23,318	11,003
Ring-billed Gull	0	2	1	1	3	5	2
Common Gull	2,640	1,857	1,853	1,750	10,212	6,102	2,130
Lesser Black-backed Gull	7,045	9,679	536	1,618	2,458	1,394	593
Herring Gull	4,046	3,053	163	1,757	5,690	4,538	2,933
Iceland Gull	0	0	0	0	0	0	1
Glaucous Gull	0	0	0	3	3	2	2
Great Black-backed Gull	1,801	991	84	1,144	2,234	1,518	942
Kittiwake	5,516	0	2	473	127	1,064	86
Total gulls	39,657	28,397	8,230	21,577	58,950	37,954	17,721
Sandwich Tern	638	0	3	0	0	0	6
Common Tern	23	1	0	0	0	0	2
Little Tern	3	0	0	0	0	0	0
Black Tern	1	0	0	0	0	0	0
Total terns	665	1	3	0	0	0	8
Kingfisher	17	9	1	11	14	3	3
TOTALWATERFOWL	173,378	249,491	109,252	284,418	603,683	281,837	157,881

RED-THROATED DIVER
Gavia stellata

International threshold: 750
Great Britain threshold: 50
All-Ireland threshold: 10*
* 50 is normally used as a minimum threshold

GB max:	571	Dec
NI max:	42	Mar

Unlike the previous two winters, there was no exceptionally high peak count of Red-throated Divers in 1997-98, and numbers in both Great Britain and Northern Ireland were within the normal range for recent years. Of particular note was the July total of 140, including 104 birds on the Don and Ythan, greatly exceeding previous counts in summer months.

Determining trends for this species is not possible at a national level due to the small proportion of the population counted by WeBS and difficulties in detecting birds present even at those sites that are covered. Nevertheless, comparison of average counts for sites in the table below with those calculated at the end of the 1993-94 winter shows very similar figures for most, perhaps suggesting little change overall.

A notable difference in the table to that in 1993-94 is the absence of Minsmere. Since counts of over 200 birds in the early 1990s, none have exceeded 50 at this site on WeBS dates. However, the continuing importance of the Suffolk coast for this species is illustrated by the large movements noted regularly by bird-watchers and an exceptional count of 1,500 off Minsmere in December 1997 (Rafe 1998).

The large number off the Aberdeenshire coast resulted in Don Mouth to Ythan Mouth attaining national importance. Durham Coast also attained this status following the high count in 1996-97, although this was not noted in the previous WeBS report.

	93-94	94-95	95-96	96-97	97-98	Mon	Mean
Great Britain							
Cardigan Bay	[1]740	[1]252	900	528	[1]536	Oct	591
Moray Firth	[2]411	[2]385	(72)	(52)	[3]284	Dec	360
Clyde Est.	11	50	126	195	136	Feb	104
Dengie Flats	89	143	41	96	100	Mar	94
Forth Est.	83	72	98	124	75	Nov	90
Don Mouth to Ythan Mouth	27	58	11	35	166	Sep	59 ▲
Durham Coast	9	63	81	103	24	Feb	56
North Norfolk Coast	67	26	71	47	43	Dec	51
Northern Ireland							
Lough Foyle	15	[4]40	83	18	4	Nov	28
Belfast Lough	20	28	10	11	41	Mar	22
Craigalea to Newcastle	-	-	13	-	-		13

Sites no longer meeting table qualifying levels
Minsmere Levels

Other sites surpassing table qualifying levels in 1997-98
Girvan to Turnberry	79	Nov
Solway Estuary	59	Nov

1   Data from Friends of Cardigan Bay, e.g. Green & Elliott (1993)
2   RSPB/BP studies, e.g. Stenning (1994)
3   Stenning (1998) RSPB report to Talisman Energy
4   unpubl. data

BLACK-THROATED DIVER
Gavia arctica

International threshold: 1,200
Great Britain threshold: 7*
All-Ireland threshold: 1*
* 50 is normally used as a minimum threshold

GB max:	39	Nov
NI max:	0	

Monthly UK totals in 1997-98 were about average for recent years: British maxima have generally fluctuated between 20 and 50 birds, whilst birds have only been recorded in Northern Ireland in three winters since counts of divers began in 1991-92. Their occurrence at little visited sites, particularly in western Scotland, means that counts are often sporadic, and even at regularly visited sites, counts can vary considerably: the highest in 1997-98 was at a site that had not previously featured in the table below.

Black-throated Divers were recorded at 37 sites in 1997-98, with concentrations in Essex, Northumberland, Glamorgan, Ayrshire to SE Argyll, SE Scotland and the Moray. Additional counts of divers in Bay of Sandoyne/Holm Sound, Orkney, recorded a peak of 23 Black-throats in December (K. Hague *in litt.*).

	93-94	94-95	95-96	96-97	97-98	Mon	Mean
Great Britain							
Moray Firth	[1]53	[1]35	(5)	(5)	[2]22	Dec	37
Loch Ewe: Aultbea	-	14	-	-	-		14
Forth Est.	9	9	19	7	8	Oct	10
Loch Indaal	0	31	11	1	1	several	9
Girvan to Turnberry	0	3	6	8	23	Nov	8 ▲
Northern Ireland							
Belfast Lough	0	1	2	2	0		1

Sites no longer meeting table qualifying levels
Arran

Other sites surpassing table qualifying levels in 1997-98
Traigh Luskentyre   12   Oct
North Norfolk Marshes   7   Nov

1   *RSPB/BP studies, e.g. Stenning (1994)*
2   *Stenning (1998) RSPB report to Talisman Energy*

GREAT NORTHERN DIVER
Gavia immer

International threshold:	50
Great Britain threshold:	30*[†]
All-Ireland threshold:	?[†]

* 50 is normally used as a minimum threshold

GB max: 73   Dec
NI max: 6   Jan

Monthly totals in 1997-98 were around normal, although the December peak was the highest yet recorded by WeBS, just exceeding that of 72 in 1991-92. Counts in Northern Ireland were much lower than the usual peak of between 20 and 40. Birds were recorded at 71 sites widely spread around the UK, though with notable concentrations off Co Down, southwest England, Essex, southwest Scotland, especially around the Clyde, and particularly off the Hebrides and northern isles. No WeBS counts have exceeded

the threshold for British importance, though dedicated counts of key sites, particularly more remote Scottish coastlines, demonstrate the need for such monitoring: 54 were recorded in the Moray, with the outer Dornoch Firth being of particular importance (Stenning 1998); additional counts off Tankerness, Orkney, recorded up to 58 birds during winter, with remarkable concentrations during spring passage of 393 in 1997 and 330 in 1998 (K. Hague *in litt.*, Corse *et al.* 1998).

	93-94	94-95	95-96	96-97	97-98	Mon	Mean
Great Britain							
Moray Firth	[1]17	[1]14	(2)	(8)	[2]54	Dec	28
Traigh Luskentyre	-	3	12	39	8	Dec	16
Loch Indaal	13	16	14	11	16	Sep	14
Lo. Beg/Scridain	5	4	6	6	4	Mar	5
Northern Ireland							
Tyrella	-	-	12	-	-		12
Lough Foyle	3	[3]20	15	9	3	Jan	10
Carlingford Lough	1	12	26	1	2	Jan	8
Kilkeel to Lee Stone Point	-	-	8	-	-		8
Craigalea to Newcastle	-	-	5	-	-		5

Sites no longer meeting table qualifying levels
Arran
Dundrum Bay

Internationally or nationally important sites not counted in last five years
Sound of Taransay

Other sites surpassing table qualifying levels in 1997-98
Fal Complex   8   Dec
Poole Harbour   6   Jan

† as no British site is of national importance for Great Northern Diver and as no all-Ireland threshold has been set, a qualifying level of five has
   been chosen to select sites for presentation in this report
1   RSPB/BP studies (e.g. Stenning 1994)
2   Stenning (1998) RSPB report to Talisman Energy
3   unpublished data

PIED-BILLED GREBE
Podilymbus podiceps

Vagrant
Native range: North America

One was recorded at Skelton Lake in June, an
unusual date for this transatlantic vagrant.

LITTLE GREBE
Tachybaptus ruficollis

International threshold:	**?**
Great Britain threshold:	**30***
All-Ireland threshold:	**?†**

* 50 is normally used as a minimum threshold

GB max: 3,321   **Sep**
NI max: 529   **Nov**

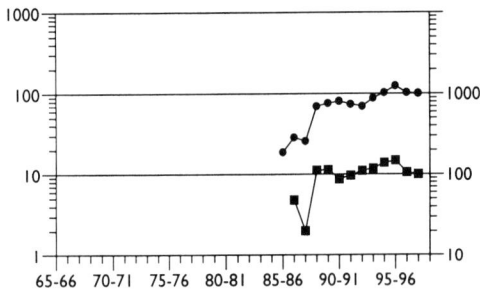

Figure 3. Annual indices for Little Grebe in GB (circles, left axis)
and NI (squares, right axis)

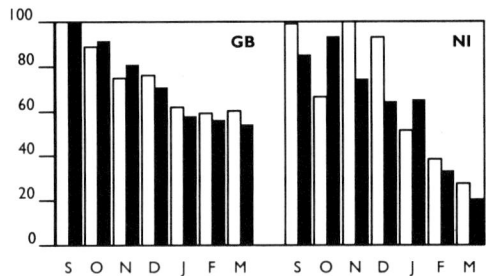

Figure 4. Monthly indices for Little Grebe in GB and NI (white
bars 1997-98, black bars 1992-93 to 1996-97)

The peak count in Britain fell markedly for the
second year in succession to a level lower than
that of 1994-95, a picture mirrored closely by the
annual indices. Since the previous winter was
very cold, this may have resulted from increased
mortality at the end of 1996-97, reflected in the
lower monthly index values that year, whilst an
extremely wet June in 1997 is likely to have
resulted in nests being flooded and may have
lowered breeding success. Similar reasons may
be responsible for the decline in Northern
Ireland, where the changes in annual index
values correspond closely with those for Great
Britain and the peak was the lowest since counts
were first made in the mid 1980s (apart from the

first two years, when coverage was much
poorer). The relatively mild conditions in 1997-
98, with correspondingly higher monthly indices
in late winter at least in Britain, may allow
numbers to recover.

Counts at several British sites show a five
year pattern matching the national trend, with
numbers increasing to a peak in 1995-96,
followed by a steady decline. This is particularly
evident on the Thames Estuary and the Wash,
where the 1997-98 peak was less than half the
five year mean, and also at Pitsford Reservoir and
Hampton & Kempton Reservoirs, although
numbers involved here are much lower. Only at
Hogganfield Loch have numbers shown a

sustained increase over the period. In Northern Ireland, numbers at the two key sites have also shown marked decreases after highs in 1995-96, though the 1997-98 peak is closer to the long term average in both cases.

Great Britain	93-94	94-95	95-96	96-97	97-98	Mon	Mean	
Thames Estuary	160	328	477	255	124	Dec	269	
Swale Estuary	77	202	195	213	244	Nov	186	
Holme Pierrepont GP	127	105	162	80	100	Sep	115	
Chew Valley Lake	75	106	122	152	-		114	
Wash	92	120	146	53	29	Sep	88	
Avon Valley (Mid)	67	81	86	68	77	Oct	76	
North Norfolk Marshes	58	56	93	51	87	Sep	69	
Deben Estuary	87	66	49	63	78	Dec	69	
Cleddau Estuary	27	49	75	91	72	Nov	63	
Chichester Harbour	35	50	100	52	72	Dec	62	
Rutland Water	68	60	83	35	62	Oct	62	
Eyebrook Reservoir	27	43	70	76	56	Oct	54	▲
R. Test: Fullerton to Stockbridge	43	55	62	52	52	Mar	53	
Cameron Reservoir	40	63	70	33	56	Sep	52	▲
Sutton/Lound Gravel Pits	-	17	72	39	72	Aug	50	▲
Middle Tame Valley GP	40	25	52	53	68	Sep	48	
Tees Estuary	47	53	42	47	52	Sep	48	
Blackwater Estuary	29	52	59	44	47	Sep	46	
Medway Estuary	51	54	60	42	18	Dec	45	▼
Lee Valley Gravel Pits	44	27	45	39	56	Oct	42	
Kilconquhar Loch	20	36	52	42	49	Sep	40	
Bewl Water	14	47	57	44	36	Sep	40	
Hamford Water ˙	52	28	72	18	26	Dec	39	
Somerset Levels	14	34	37	55	47	Oct	37	
Fleet/Wey	46	37	37	30	34	Dec	37	
Southampton Water	42	26	37	42	(14)	Nov	37	
Alde Complex	9	37	51	38	44	Feb	36	
Orwell Estuary	26	37	36	45	34	Dec	36	
Blagdon Lake	26	39	59	23	31	Aug	36	
Kings Mill Reservoir	68	40	23	29	14	Sep	35	
Pirton Pool	-	29	37	41	32	Aug	35	
Hogganfield Loch	19	22	31	45	56	Sep	35	▲
Cemlyn Bay	32	33	40	33	32	Jan	34	
Portsmouth Harbour	32	36	36	30	35	Dec	34	
Pitsford Reservoir	9	53	64	32	10	Oct	34	
Hampton & Kempton Reservoirs	26	43	54	28	16	Feb	33	▲
Rye Harbour/Pett Level	24	26	46	28	37	Sep	32	▲
Morecambe Bay	46	32	27	31	22	Jan	32	
Barleycroft Gravel Pit	-	-	54	23	15	Sep	31	
Wraysbury Gravel Pits	33	32	27	32	27	Jan	30	
Northern Ireland[†]								
Lo. Neagh/Beg	399	535	626	376	330	Sep/Nov	453	
Strangford Lough	123	102	169	140	101	Dec	127	
Upper Lough Erne	54	84	62	73	50	Feb	65	
Lough Money	26	21	33	35	51	Dec	33	▲

Sites no longer meeting table qualifying levels
Fisherwick/Elford Gravel Pits
Hanningfield Reservoir

Internationally or nationally important sites not counted in last five years
R. Soar: Leicester

Other sites surpassing table qualifying levels in 1997-98

Barton Gravel Pits	44	Sep	Hilfield Park Reservoir	34	Oct
Lower Derwent Valley	42	Oct/Mar	R. Clyde: Lamington	31	Nov
Kirkby-on-Bain Gravel Pits	40	Sep	King's Dyke Pits	31	Sep
Langstone Harbour	37	Nov	Hanningfield Reservoir	30	Sep
Duddon Estuary	35	Nov			

† as no all-Ireland threshold has been set for Little Grebe, a qualifying level of 30 has been chosen to select sites for presentation in this report

GREAT CRESTED GREBE
Podiceps cristatus

International threshold: 1,500
Great Britain threshold: 100
All-Ireland threshold: *30

* 50 is normally used as a minimum threshold

GB max: 7,991 Nov
NI max: 2,994 Mar

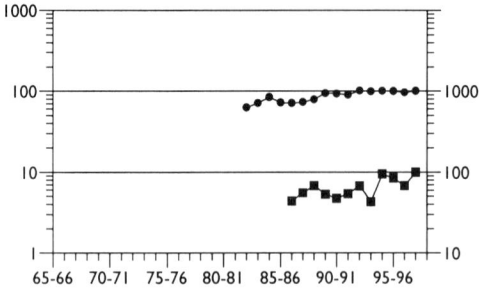

Figure 5. Annual indices for Great Crested Grebe in GB (circles, left axis) and NI (squares, right axis)

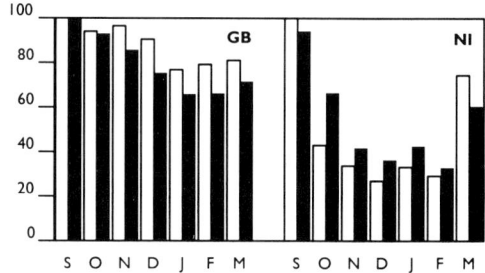

Figure 6. Monthly indices for Great Crested Grebe in GB and NI (white bars 1997-98; black bars 1992-93 to 1996-97)

Maximum counts of Great Crested Grebes in Britain have declined fairly steadily from a peak of 9,580 in 1992-93; the most recent total was the lowest since 1988-89. Annual indices, however, suggest that numbers have been very stable over this period, and are around a third higher than during the 1980s. Despite the low peak, monthly totals in 1997-98 remained remarkably consistent, and monthly indices were higher than normal in mid and late winter.

Maxima in Northern Ireland have varied between 1,500 and 4,000 in recent years, with the 1997-98 peak about average, though, unusually, in March. Monthly indices show an upturn in numbers in late winter is normal, and, whilst that in 1997-98 was particularly pronounced, it should be noted that these figures exclude data from Belfast Lough which was not counted in all months. Consequently, the real March figure is undoubtedly much higher still. Annual indices for the province fluctuate considerably, but there appears to be a pattern of general increase since the mid 1980s, with the 1997-98 value the highest yet.

Counts at Loughs Neagh & Beg regularly decline from a late summer peak, and these birds are thought to move to Belfast Lough as winter progresses. With very low counts at the former throughout 1997-98, it appears that a greater proportion of birds may have switched between the sites. However, there is some suggestion from summer breeding surveys that land-based WeBS counts at Loughs Neagh & Beg are missing some of the significant concentrations of waterfowl (Forster 1998). The higher numbers at Belfast Lough were sufficient to elevate it's status to that of international importance for Great Crested Grebe.

There were low counts at a number of key sites in 1997-98, notably Queen Mary Reservoir, Grafham Water, King George VI Reservoir and Lough Foyle. Low counts for the second year in succession at Wraysbury Reservoir are contrasted by higher counts on Wraysbury Gravel Pits. There is known movement of many waterfowl between these adjacent sites as a result of disturbance, though indications are that disturbance has been greatest on the gravel pit complex in recent years. Numbers were particularly low off the Thanet Coast, though, as with nearby Pegwell Bay, counts at such sites are likely to fluctuate as a result of conditions and perhaps local movements of birds. Notable high counts in 1997-98 were made on the Solway and Mersey Estuaries and on Upper Lough Erne.

	93-94	94-95	95-96	96-97	97-98	Mon	Mean	
International								
Lo. Neagh/Beg	571	2,533	2,440	1,537	863	Aug	1,589	
Belfast Lough	1,318	1,650	1,350	1,200	2,403	Nov	1,584	▲
Great Britain								
Rutland Water	894	741	579	378	767	Nov	672	
Chew Valley Lake	675	600	615	645	-		634	
Forth Est.	671	627	411	597	491	Sep	559	
Lade Sands	580	-	[1]277	(7)	425	Jan	427	
Queen Mary Reservoir	411	307	298	593	98	Aug	341	
Lavan Sands[2]	275	508	283	244	360	Sep	334	
Morecambe Bay	348	277	296	286	282	Nov	298	
Grafham Water	181	175	377	506	197	Nov	287	
Stour Estuary	250	260	312	261	185	Oct	254	
Cardigan Bay[3]	229	341	176	311	177	Jan	247	
Thanet Coast	250	504	-	166	15	Dec	234	
Pitsford Reservoir	172	215	188	304	147	Dec	205	
Wraysbury Gravel Pits	178	167	167	263	246	Nov	204	
Cotswold WP West	214	233	189	181	175	Mar	198	
Solway Estuary	96	113	36	205	430	Feb	176	
Hanningfield Reservoir	298	185	124	59	123	Sep	158	
Abberton Reservoir	55	59	238	248	149	Nov	150	
Queen Elizabeth II Reservoir	88	105	258	118	168	Aug	147	
Pegwell Bay	44	450	82	8	137	Feb	144	
Loch Ryan	(42)	[4]258	[4]201	(15)	54	Sep	139	
Lee Valley Gravel Pits	44	157	132	170	190	Oct	139	
Attenborough Gravel Pits	134	137	120	155	135	Oct	136	
Mersey Estuary	139	95	61	169	214	Dec	136	
Dee Estuary (Eng/Wal)	140	147	110	205	73	Mar	135	
Blithfield Reservoir	153	155	70	169	105	Oct/Nov	130	
King George VI Reservoir	47	123	401	41	16	Jul	126	
Blackwater Estuary	84	145	171	118	99	Mar	123	
Alton Water	107	183	120	109	73	Jan	118	
Wraysbury Reservoir	114	112	265	52	43	Oct	117	
Ardleigh Reservoir	112	123	82	84	171	Feb	114	
Eyebrook Reservoir	38	99	167	155	103	Nov	112	
Blagdon Lake	62	87	67	270	73	Nov	112	
Loch Leven	33	102	210	98	112	Sep	111	▲
Southampton Water	60	68	169	94	117	Dec	102	▲
Northern Ireland								
Carlingford Lough	101	295	143	364	201	Dec	221	
Lough Foyle	80	[5]480	488	116	86	Oct	195	
Upper Lough Erne	164	111	90	276	304	Feb	189	
Larne Lough	110	122	147	124	76	Oct	116	
Strangford Lough	95	40	182	83	64	Dec	93	
Craigalea to Newcastle	-	-	35	-	-		35	

Sites no longer meeting table qualifying levels
Colne Estuary
Medway Estuary
Wash

Other sites surpassing table qualifying levels in 1997-98

Sth Muskham/Nth Newark GP	130	Nov	Clyde Estuary	109	Jan
Swansea Bay	128	Dec	Holme Pierrepont Gravel Plts	100	Aug
Colne Estuary	118	Feb	Minsmere Levels	100	Jan
Bewl Water	111	Sep			

1   D. Walker (in litt.)
2   data from CCW
3   data from Friends of Cardigan Bay (e.g. Green & Elliott 1993)
4   P. Collin (in litt.)
5   unpublished data

RED-NECKED GREBE
Podiceps grisegena

GB max:	70	Aug
NI max:	0	

International threshold: 330
Great Britain threshold: 1*
All-Ireland threshold: ?
* 50 is normally used as a minimum threshold

The 1997-98 maximum was about average for the last five years, having varied between 50 and 100 birds during that time. Although the peak has occurred in August previously, the timing is curious given the absence of any local breeding populations, particularly since this largely comprises birds on the Forth Estuary, many hundreds of miles from the nearest breeding grounds. Whilst the Forth is clearly a traditional site for Red-necked Grebes, the variation in the timing of peak numbers, which has occurred in mid and even late winter in recent years, means our understanding of how and why birds use this site is far from clear.

Red-necked Grebes were recorded at 35 sites in 1997-98 in addition to those in the table below, mostly off coasts in southwest England and in the southeast from Sussex to Norfolk. There was only one record in Wales, whilst, surprisingly, none was recorded in Scotland away from the Forth. The presence of six and 10 birds in January and March, respectively, at the very northerly location of Bay of Sandoyne/Holm Sound, Orkney, during additional counts of divers and grebes (K. Hague *in litt.*) is thus noteworthy.

	93-94	94-95	95-96	96-97	97-98	Mon	Mean
Great Britain[†]							
Forth Est.	44	89	¹52	44	64	Aug	59
North Norfolk Marshes	0	4	19	2	17	Feb	8
Other sites surpassing table qualifying levels in 1997-98							
Alde Complex			5	Dec			

[†] *as the British threshold for national importance is so small, a qualifying level of five has been chosen to select sites for presentation in this report*
[1] *SNH funded surveys in SE Scotland, WWT unpubl. data*

SLAVONIAN GREBE
Podiceps auritus

GB max:	315	Dec
NI max:	8	Jan

International threshold: 50
Great Britain threshold: 4*
All-Ireland threshold: ?
* 50 is normally used as a minimum threshold

The British peak in 1997-98 was some 20% greater than the previous highest WeBS total, following three years in which numbers had stabilised at around 250. Perhaps more impressive was that this occurred in mid winter, rather than, as is usual, late winter or early spring when passage birds gather at key sites. This was influenced by a particularly large count of 88 birds between the Dornoch and Loch Fleet, birds which were also recorded during dedicated surveys of the Moray Firth, though numbers in the months both immediately prior to and after this count were much lower (Stenning 1988). Notable counts in the table below are the continued high numbers in the Clyde Estuary. A peak of 33 birds, in January, was recorded in Bay of Sandoyne/Holm Sound, Orkney (K. Hague *in litt.*).

Slavonian Grebes were recorded at 103 UK sites by WeBS in 1997-98. In December and January alone, the number of British sites supporting this species has risen more or less steadily from 36 in 1993-94 to 67 in 1997-98. Whilst this suggests a trend of either greater dispersal or improved coverage of the Slavonian Grebe site network by WeBS in recent years, the presence of average numbers at nearly all key sites in 1997-98 suggests that the increased British total results, presumably, from better detection of birds off the Dornoch rather than a general influx.

International	93-94	94-95	95-96	96-97	97-98	Mon	Mean	
Moray Firth	[1]53	[1]66	(8)	(22)	[2]163	Dec	94	
Forth Est.	53	78	[3]108	107	75	Nov	84	
Lough Foyle	(3)	[4]71	103	20	(6)	Jan	65	
Great Britain								
Pagham Harbour	14	75	23	29	39	Dec	36	
Loch Indaal	22	37	20	13	32	Oct	25	
North Norfolk Marshes	2	6	77	17	9	Feb	22	
Loch of Harray	9	36	31	6	14	Nov	19	
Clyde Est.	1	8	25	32	25	Dec	18	
Blackwater Estuary	8	13	22	14	18	Mar	15	
Studland Bay	8	17	16	-	-		14	
Traigh Luskentyre	-	9	24	13	8	Nov/Dec	14	
Poole Harbour	8	15	13	10	9	Mar	11	
Chichester Harbour	5	10	3	13	8	Jan	8	
North West Solent	6	5	13	12	2	Dec	8	
Lindisfarne	4	3	15	2	12	Mar	7	
Loch Ryan	0	[5]6	[5]19	0	11	Feb	7	
Exe Estuary	11	5	6	2	11	Jan	7	
Loch of Swannay	4	4	8	10	5	Oct	6	
Tamar Complex	1	2	9	7	5	Dec	5	
Langstone Harbour	5	4	5	3	3	Dec	4	▲

Internationally or nationally important sites not counted in last five years
Sound of Taransay

Other sites surpassing table qualifying levels in 1997-98

Dengie Flats	8	Mar	Fal Complex	4	Dec
Medway Estuary	8	Nov	Ryde Pier to Puckpool Point	4	Jan
Beaulieu Estuary	7	Nov	St Andrews Bay	4	Mar
Cleddau Estuary	5	Dec			

1   RSPB/BP studies (e.g. Stenning 1994)
2   Stenning (1998) RSPB report to Talisman Energy
3   SNH funded surveys in SE Scotland (WWT, unpubl. data)
4   unpublished data
5   P. Collin (in litt.)

BLACK-NECKED GREBE
Podiceps nigricollis

International threshold: 1,000
Great Britain threshold: 1*[†]
All-Ireland threshold: ?

GB max: 48   Jan
NI max: 1   Oct

* 50 is normally used as a minimum threshold

The peak British count closely matched those of recent years. One at Belfast Lough in October was the first to be recorded by WeBS in Northern Ireland, and compares with annual totals of 10 or less for the whole of Ireland (e.g. Milne & O'Sullivan 1998). Birds were recorded at 54 British sites during 1997-98, mostly along the south coast and in southeast England, but with marked concentrations in the east midlands and southeast Scotland also. The peak on the Fal in December was the highest at an individual site recorded by WeBS to date. In view of the size of counts at this site in recent years, earlier nil counts have now been regarded as incomplete and the Fal is elevated to the key UK site for Black-necked Grebes.

Great Britain[†]	93-94	94-95	95-96	96-97	97-98	Mon	Mean	
Fal Complex	(0)	(0)	24	23	33	Dec	27	▲
Langstone Harbour	26	21	24	19	9	Nov	20	
Studland Bay	11	14	12	-	-		12	
Poole Harbour	3	16	15	7	12	Jan	11	▲

Other sites surpassing table qualifying levels in 1997-98

Swithland Reservoir	7	Sep/Oct
Tamar Complex	6	Dec

†   as the British threshold for national importance is so small, a qualifying level of 10 has been chosen to select sites for presentation in this report

CORMORANT
Phalacrocorax carbo

International threshold: 1,200
Great Britain threshold: 130
All-Ireland threshold: ?[†]

GB max: 13,658 Oct
NI max: 1,977 Oct

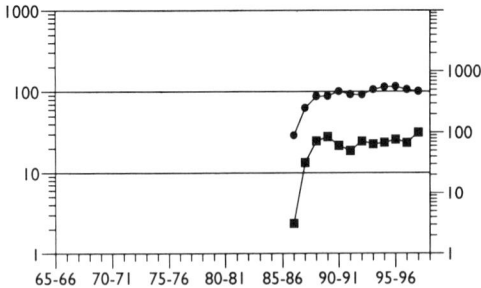

Figure 7. Annual indices for Comorant in GB (circles, left axis) and NI (squares, right axis)

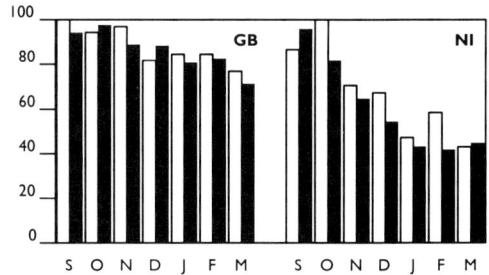

Figure 8. Monthly indices for Cormorant in GB and NI (white bars 1997-98; black bars 1992-93 to 1996-97)

The peak British count of Cormorants in 1997-98 fell to its lowest level since 1991-92, markedly below the count of 16,266 two years previously. This is partly reflected in the annual indices which returned to early 1990s levels. By contrast, numbers in Northern Ireland reached their highest level since 1989-90. Annual index values suggest that numbers in the province in 1997-98 were the highest to date by a considerable margin, but the count total only just exceeded that of 1,900 in 1995-96 and was well below the peak of 2,300 in 1989-90. This will have resulted, in part, from the higher than normal monthly index values for Northern Ireland in nearly all months except the usual peak in September.

The low national total was matched by lower than normal peak counts at many key sites: sharp decreases were recorded at Abberton Reservoir, North Norfolk Marshes and the Ouse Washes; numbers at several sites returned to low levels following recent peaks, notably the Tees Estuary

and Wraysbury Gravel Pits; whilst counts show continuing declines on the Thames Estuary, Draycote Water, Colne Estuary, and especially the Inner Moray Firth, where counts have failed to exceed the national threshold in the two most recent years. Above average counts in 1997-98 were thus notable, particularly those on the Inner Clyde, Dungeness Gravel Pits and Sonning Gravel Pits, the last two comprising some of the five sites elevated to national importance for Cormorant in 1997-98. The largest count during the year, on Loughs Neagh & Beg, continues the growth at this site in recent years. A notable feature of the table below is the large number of waterbodies in the London area, e.g. Queen Mary, Queen Elizabeth II, Queen Mother, Wraysbury and King George VI Reservoirs, and the great variability in numbers at these sites between years. Fish stocking data may shed some light on these patterns.

	93-94	94-95	95-96	96-97	97-98	Mon	Mean
Great Britain							
Morecambe Bay	895	793	1,115	977	1,099	Oct	976
Abberton Reservoir	800	722	800	900	410	Mar	726
Forth Est.	622	579	806	657	632	Jan	659
Inner Moray Firth	1,945	624	388	118	99	Oct	635
Rutland Water	800	661	655	391	385	Oct	578
Solway Estuary	682	450	639	457	510	Aug	548
Queen Mary Reservoir	407	137	387	1050	(48)	Aug	495
Clyde Est.	377	459	464	404	610	Oct	463
Alt Estuary	455	447	285	514	397	Nov	420
Tees Estuary	181	396	676	471	320	Aug	409
North Norfolk Marshes	[R]426	398	463	492	224	Sep	401
Loch Leven[R]	297	442	410	405	400	Dec	391
Poole Harbour	368	284	471	375	400	Oct	380
Dee Estuary (Eng/Wal)	431	354	460	253	374	Sep	374
Grafham Water	470	170	310	610	297	Nov	371

	93-94	94-95	95-96	96-97	97-98	Mon	Mean	
Ranworth/Cockshoot Broads	259	462	295	254	405	Nov	335	
Wash	297	394	348	337	295	Sep	334	
Blackwater Estuary	501	269	249	348	273	Feb	328	
Walthamstow Reservoirs	90	400	300	450	-		310	
Rostherne Mere	369	273	244	229	270	Mar	277	
Ouse Washes	335	244	285	391	125	Mar	276	
Chichester Gravel Pits[R]	308	222	265	346	213	Nov	271	
Hanningfield Reservoir	240	283	211	223	272	Jan	246	
Thames Estuary	399	246	205	164	150	Aug	233	
Irvine to Saltcoats	190	197	250	230	-		217	
Medway Estuary	212	212	310	154	179	Jan	213	
Draycote Water	152	347	292	130	125	Nov	209	
Chew Valley Lake	220	195	250	170	-		209	
Queen Elizabeth II Reservoir	98	118	169	380	268	Sep	207	
Colne Estuary	676	181	43	65	59	Mar	205	
Dysynni Est.	245	141	248	214	173	Aug	204	
Besthorpe/Girton Gravel Pits	79	176	255	262	236	Jun	202	
Swale Estuary	236	208	174	200	187	Jan	201	
William Girling Reservoir	132	400	200	91	180	Aug	201	
Pagham Harbour	199	158	204	246	183	Oct	198	
Queen Mother Reservoir	45	180	105	600	46	Oct	195	
Wraysbury Reservoir	69	43	241	142	479	Oct	195	
Dungeness Gravel Pits	145	161	186	144	330	Jul	193	▲
Lee Valley Gravel Pits	77	156	231	210	254	Nov	186	
South Stoke[R]	-	118	105	332	187	Dec	186	▲
Ribble Estuary	175	167	191	179	123	Feb	167	
Sonning GP	-	130	72	150	312	Jan	166	▲
Farmoor Reservoirs	183	97	225	185	120	Dec	162	
Windermere	186	167	137	142	-		158	
Wraysbury Gravel Pits	70	217	206	169	105	Nov	153	
Breydon Water & Berney Marshes	113	187	198	132	129	Aug	152	
Carmarthen Bay	237	249	60	129	77	Nov	150	
Tay Estuary	96	95	245	212	98	Sep	149	
Herne Bay[R]	150	140	-	-	-		145	
Stour Estuary	93	169	157	153	137	Aug	142	
Deeping St James	233	91	93	-	-		139	
Blithfield Reservoir	97	90	88	323	77	Dec	135	
Staines Reservoirs	226	6	194	32	216	Oct	135	▲
Attenborough Gravel Pits	116	115	121	181	137	Feb	134	▲
Rye Harbour/Pett Level	147	152	131	61	179	Sep	134	▲
Coombe Pool	119	233	44	-	-		132	

Northern Ireland[†]
	93-94	94-95	95-96	96-97	97-98	Mon	Mean
Lo. Neagh/Beg	718	631	951	927	1,184	Oct	882
Belfast Lough	483	401	536	352	514	Nov	457
Strangford Lough	259	165	180	167	164	Sep	187
Carlingford Lough	130	101	244	187	133	Dec	159
Outer Ards	100	177	147	152	158	Dec	147

Sites no longer meeting table qualifying levels

Dengie Flats

Ayr to Troon

Clwyd Estuary

Exe Estuary

Durham Coast

Southampton Water

Humber Estuary

Other sites surpassing table qualifying levels in 1997-98

King George VI Reservoir	272	Oct	Middle Tame Valley Gravel Pits	150	Jan
Dengie Flats	201	Mar	Eyebrook Reservoir	148	Nov
Marsh Lane Gravel Pits	[R]194	Dec			

† *as no all-Ireland threshold has been set for Cormorant, a qualifying level of 130 has been chosen to select sites for presentation in this report*

BITTERN
Botaurus stellaris

International threshold: ?
Great Britain threshold: ?
All-Ireland threshold: ?

GB max: 14 Dec
NI max: 0

Numbers returned to more normal levels in 1997-98, following the highs of the previous year. However, given the normal requirement to wait patiently for a considerable time in order to see this secretive species, it is perhaps surprising that any Bitterns are seen at all during WeBS counts. The larger numbers in mid winter presumably reflect both the arrival of birds from the continent and also their greater visibility during cold weather.

Birds were noted at 20 sites in total, all in England except for one each in the Channel Islands and Wales. Most, as usual, were in the south, but a handful of records from the north and northwest presumably relate to wandering birds from the native population there.

Sites with two or more birds in 1997-98

Leighton Moss	7	Dec	Minsmere	2	Mar
Walland Marsh	5	Oct	Middle Yare Valley	2	Jan
Middle Thame Valley GP	3	Feb	Rostherne Mere	2	Nov
Fen Drayton Gravel Pit	2	Jan	Marton Mere	2	Feb

LITTLE BITTERN
Ixobrychus minutus

Vagrant
Native range: Europe, Africa and S Asia

One was seen at Marton Mere in June.

LITTLE EGRET
Egretta garzetta

International threshold: 1,250
Great Britain threshold: ?[†]
All-Ireland threshold: ?[†]

GB max: 430 Aug
NI max: 0

Following the peak of 733 birds in 1995-96, the national total has declined in both subsequent years. Counts at key sites in 1997-98, however, were generally similar to peak counts in recent years and, with records from 68 sites in total, there is no reason to suspect that the increase in recent years will not continue. This seems all the more likely, given the successful breeding of Little Egrets at two sites in Britain in 1997, for the second successive year (Lock & Cook 1998): on Brownsea Island in Poole Harbour, one pair raised three young in 1996 and five pairs reared 12 young in 1997; and at the second site, in southwest Britain, one pair raised two young in 1996, and two pairs raised two young in 1997.

Most birds arrive in late summer as part of the post-breeding dispersal from the continent, numbers falling slightly as the winter progresses, perhaps as birds disperse to smaller sites from the key arrival points on the south coast.

Roost counts at the key sites normally reveal markedly higher numbers than Core Counts, e.g. a peak of 137 birds at Chichester Harbour in August (Holloway 1998), whilst a pilot survey of key roosts in autumn 1997 estimated around 750 birds (WeBS unpubl. data). A full national survey, scheduled for 1999-2000, will provide a benchmark against which to measure, as most anticipate, the growth of this population in the UK.

	93-94	94-95	95-96	96-97	97-98	Mon	Mean
Great Britain[†]							
Longueville Marsh	(0)	(90)	(82)	130	(98)	Nov	130
Chichester Harbour	(44)	55	99	74	90	Aug	80
Tamar Complex	(48)	45	83	69	42	Aug	60
Poole Harbour	24	42	58	57	60	Sep	48

	93-94	94-95	95-96	96-97	97-98	Mon	Mean
Camel Estuary	(2)	29	49	46	(33)	Sep	41
Kingsbridge Estuary	27	23	48	47	45	Oct	38
North West Solent	9	16	86	16	(3)	Jan	32
Exe Estuary	(7)	11	38	34	37	Oct	30
Langstone Harbour	(10)	(14)	36	32	19	Nov	29
Fowey Estuary	13	14	30	35	27	Sep	24
Pagham Harbour	(6)	20	19	29	27	Oct	24
Newtown Estuary	6	16	(34)	21	34	Aug	22
Fal Complex	7	20	16	24	21	Aug	18
Taw/Torridge Estuary	(11)	9	22	23	19	Mar	18
Medway Estuary	(1)	(0)	(30)	(17)	8	Nov	18
Guernsey Shore	(10)	13	13	18	-		15
Burry Inlet	(7)	10	23	9	14	Oct	14
Cleddau Estuary	7	(11)	9	14	21	Dec	13
Erme Estuary	9	8	17	13	13	Aug	12
Portsmouth Harbour	(0)	(0)	10	(0)	14	Mar	12
Fleet/Wey	(1)	1	8	18	(13)	Feb	10
Beaulieu Estuary	2	5	14	21	9	Oct	10

Other sites surpassing table qualifying levels in 1997-98

Fowey Estuary	27	Sep
Teign Estuary	13	Sep
Avon Estuary	10	Aug

† as no British threshold for national importance has been set, a qualifying level of 10 has been chosen to select sites for presentation in this report

---

GREAT WHITE EGRET
Vagrant

Ardea alba

Native range: S Europe, Africa, Asia, North and C America

Singles were reported from River Avon: Britford in April, on the North Norfolk Marshes in January and March, and from Earls Barton Gravel Pits, also in January.

GREY HERON

Ardea cinerea

International threshold:	4,500
Great Britain threshold:	?†
All-Ireland threshold:	?†

GB max:	3,684	Oct
NI max:	334	Sep

With the exception of the 4,000 birds recorded in 1995-96 and the slightly lower numbers in the early 1990s, when Grey Heron was first included in WeBS, numbers in both Great Britain and Northern Ireland have remained remarkably consistent between years.

This consistency is reflected in counts at individual sites also. The most notable departures from long-term averages were recorded on the Ribble Estuary, where counts exceeded 90 for the second year in succession, and the Inner Clyde.

Given totals of over 10,000 nests in the early 1990s (Marquiss 1993), and the continuing increase in number recorded by the BTO's Heronry Census since that time (*BTO News* 216/217), it seems that the WeBS peak count represents at most one tenth of the post-breeding population in Great Britain.

	93-94	94-95	95-96	96-97	97-98	Mon	Mean
Great Britain							
Walthamstow Reservoir	100	310	200	300	-	Aug	228
Somerset Levels	99	142	100	115	119	Mar	115
Deeping St James GP	100	110	-	-	-		105
Thames Estuary	70	(95)	119	98	84	Sep	93
Taw/Torridge Estuary	(64)	68	78	125	94	Jul	91

	93-94	94-95	95-96	96-97	97-98	Mon	Mean	
Tamar Complex	(91)	114	87	64	75	Sep	86	
Morecambe Bay	45	72	87	70	88	Oct	72	
Coombe Pool	120	60	31	-	-		70	
Dee Estuary (Eng/Wales)	86	42	73	58	76	Sep	67	
Wash	63	84	55	(35)	(45)	Sep	67	
Severn Estuary	47	41	121	54	59	Feb	64	
Montrose Basin	44	86	(74)	71	42	Aug	63	
Ribble Estuary	39	42	40	99	95	Oct	63	
R. Avon: Britford	50	(68)	56	70	65	Jan	62	
Ouse Washes	46	66	75	63	57	Feb	61	
Burry Inlet	(33)	67	57	50	64	Sep	59	
Clyde Est.	(26)	(58)	40	46	86	Oct	58	▲
Tees Estuary	71	70	(43)	38	38	Jul	54	
Poole Harbour	62	57	55	34	47	Oct	51	
Colne Valley Gravel Pits	58	136	18	12	27	Sep	50	▲
Northern Ireland								
Lo. Neagh/Beg	200	123	207	198	217	Aug	189	
Strangford Lough	73	69	87	79	87	Oct	79	

Sites no longer meeting table qualifying levels
Tring Reservoir
Avon Valley (Mid)

Other sites suprassing table qualifying levels in 1997-98

Kentra Moss/Lower Loch Shiel	69	Jan	Aldford Brook & Eaton Park	50	Jan
Timsbury Gravel Pits	65	Dec	Alde Complex	50	Oct
Durham Coast	58	Sep			

† as no British threshold for national importance has been set, a qualifying level of 50 has been chosen to select sites for presentation in this report

---

WHITE STORK
Ciconia ciconia

Vagrant and escape
Native range: Europe, Africa and Asia

Up to three known escapes were seen at Harewood Park throughout the year.

SPOONBILL
Platalea leucorodia

Scarce

Birds were reported from nine sites, mostly in the south, southwest and northwest. Many were long-staying birds, and several involved multiple sightings.

Sites with two or more birds in 1997-98

Taw/Torridge	5	Dec	Ribble Estuary	2	May-Aug
Dee Estuary (Eng/Wales)	4	Jan/Feb	Tamar Estuary	2	Dec/Feb

GREATER FLAMINGO
Phoenicopterus ruber

Escape
Native range: S Europe, Africa and Central America

The long-staying bird remained on Thames Estuary throughout the year and it or another paid a brief visit to Livermere in November.

CHILEAN FLAMINGO
Phoenicopterus chilensis

Escape
Native range: South America

A single bird was seen on the Thames Estuary's Cliffe Pools twice during the summer, on both occasions forming a mixed flock with a Greater Flamingo!

LESSER FLAMINGO
Phoenicopterus minor

Escape
Native range: Africa and S Asia

The regular bird was seen in most winter months on the Mersey Estuary.

FULVOUS WHISTLING DUCK
Dendrocygna bicolor

Escape
Native range: C & S America, Africa and S Asia

Two were at Thrapston Gravel Pit in September and a single was seen in Poole Harbour in March.

MUTE SWAN
Cygnus olor

International threshold: **2,400**
Great Britain threshold: **260**
All-Ireland threshold: **55**

GB max: 18,170 Nov
NI max: 2,133 Nov

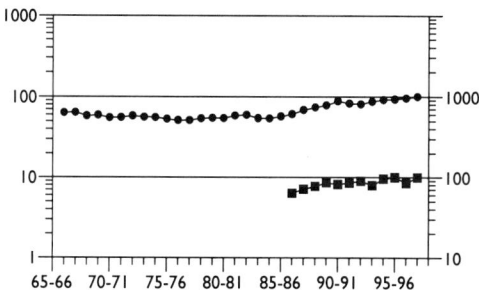

Figure 9. Annual indices for Mute Swan in GB (circles, left axis) and squares (right axis)

The rise in recorded totals in Great Britain continued in 1997-98, with numbers surpassing 18,000 for the first time. Annual indices were correspondingly at their highest level and are 75% higher than when the ban on the sale of lead shot for fishing was introduced in 1986. In Northern Ireland, peak counts were the lowest for three years, although annual index values, which equalled their highest ever figure, suggest that this may result from a reduction in site coverage rather than a true fall in numbers.

Loch of Harray rejoined the list of nationally important sites following an increase in 1997-98, though numbers remain much lower than in the late 1980s and early 1990s, when up to 1,200 birds exploited a flush of Canadian pondweed *Elodea canadensis* at the site (Meek 1993). Rutland Water recorded its highest ever count of Mute Swans; numbers on the site have grown steadily since it was created in the mid 1970s. In Northern Ireland, Loughs Neagh & Beg hold a large percentage of the population in all months of the year.

Great Britain	93-94	94-95	95-96	96-97	97-98	Mon	Mean
Fleet/Wey	1,196	1,227	1,151	1,185	1,313	Nov	1,214
Somerset Levels	511	687	608	731	734	Jan	654
Tweed Estuary[R]	720	593	450	664	544	Jul	594
Ouse Washes	923	726	427	364	432	Jan	574
Abberton Reservoir	572	624	538	480	428	Aug	528
Avon Valley (Mid)	327	438	476	368	350	Jan	392
Rutland Water	342	280	295	396	485	Nov	360
Montrose Basin	291	297	299	356	315	Jul	312
Morecambe Bay	250	330	285	281	237	Feb	277
Loch of Harray	275	211	219	249	413	Nov	273 ▲

Northern Ireland

Lo. Neagh/Beg	1,170	1,683	2,179	1,844	1,612	Nov	1,698	
Upper Lough Erne	413	456	456	590	377	Feb	458	
Strangford Lough	213	133	98	83	96	Sep	125	
Castlecaldwell Refuge Area	-	-	-	-	116	Dec	116	▲
Lough Foyle	80	102	104	130	110	Dec	105	
Upper Quoile	38	114	73	104	116	Oct	89	▲
Broad Water Canal	175	26	-	78	66	Nov	86	
Dundrum Bay	145	80	59	67	76	Dec	85	

Sites no longer meeting table qualifying levels
Loch of Skene

Internationally or nationally important sites not counted in last five years
Ballyroney Lake

Other sites surpassing table qualifying levels in 1997-98

Tring Reservoirs	329	Nov	Severn Estuary	302	Feb
Stodmarsh	320	Aug	Fen Drayton Gravel Pit	264	Jul
Stour Estuary	307	Dec	Lough Aghery	67	Oct

BLACK SWAN
Cygnus atratus

Escape
Native range: Australia

GB max: 31   Dec
NI max: 0

Although the number of sites with this species fell from 44 in the previous winter to 37 in 1997-98, summed site maxima increased from 62 to 67. Birds were distributed widely, recorded at 20 sites in England, five in Scotland and two in Wales. This species has yet to be recorded by WeBS in Northern Ireland.

Sites with three or more birds in 1997-98

Woburn Park Lakes	8	Nov	Avon Valley (Mid)	3	Sep/Mar
Deene Lake	6	Dec	Lindisfarne	3	Mar
Poole Harbour	4	Dec			

BEWICK'S SWAN
Cygnus columbianus

International threshold:	170
Great Britain threshold:	70
All-Ireland threshold:	25*

GB max: 5,441   Dec
NI max: 133   Jan

* 50 is normally used as a minimum threshold

% young: 4.1-6.7   brood size: n/a

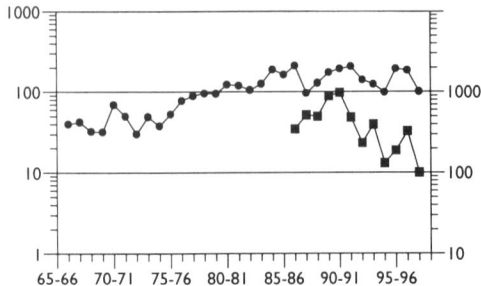

Figure 10. Annual indices for Bewick's Swan in GB (circles, left axis) and NI (squares, right axis)

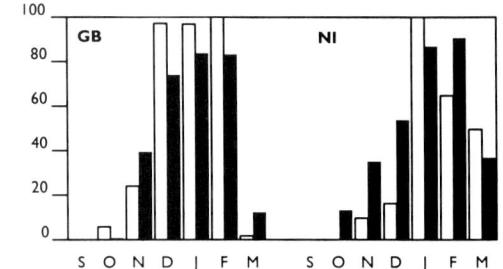

Figure 11. Monthly indices for Bewick's Swan in GB and NI (white bars 1997-98; black bars 1992-93 to 1996-97)

Following high counts in the previous two winters, 1997-98 saw numbers in Great Britain fall to their lowest levels for ten years. However, no count of the roosting birds on the Ouse Washes was possible in January. Adding December or February counts at this site to recorded January totals brings the peak national total to between 6,600 and 8,000 birds, around average for recent winters. Annual indices, however, also suggest a considerable fall since 1996-97 (-46%). Mild weather on the continent and exceptionally low productivity, with only 4.1% young birds in flocks at WWT Slimbridge, 6.7% at WWT Martin Mere and 5.2 at WWT Welney (WWT, unpubl. data) will have contributed to the decline.

In Northern Ireland, numbers were well below average in all months, with the peak count being the lowest since co-ordinated wildfowl counts commenced. Annual indices, whilst fluctuating, have shown a steady decline since the late 1980s, and it seems increasingly apparent that the relative importance of Northern Ireland has fallen in recent years. Undoubtedly the run of relatively mild winters will have been a major influence.

Declines in numbers of Bewick's at individual WWT Centres in recent years contrasts with increasing Whooper Swan numbers (Rees & Bowler 1997), although there is, as yet, no direct evidence to suggest that the two are linked.

Examination of site use by marked individuals suggests that the site fidelity exhibited during the winter also applies during migration (Rees & Bacon 1996). Paired birds, particularly those with families, showed the greatest tendency to use the same sites in successive years, presumably constrained by the greater food requirements of developing young or of females in preparation for the breeding season; single birds or young used a greater number of sites. As a result of this study, it might be speculated that declines in numbers at more northerly sites, including those in Northern Ireland, represent a decline in a sub-group of 'northern' birds or a tendency to winter further east.

The small number of juveniles and mild winter will have contributed to low counts at many sites, although the size of the decrease at some was particularly large, e.g. Somerset Levels (70% below average), Loughs Neagh & Beg (-67%), St Benets Levels (-44%), Breydon Water & Berney Marshes (-43%). The record count on the Nene Washes was thus all the more impressive, as was the addition of two haunts to the list of nationally important sites in Great Britain.

	93-94	94-95	95-96	96-97	97-98	Mon	Mean	
International								
Ouse Washes	4,172	3,920	4,830	[1]4,977	[1]4,257	Dec	4,431	
Nene Washes	1,922	1,913	1,025	863	2,585	Jan	1,662	
Martin Mere/Ribble Est.	[1]582	[1]548	[1]350	[1]669	368	Dec	503	
Breydon Water & Berney Marshes	331	209	752	476	231	Feb	400	
Severn Estuary	[1]313	253	[1]370	555	[1]393	Feb	377	
Walland Marsh	[2]288	-	[2]327	324	306	Jan	315	
St Benet's Levels	179	404	391	286	161	Feb	284	
Lo. Neagh/Beg	703	90	80	117	77	Jan	213	
Somerset Levels[R]	195	119	345	285	68	Dec	202	
Great Britain								
Avon Valley (Mid)	90	81	118	137	91	Feb	103	
Walmore Common	127	75	106	135	68	Jan	102	
Alde Complex	18	18	178	52	165	Dec	86	▲
Arun Valley	59	68	133	68	98	Feb	85	
Lower Derwent Valley	35	74	30	139	81	Feb	72	▲
Northern Ireland								
Lough Foyle	92	37	94	90	14	Jan	65	
Canary Road	59	-	43	-	26	Jan	43	▼
R. Lagan: Flatfield	84	18	32	17	38	Nov	38	
Strangford Lough	133	0	0	10	2	Jan	29	

Other sites surpassing table qualifying levels in 1997-98
Dee Estuary (Eng/Wal)   79   Jan

1   *from WWT annual swan reports (e.g. Bowler et al. 1994)*
2   *D. Walker (in litt.)*

TRUMPETER SWAN
Cygnus buccinator

Escape
Native range: North America

The three regular birds were recorded at Tansor Gravel Pits during most winter months. These derive from a dozen which escaped from Apethorpe Hall, Northamptonshire, in late 1989. Three pinioned adults settled at Tansor, although one died and just two remained at the end of 1996-97. Three birds were again seen, however, when a juvenile was observed in September 1997, the first wild breeding of this introduced species in Great Britain (Stroud 1998).

This highlights the problems posed by naturalised introductions, even of pinioned birds, and the need for more effective management of wildfowl collections.

WHOOPER SWAN
Cygnus cygnus

International threshold:	160
Great Britain threshold:	55
All-Ireland threshold:	100

GB max:	3,781	Feb
NI max:	2,127	Feb

% young: 13.8-16.5   brood size: n/a

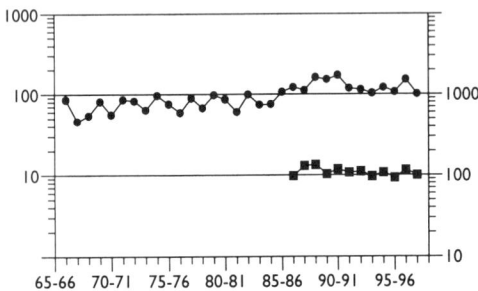

Figure 12. Annual indices for Whooper Swan in GB (circles, left axis) and NI (squares, right axis)

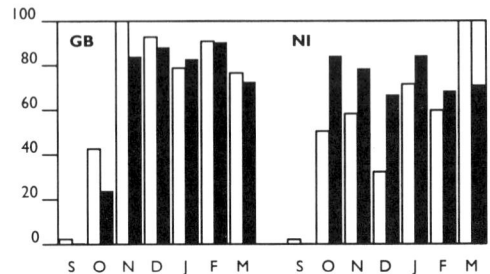

Figure 13. Monthly indices for Whooper Swan in GB and NI (white bars 1997-98; black bars 1992-93 to 1996-97)

Because of their widespread distribution across northern Britain, WeBS records only 70-75% of Whooper Swan numbers in Great Britain, compared with a much higher proportion of Bewick's, found mainly in southeast England. The peak in 1997-98 was around average for recent years although, as for Bewick's Swan, the absence of data for the Ouse Washes in January significantly affected national totals in this month. Annual indices revealed a 34% fall in numbers, though such a drop is not unprecedented. Whilst a relatively poor breeding season in 1997, with only 13.8-16.5% young at WWT centres (WWT, unpubl. data), will have contributed to this, the decline is particularly dramatic for a population which winters wholly within Britain and Ireland. Mild and, especially, wet weather will have meant a larger proportion of birds used non WeBS sites, and perhaps enabled more birds to remain further north, where they are also less likely to have been detected by the WeBS network. Despite this drop, monthly indices (which, because they use data only from sites counted in all months from September to March,

exclude the Ouse Washes in 1997-98) show a similar pattern to previous years.

Numbers in Northern Ireland are relatively well monitored, with almost all of the key sites counted on a regular basis. Peak totals for 1997-98 were around 30% below the exceptionally high counts of the previous winter although, because of the large scale movements of birds through Northern Ireland, monthly counts, and hence yearly peaks, can fluctuate quite widely. Despite this variation, annual indices for Whooper Swans are more stable than for many other wildfowl species in Northern Ireland, varying by approximately ±15% over the 12 years for which data are available.

The species' northerly distribution is amply reflected below, including only four sites outside Scotland and Northern Ireland. Numbers on two, the Ouse Washes and Martin Mere/Ribble Estuary, reached an all time high, having doubled at both sites over the last ten years or so. A detailed study at Black Cart Water collated additional data which shows this site to be of international importance (Rees & White 1998).

	93-94	94-95	95-96	96-97	97-98	Mon	Mean	
International								
Ouse Washes	[1]986	[1]1,142	[1]1,288	[1]1,211	1,299	Feb	1,185	
Lo. Neagh/Beg	740	1,102	906	1,169	1,113	Mar	1,006	
Upper Lough Erne	721	756	980	1,094	799	Feb	870	
Lough Foyle	569	596	1,521	671	566	Mar	785	
Martin Mere/Ribble Estuary	[1]650	[1]738	[1]740	[1]827	[1]1,041	Dec	799	
R. Foyle: Grange	297	-	266	380	150	Jan	273	
Solway Estuary	[1]175	[1]176	220	350	[1]221	Oct	228	
Loch of Strathbeg	302	75	221	158	310	Nov	213	
Black Cart Water[2]	262	250	149	163	180	Oct	201	
Great Britain								
Loch of Spiggie	-	84	180	-	-		132	
Loch of Wester	187	49	-	98	114	Nov	112	
Loch Insh & Spey Marshes	(0)	200	115	82	-		132	
R Clyde: Carstairs Junction	-	-	-	60	157	Mar	109	▲
Loch Eye/Cromarty Firth	72	191	89	120	52	Oct	105	▼
R Tweed: Kelso to Coldstream	137	75	88	48	138	Jan	97	
Loch Leven	99	96	94	97	98	Jan	97	
R. Nith: Keltonbank to Nutholm	-	-	-	75	115	Nov	95	▲
Rutherford	102	-	110	36	-		83	
Wigtown Bay	75	98	72	59	75	Mar	76	
Islesteps	74	-	-	-	-		74	
Merryton Ponds	72	72	67	72	74	Dec	71	
R. Tweed: Magdalenehall	-	70	-	-	-		70	
Loch of Skaill	21	104	95	78	51	Nov	70	
R. Frome: Wareham to Wool	0	137	-	-	-		69	
Milldam & Balfour Mains Pools	60	57	46	87	76	Jan	65	
Loch of Lintrathen	24	136	1	67	77	Dec	61	
Loch Heilen	17	110	-	15	99	Oct	60	▲
Lower Derwent Valley	22	73	42	96	61	Feb	59	
Loch of Skene	243	0	8	2	26	Oct	56	▼

Sites no longer meeting table qualifying levels
R. Teviot: Nisbet to Kalemouth
R. Lagan: Flatfield
East Fortune Ponds

Internationally or nationally important sites not counted in last five years
R. Teviot: Kalemouth to Roxburgh
Easterloch /Uycasound

Other sites surpassing table qualifying levels in 1997-98

R. Lagan: Flatfield	152	Nov	Lindisfarne	58	Feb
Loch of Isbister	85	Oct	Ouse/Lairo Water	58	Feb
Threave Estate	85	Feb	Loch of Harray	56	Nov
Dornoch Firth	73	Oct	Bush River: Deepstown	114	Jan
Barons Haugh	64	Jan	Strangford Lough	100	Nov

1   from WWT annual swan reports (e.g. Bowler et al. 1994) or WWT unpubl. data
2   Rees & White (1998)

SWAN GOOSE
Anser cygnoides

Escape
Native range: Eastern Asia

Although recorded at just four sites in 1997-98, compared with double that number in the previous winter, the peak count jumped from 26 to 38, with maxima of 16 birds at Etherow Country Park, 15 at Esthwaite Water and eight at Nafferton Mere.

BEAN GOOSE
Anser fabalis

International threshold (*fabalis*): **800**
Great Britain threshold: **4*†**
All-Ireland threshold: **+***

* 50 is normally used as a minimum threshold

GB max: 164 Nov
NI max: 1 Oct

Although never common, numbers of this regular visitor recorded by WeBS vary greatly, both monthly and annually, since significant numbers occur regularly at only two sites and birds may be dispersed over a wide area. Indeed, the Yare Valley flock was not detected during any of the WeBS Core Counts. There was a marked increase in the number of birds in the Slamannan area.

At the very edge of the species' distribution, only in cold winters with prolonged periods of easterly winds are numbers supplemented by arrivals from continental Europe, generally to east coast localities. With mild weather in 1997-98, counts away from the two regular wintering sites were considerably lower than in recent winters: only 12 sites recorded flocks of four or more birds (*cf.* 19 and 20 in the colder winters of 1995-96 and 1996-97 respectively). The most noteworthy count was of 28 birds at Holland Haven in December. A single bird in the Myroe area of Lough Foyle was the first WeBS record of this species in Northern Ireland.

	93-94	94-95	95-96	96-97	97-98	Mon	Mean
Great Britain[†]							
Middle Yare Valley[1]	305	310	195	224	266	Dec	260
Slamannan Plateau[2]	135	132	123	127	157	Jan	135
Heigham Holmes	365	8	103	0	0		95
North Warren/Thorpness Mere	0	13	48	36	12	Feb	22
Ouse Washes	25	1	2	34	8	Feb	14

Other sites surpassing table qualifying levels in 1997-98
Holland Haven   28   Dec

† as the British threshold for national importance is so small, a qualifying level of 10 has been chosen to select sites for presentation in this report
1 RSPB pers. comm.
2 data from Bean Goose Working Group annual reports, e.g. Smith et al. (1994), Simpson & Maciver (1997)

PINK-FOOTED GOOSE
Anser brachyrhynchus

International threshold: **2,250**
Great Britain threshold: **2,250**
All-Ireland threshold:**+**

GB max: 235,559 Oct
NI max: 30 Sep

% young: **15.5**   brood size: **2.3**

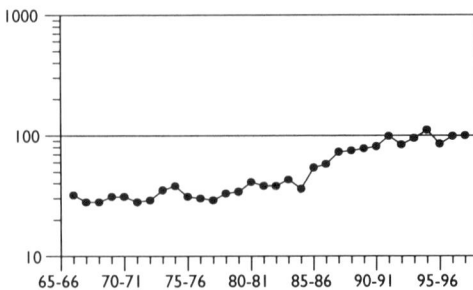

Figure 14. Annual indices for Pink-footed Goose in GB

The 38th national grey goose census in 1997 (Hearn 1998) recorded only a very modest increase in numbers of Pink-footed Geese compared with the previous winter, the population having remained at around the same level since the early 1990s. This accords reasonably with moderate breeding success in 1997.

A notable feature of 1997-98 was the relatively slow dispersal of birds from the main arrival area in east-central Scotland (Hearn 1998), perhaps a result of much spilled grain after the harvest. Counts at many sites fluctuate considerably between years, although this will be partly due to the rapid movement of birds through the region; consequently, the peak may be missed if the count is not made during the critical period which may last just two or three days. The slow departure in 1997-98 may have accounted for elevated totals at several sites in

52

the key arrival and early staging areas, e.g. at West Water Reservoir, Montrose Basin, Loch Tullybelton and Fala Flow. This corresponds with relatively low numbers in Lancashire in early winter.

By contrast, the increasing use of Norfolk by larger numbers earlier in the winter continued: the 1995-96 peak was of 54,760 birds on 12 January; in 1996-97, 55,500 were recorded on 20 December; and in 1997-98, 76,170 were present by 1 December. Birds are also moving further east in the county, with remarkable counts of 5,500 at Breydon Water & Berney Marshes and 10,000 at nearby Heigham Holmes in February (P. Allard *in litt.*).

Low counts were notable on the Solway Estuary and on the Tay/Isla Valley, whilst the roost at Glenfarg appears now to have been abandoned.

	93-94	94-95	95-96	96-97	97-98	Mon	Mean	
International								
Lo. of Strathbeg	38,970	58,150	48,500	32,000	33,556	Oct	42,235	
Dupplin Lo.	36,500	62,000	35,000	40,500	29,850	Oct	40,770	
Snettisham[1]	45,925	31,038	39,130	35,930	40,350	Dec	38,475	
West Water Rsr	40,000	26,500	31,500	25,500	38,700	Oct	32,440	
SW Lancashire[2]	27,260	31,000	28,850	41,680	28,960	Dec	31,550	
Montrose Basin	41,210	36,000	18,500	17,150	35,000	Nov	29,572	
Holkham[1]	26,760	16,000	19,230	26,000	33,700	Dec	24,338	
Slains Lo./Ythan Est.	23,880	21,400	25,000	17,400	12,200	Oct	19,976	
Lo. Leven	18,870	16,154	17,900	18,150	14,740	Oct	17,163	
Scolt Head[1]	16,860	13,150	15,635	17,900	15,890	Dec	15,887	
Solway Est.	17,470	20,202	20,523	11,546	7,770	Mar	15,502	
Hule Moss	14,100	8,100	15,200	19,400	19,675	Oct	15,295	
Cameron Rsr	27,300	14,860	11,260	3,460	11,280	Oct	13,632	
Carsebreck/Rhynd Lo.	7,120	14,500	13,500	12,000	13,560	Oct	12,136	
Aberlady Bay	26,000	5,750	11,320	4,650	6,540	Nov	10,852	
R Clyde: Carstairs Junction	-	-	-	-	(8,000)	Oct	(8,000)	▲
Wigtown Bay	3,530	5,912	7,229	7,280	5,234	Mar	5,837	
Tay Est.	(300)	1,938	6,117	8,897	3,765	Nov	5,179	
Cowgill Rsr	5,400	3,820	4,560	6,060	6,000	Nov	5,168	
Fala Flow	6,450	3,500	2,437	5,000	7,500	Oct	4,977	
Alloa Inch	-	2,300	6,700	-	-		4,500	
Gladhouse Rsr	2,500	4,550	3,290	6,200	5,000	Oct	4,308	
Forth/Teith Valley	360	7,780	-	-	-		4,070	
Morecambe Bay	2,229	687	5,503	8,671	3,000	Oct	4,018	
Lo. Tullybelton	4,100	1,800	1,395	4,658	8,000	Nov	3,991	
Lo. of Kinnordy	9,195	3,420	434	2,730	(84)	Nov	3,945	
Lo. Eye/Cromarty Fth	2,797	5,816	7,150	1,500	465	Oct	3,546	
Whitton Loch	-	3500	-	-	-		3,500	
Drummond Pond	2,550	2,250	110	7,000	3,300	Oct	3,042	
Crombie Rsr	3,000	-	-	-	-		3,000	
Tay/Isla Valley	3,820	3,202	2,785	2,911	229	Oct	2,589	
Glenfarg Rsr	3,800	9,080	0	0	0		2,576	
Loch Mahaick	600	970	600	2,700	6,465	Oct	2,267	▲

Sites no longer meeting table qualifying levels
Skinflats
Castle Loch (Lochmaben)

Other sites surpassing table qualifying levels in 1997-98

Heigham Holmes	10,000	Feb	R. Clyde: Lamington	2,600	Oct
Breydon Wtr & Berney Marshes	5,500	Feb	Loch Spynie	2,300	Nov
Alt Estuary	5,001	Dec	Loch Mullion	3,000	Oct
Holburn Moss	4,500	Feb	Loch of Lintrathen	2,800	Nov
R. Nith: Kelton to Nutholm	3,140	Feb			

1   *includes data from Paul Fisher (in litt.)*
2   *includes data from Lancashire Goose Report (e.g. Forshaw 1998)*

EUROPEAN WHITE FRONTED GOOSE
Anser albifrons albifrons

International threshold: 6,000
Great Britain threshold: 60
All-Ireland threshold: +

GB max: 5,345   Feb
NI max: 0

% young: 14.8   brood size: n/a

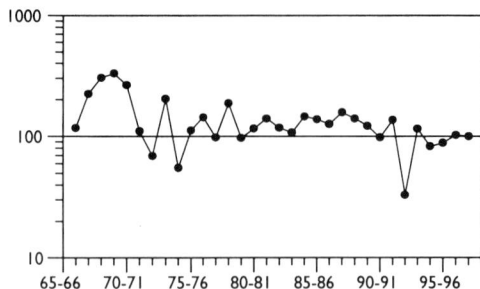

Figure 15. Annual indices for European White-fronted Goose in GB

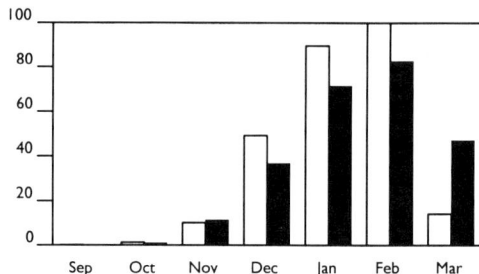

Figure 16. Monthly indices for European White-fronted Goose in GB (white bars 1997-98; black bars 1992-93 to 1996-97)

Numbers in Great Britain represent only a small proportion of the European population which had risen dramatically from 60,000 birds in the 1960s to around 10 times that number in the late 1980s (Madsen 1991). In contrast, numbers wintering in the UK have fallen steadily during this period and, though the peak in 1997-98 was typical for recent years, national counts of double this figure were not uncommon thirty years ago (Owen *et al.* 1986). Breeding success, however, was relatively poor in 1997, with only 14.8% young observed in flocks at Slimbridge.

Depending on weather conditions, birds may continue to arrive in Great Britain through to February. Monthly counts and indices show that in 1997-98 most of the wintering birds had arrived by January, and, with predominant southwesterly winds in the first two weeks of March, few remained by the March count date.

Although only the WWT Slimbridge reserve on the Severn Estuary and Elmley Marshes on the Swale regularly support more than 1,000 birds, numbers have risen sharply at Heigham Holmes in recent years. With large counts of other species at this site as a result of sympathetic ESA management, this new 'goose mecca' is likely to become one of the key UK sites.

Decreases, such as that on the Alde and the Ouse Washes, were to be expected given the contrasting weather in the last three winters. The well above average counts for the second successive year at both Minsmere Levels (+168%) and Dungeness Gravel (+118%) were thus noteworthy, whilst the continuing use of the most northerly site, Lower Derwent Valley, by large numbers is particularly remarkable.

Great Britain	93-94	94-95	95-96	96-97	97-98	Mon	Mean
Severn Estuary	3,000	2,200	2,170	2,780	2,501	Feb	2,530
Swale Estuary	1,703	1,681	2,088	1,604	1,402	Feb	1,696
Heigham Holmes	163	185	1,043	640	475	Feb	501
North Norfolk Marshes	316	248	476	491	290	Feb	364
Walland Marsh	-	-	[1]300	328	198	Jan	275
North Warren/Thorpness Mere	120	47	450	302	220	Feb	228
Dungeness Gravel Pits	174	0	8	355	320	Feb	171
Alde Complex	19	0	427	317	60	Jan	165
Middle Yare Valley	[2]265	189	180	47	107	Jan	158
Minsmere Levels	69	64	83	215	236	Jan	133
Wash	483	0	38	21	3	Jan	109
Lower Derwent Valley	7	1	244	114	152	Feb	104
Thames Estuary	103	107	59	146	69	Feb	97
Breydon Water & Berney Marshes	80	88	64	69	90	Jan	78
Ouse Washes	106	16	88	76	18	Dec	61

Internationally or nationally important sites not counted in last five years
Kessingland Levels

1   *D. Walker (in litt.)*
2   *RSPB pers. comm.*

54

GREENLAND WHITE-FRONTED GOOSE
Anser albifrons flavirostris

International threshold:	**300**
Great Britain threshold:	**140**
All-Ireland threshold:	**140**

GB max: **20,654** Nov
NI max: **111** Oct

% young: **14.7**   brood size: **3.3**

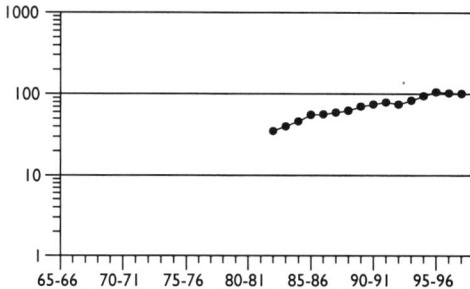

Figure 17. Annual indices for Greenland White-fronted Goose in GB

The peak count, obtained during the autumn national census by the Greenland White-fronted Goose Study (Fox & Francis 1998, AD Fox pers. comm.), represents the second consecutive, albeit small, annual decline in numbers for this population. This apparent stabilising of numbers is consistent with moderate breeding success in 1997, yet the rather low count on Islay during the census, compared with higher counts before and after, suggests that some birds may have been missed (13,400 birds were recorded on the census, *cf.* 14,900 and 15,100 either side of this; however, both of the higher totals were obtained

during counts made over two days which are therefore deemed unsatisfactory due to the possibility of double-counting or missing birds).

Despite the relatively low national total, 1997-98 maxima at all but two of the internationally and three of the nationally important sites were higher than their respective means; the count at Appin/Eriska/Benderloch was sufficient to elevate its status to internationally important. The sudden increase on Tiree in 1995-96, of around 1,000 birds, was maintained in 1997-98 also. Clachan features in the table below for the first time, although it was inadvertently omitted from the previous report.

Preliminary results of research in western Greenland suggest that the rapid expansion of the Canada Goose's breeding range into the area (Fox *et al.* 1996) may be having a detrimental effect upon White-fronted Geese (Fox & Francis 1998). Moulting flocks tended to consist of just one species, and at those where both occurred, the more aggressive and larger Canadas dominated the best feeding areas. Whilst Canada Goose numbers have continued to increase in the area, White-front numbers have declined rapidly in the last three years.

	93-94	94-95	95-96	96-97	97-98	Mon	Mean
International[†]							
Islay[1]	11,368	12,350	14,495	12,964	13,414	Dec	12,918
Rhunahaorine	1,050	1,361	1,360	1,272	1,193	Nov	1,247
Machrihanish	1,103	1,044	1,339	1,629	931	Nov	1,209
Tiree	499	512	1,387	1,455	1,464	Mar	1,063
Coll	896	1,026	962	1,047	1,052	Mar	997
Stranraer Lochs	565	565	550	535	680	Dec	579
Danna/Keills	308	381	414	333	441	Mar	375
Loch Ken	325	293	360	318	450	Mar	354
Appin/Eriska/Benderloch	323	336	376	217	318	Mar	314 ▲
Great Britain[†]							
Westfield Marshes	196	206	352	210	206	Nov	234
Lo. Lomond: Endrick Mouth	137	230	230	245	261	Nov	221
Bute	213	226	210	224	223	Mar	219
Lo. Heilen/Lo. Mey	180	196	258	199	217	Nov	210
Colonsay/Oronsay	150	185	206	169	288	Mar	197
Clachan	-	-	191	184	203	Mar	193
Jura	-	148	160	140	-		149
Dyfi Est.	160	155	147	125	110	Apr 98	139

Sites no longer meeting table qualifying levels
Linnhe Mhuirich/Lo. na Cille

Other sites surpassing table qualifying levels in 1997-98
Loch Calder   160   Jan

† counts based largely on data from GWGS reports (e.g. Fox & Francis 1998)
1 data represent SNH 'adopted' counts: whole-island counts are made on two consecutive days and the average taken, unless one count is deemed inaccurate due to operational difficulties.

LESSER WHITE-FRONTED GOOSE
Anser erythropus

Singles were recorded on the Severn Estuary in February and Alton Water in June, whilst one moved between Ogden, Holden Wood and Calf Hey Reservoirs from July to February.

GREYLAG GOOSE
Anser anser

ICELANDIC POPULATION

GB max:	79,477	Nov
NI max:	597	Dec

International threshold: **1,000**
Great Britain threshold: **1,000**
All-Ireland threshold: **40***

* 50 is normally used as a minimum threshold

% young:	**13.5**	brood size:	**2.5**

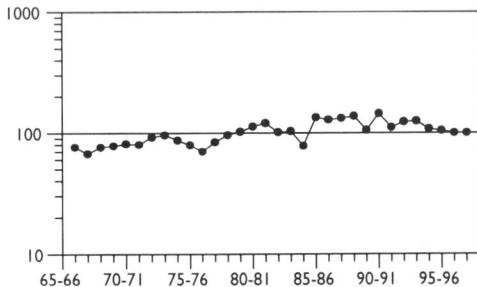

Figure 18. Annual indices for Icelandic Greylag Geese in GB

The census total in 1997-98 (Hearn 1998), although only fractionally below that of the previous winter, now becomes the lowest since 1978-79 (excluding that in 1984, known to be an undercount). This is in stark contrast to the other major goose populations wintering in the UK: both Pink-feet and Greenland White-fronts have tripled in number over the last 20 years whilst Dark-bellied Brents have increased by 50%. Although poorer than normal breeding success in 1997 will have contributed to the most recent low total, breeding success in this population has been virtually identical to the sympatric Pink-foot over the last 10 years. Using five year running means, the Greylag population has shown a sustained decline over this period, whereas Pink-feet have continued to grow steadily. The continued large bag of Greylag Geese in Iceland has prompted an education programme by the Iceland Institute of Natural History to discourage the hunting of Greylags on a voluntary basis (A. Sigfússon *in litt.*).

Peak counts in 1997-98 at a large number of sites were markedly different from their respective averages. One of the most notable was on Orkney, having risen from ninth position in the table following the 1993-94 winter, and likely to usurp Loch Eye/Cromarty Firth as the third most important site in the near future if low counts continue at the latter site. Numbers dropped sharply at Haddo House Lakes, on the Tay/Isla Valley, Holborn Moss and the Eden Estuary, and remained low at both Stranraer Lochs and the Beauly Firth. The fall in numbers at Lower Bogrotten coincides with an expansion of reeds and other successional species at the site and it may be that this roost is soon lost. There were two very marked increases, at Loch of Lintrathen and on the Dornoch Firth, suggesting a redistribution of birds in the Moray area in autumn 1997. It is notable that, following 1997-98, a further five sites are no longer of international importance.

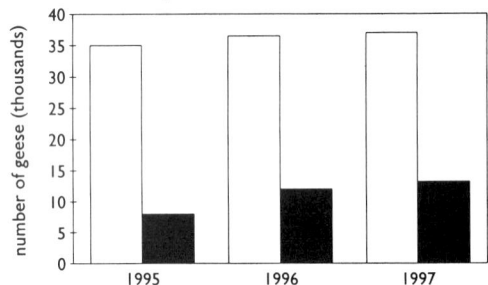

Figure 19. Numbers of Greylag (white bars) and Pink-footed Geese (white) shot in Iceland.

Sightings of darvic-ringed birds have shown a number in Ireland to belong to the Icelandic population (B. Swann pers. comm.). It is thought that perhaps 1,000 birds use sites here, but the picture is clouded by the presence of introduced birds. Special effort was made during the census in autumn 1998 to assess Greylags in Ireland in an attempt to clarify the situation.

International	93-94	94-95	95-96	96-97	97-98	Mon	Mean	
Dinnet Lo./R. Dee	27,173	33,119	36,525	26,185	24,346	Oct	29,470	
Lo. of Skene	14,000	8,500	12,300	12,876	11,200	Nov	11,775	
Lo. Eye/Cromarty Fth	14,842	11,714	8,550	4,433	5,416	Nov	8,991	
Orkney	4,112	2,702	¹9,880	9,338	13,361	Nov	7,879	
Caithness Lo.	5,433	5,563	12,376	5,378	7,200	Oct	7,190	
Lo. Spynie	5,000	7,000	5,500	5,500	3,000	Nov	5,200	
Haddo House Lo.	4,600	4,900	4,900	4,360	1,110	Nov	3,974	
Lo. of Lintrathen	4,100	1,240	2,300	960	7,200	Nov	3,160	
Bridge of Earn	-	-	3,000	-	-		3,000	
Lower Bogrotten	5,620	5,180	3,000	850	0		2,858	
Findhorn Bay	2,640	3,065	3,150	1,860	2,350	Nov	2,613	
Killimster Loch	-	-	-	2,500	-		2,500	
Drummond Pond	4,000	3,430	1,680	1,021	1,834	Nov	2,393	
Tay/Isla Valley	3,877	3,064	1,661	2,096	857	Jan	2,311	
Bute	1,500	2,370	4,280	1,797	1,200	Dec	2,229	
R. Eamont: Watersmeet								
- Pooley Bridge	2,150	-	-	-	-		2,150	
Beauly Firth	6,300	2,510	194	520	400	Oct	1,985	
Dornoch Fth	692	1,975	1,937	1,132	3,211	Oct	1,789	
Stranraer Lo.	2,500	2,500	760	-	645	Nov	1,601	
Lindisfarne	2,000	2,600	750	1,000	900	Nov	1,450	
R. Tay: Dunkeld	-	1,400	-	-	-		1,400	
Holburn Moss	1,500	2,000	2,000	1,200	200	Oct	1,380	
Strathearn	-	-	2,665	0	-		1,333	
Kilconquhar Loch	1,844	918	1,135	1,300	1,216	Feb	1,283	▲
Lo. Garten/Mallachie	1,550	1,482	1,987	587	735	Nov	1,268	
Lo. Fleet	1,500	1,300	960	1,200	843	Oct	1,161	
Eden Est.	2,020	1,071	1,520	1,070	68	Nov	1,150	
Upper Tay	534	2,030	746	971	1,333	Nov	1,123	▲
R. Spey: Boat of Balliefirth	1,115	-	-	-	-		1,115	
Corby Lo.	1,080	-	-	-	-		1,080	

Internationally or nationally important sites not counted in last five years
Fincastle Loch

Sites no longer meeting table qualifying levels
Lo. of the Lowes   Lo. Clunie
R. Tay: Scone

Other sites surpassing table qualifying levels in 1997-98

Dowlaw Dam	1,600	Nov	Threave Estate	1,089	Dec
R. Forth: West Carse Farm			Lowbank Gravel Pit	1,075	Dec
to R. Teith Confluence	1,250	Mar	R. Eamont & Eden:		
Summerston	1,150	Oct	Honeypot to Edenhall	1,023	Mar

1   *Orkney Bird Report*

NORTHWEST SCOTTISH POPULATION

International threshold: **50**
Great Britain threshold: **50**

GB max: **9,793**   **Aug**   **% young:**   n/a   **brood size:**   n/a

A full census of all known and suspected haunts of this population was undertaken in autumn 1997.   Provisional results suggest a total population size of just under 10,000 birds (WWT unpubl. data), double the previous estimate. The traditional strongholds remained as before, with the Outer Hebrides and Coll and Tiree supporting a high proportion of the total. In excess of 1,200 birds were also found in Sutherland and on the Orkney Islands, whilst small numbers were found at an additional 18 sites on the Inner Hebrides and along the west coast, particularly Wester

Ross, and at two sites in the Shetland Islands, indicating range expansion. Consequently, the population estimate and 1% threshold will be revised later this year.

International[†]

Tiree	1,206	1,526	1,451	2,475	2,417	Feb	1,815
North Uist	1,556	1,346	1,345	1,630	1,670	Aug	1,509
South Uist	880	752	1,157	1,270	1,046	Aug	1,021
Coll	-	792	707	1,016	953	Mar	687
Benbecula	136	156	264	440	595	Aug	318

† Counts based largely on data from Mitchell (et al. 1995), R MacDonald in litt. and SNH. Note that birds occasionally move between adjacent islands, particularly between South Uist and Benbecula, and between Coll and Tiree

NATURALISED POPULATION

GB max: 19,637   Oct
NI max: 1,159   Mar

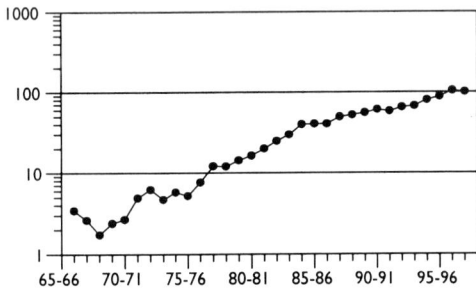

Figure 20. Annual indices for naturalised Greylag Geese in GB

Whilst the peak count of birds in Britain increased further, the annual index value dropped slightly. This is reflected in the table below, with counts at 14 sites noticeably higher than the long-term mean, around the same at seven, and noticeably lower at 12 (note that no sites are identified as having joined or fallen from the table in the most recent year, since this table was not included in the report previously). Many

Naturalised re-establishment[†]

of the increases, however, were particularly large, notably on the North Norfolk Marshes, Llyn Traffwll, Wynyard Lake and the Humber Estuary, whilst there has been an increase at Sutton/Lound Gravel Pits in each of the last four years. Notable, also, are counts in excess of 600 at three sites for which average maxima fall below 300.

In Northern Ireland, numbers were similar or slightly lower than during 1996-97, with the exception of a particularly large count in March. Although it is clear that both Iceland and naturalised birds occur in the province, separating the different populations is problematic. Nevertheless, this large peak, around one third higher than the previous maximum, perhaps indicates some sort of pre-migratory gathering of Icelandic birds. The large peak count on Lough Foyle in 1997-98 was notable, whilst counts at Loughs Neagh & Beg have also risen in each of the last four winters.

	93-94	94-95	95-96	96-97	97-98	Mon	Mean
Great Britain							
North Norfolk Marshes	1,153	1,089	1,204	1,669	2,177	Nov	1,458
Lower Derwent Valley	840	658	1,304	1,200	1,063	Feb	1,013
Bolton-on-Swale GP	1,266	841	572	955	635	Aug/Dec	854
Tophill Low Reservoirs	668	1,263	481	561	450	Dec	685
Swale Estuary	416	651	673	456	589	Feb	557
Little Paxton Gravel Pits	364	500	644	518	655	Oct	536
Alton Water	259	356	815	514	647	Oct	518
Sutton/Lound Gravel Pits	-	356	458	570	650	Nov	509
Revesby Reservoir	479	-	602	273	571	Sep	481
Baston/Langtoft Gravel Pits	905	408	270	349	320	Mar	450
Wash	160	505	511	747	314	Sep	447
Heigham Holmes	430	465	373	538	410	Oct	443
Langtoft West End Gravel Pits	300	430	550	420	490	Jan	438
Martin Mere	424	430	458	420	419	Nov	430
Tattershall Pits	245	350	-	700	340	Jun	409
Castle Howard Lake	406	-	-	-	-		406

	93-94	94-95	95-96	96-97	97-98	Mon	Mean
Llyn Traffwll	245	264	466	349	646	Sep	394
Bough Beech Reservoir	464	64	650	-	-		393
Earls Barton Gravel Pits	250	392	486	542	284	Jul	391
Thrapston Gravel Pit	340	343	305	417	520	Oct	385
Dungeness Gravel Pits	256	349	446	381	473	Aug	381
Ferry Meadows (Nene Park)	369	-	-	-	-		369
Hamford Water	182	559	576	358	168	Sep	369
Ouse Washes	145	324	305	521	453	Dec	350
Wynyard Lake	14	530	241	224	710	Sep	344
Morecambe Bay	311	342	287	370	401	Feb	342
Humber Estuary	109	160	126	459	854	Sep	342
Livermere	-	400	335	330	300	Aug/Feb	341
Hornsea Mere	428	488	-	-	92	Mar	336
Linford Gravel Pits	412	168	365	409	301	Sep	331
Thames Estuary	321	499	273	293	252	Feb	328
Medway Estuary	302	470	361	290	203	Feb	325
St Benet's Levels	350	330	336	268	268	Sep/Oct	310
Clifford Hill Gravel Pits	485	352	216	378	92	Dec	305
Northern Ireland							
Strangford Lough	461	591	173	351	379	Dec	391
Lo. Neagh/Beg	70	243	347	448	510	Mar	324
Temple Water	250	378	158	15	-		200
Lough Foyle	184	22	43	88	383	Mar	144
Belfast Lough	20	41	77	86	86	Dec	62 ▲

Other sites surpassing table qualifying levels in 1997-98

Nosterfield Gravel Quarry	771	Feb	Bardney Pits	350	Jan
Kirkby-on-Bain Gravel Pits	627	Nov	Grimsthorpe Lake	334	Feb
Emberton Gravel Pits	602	Sep	Deene Lake	328	Aug
Middle Yare Valley	481	Oct	Rutland Water	323	Sep
Derwent Reservoir	442	Mar	Fillingham Lake	322	Sep
Aldford Brook & Eaton Park	430	Nov	Severn Estuary	320	Sep
Didlington	392	Aug	Orwell Estuary	313	Jan
Lough Foyle	383	Mar	Ardleigh Reservoir	312	Aug
Leighton/Roundhill Reservoirs	370	Sep	Alaw Reservoir	312	Aug
Eccup Reservoir	368	Dec			

†   *as site designation does not occur and the 1% criterion is not applied, qualifying levels of 300 and 50 have been chosen to select sites in Great Britain and Northern Ireland, respectively, for presentation in this report*

BAR-HEADED GOOSE
Anser indicus

Escape
Native range: Southern Asia

This species was recorded at 35 sites in 1997-98, with a peak count of 16 in March, both figures slightly lower than during the previous winter. Summing site maxima suggests as many as 56 birds, though comparatively few sites held birds for extended periods and it is likely that this figure includes a degree of double-counting.

Sites with three or more birds in 1997-98

Spade Oak Gravel Pit	5	Dec	Westport Lake	3	several
Pennington Flash	5	Mar	Chatsworth Park Lakes	3	Sep
Emberton Gravel Pits	3	Sep			

SNOW GOOSE
Anser caerulescens

Escape and vagrant
Native range: North America

GB max: **70**   **Oct**
NI max: **0**

There was a marked fall in the number of records in 1997-98, with birds noted at only 32 sites (*cf.* 42 in 1996-97 and a peak of 99 birds). There was a corresponding fall in numbers at most of the key sites. Summing site maxima suggests as many as 146 birds, probably all escapes.

Sites with 10 or more birds in 1997-98

Eversley Cross/Yateley Gravel Pits	26	Jan	Blenheim Park Lake	14	several
Stratfield Saye	21	Oct	Emberton Gravel Pits	12	Mar
Bramshill Park Lake	19	Jun/Jul			

ROSS'S GOOSE
Anser rossii

Escape
Native range: North America

Birds were noted at seven sites in 1997-98, although almost certainly the same birds were involved in two cases. All records involved singles except for a count of three at St Mary's Island in August.

EMPEROR GOOSE
Anser canagicus

Escape
Native range: Alaska and NE Siberia

Occurring on widely separated sites, summing site maxima suggests 10 birds in Britain in 1997-98. All counts were of singles except for three birds at both Ramsbury Lake and Eccup Reservoir.

CANADA GOOSE
Branta canadensis

Naturalised introduction[†]
Native range: North America

GB max: **43,225**   **Nov**
NI max: **456**   **Feb**

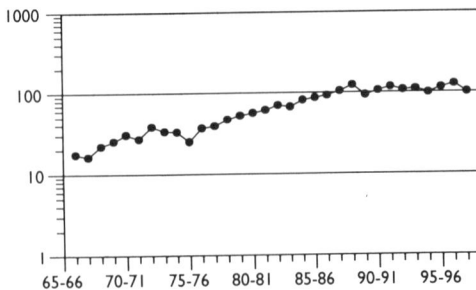

Figure 21. Annual indices for Canada Geese in GB

Following many years of rapid growth up to the late 1980s, annual indices suggest that the Canada Goose population has stabilised over the past ten years. The peak national total, however, surpassed 43,000 birds for the second consecutive year, some 35% higher than the figure recorded ten years previously. A partial explanation may be that population growth is occurring primarily on sites only recently covered by WeBS, e.g. newly created gravel pits, which cannot be included in the index calculations until there is a sufficiently long run of data. In Northern Ireland, numbers returned to normal levels after the high counts in 1996-97.

Rutland Water is the only site with counts in excess of 1,000 birds in each of the past five years, with numbers apparently continuing to grow. The largest count, however, came from the Arun Valley where numbers were almost double those of the previous winter. Numbers on the Dyfi Estuary rose sharply, one of two new sites in the table below. That six sites no longer hold averages of 600 or more birds following 1997-98 counts is perhaps surprising in view of the national increase. To what extent licensed control measures have influenced counts at these sites is uncertain.

	93-94	94-95	95-96	96-97	97-98	Mon	Mean	
Great Britain								
Rutland Water	1,025	1,137	1,282	1,266	1,395	Oct	1,221	
Arun Valley	1,012	775	868	796	1,490	Nov	988	
Fairburn Ings	580	922	953	1,030	1,340	Jun	965	
Walthamstow Reservoirs	516	1,072	1,062	1,030	816	Jun	899	
Lower Derwent Valley	919	732	831	841	1,170	Nov	899	
Chew Valley Lake	805	625	855	740	-		756	
Kedleston Park Lake	1,100	-	900	360	650	Nov	753	
Holme Pierrepont GP	552	717	648	1,001	715	Aug	727	
Stour Estuary	593	551	1,261	492	608	Feb	701	
Dee Estuary (Eng/Wal)	553	421	645	877	875	Jan	674	▲
Dyfi Estuary	453	520	681	682	1,020	Dec	671	▲
Blithfield Reservoir	484	688	342	916	850	Sep	656	
Middle Tame Valley GP	666	649	769	441	750	Sep	655	
Bewl Water	420	833	820	600	548	Nov/Dec	644	
Northern Ireland								
Upper Lough Erne	211	242	194	451	168	Feb	253	
Strangford Lough	153	297	185	257	183	Feb	215	
Drumgay Lough	205	0	265	236	95	Feb	160	
Woodford River	128	-	-	-	-		128	
Castlecaldwell Refuge Area	-	-	-	-	10	Feb/Mar	10	

Sites no longer meeting table qualifying levels

Abberton Reservoir	Alde Estuary
Stratfield Saye	Dorchester Gravel Pits
Livermere	Eversley Cross/Yateley Gravel Pits

Other sites surpassing table qualifying levels in 1997-98

Tundry Pond	840	Oct	Barton Pits	697	Aug
Arlington Reservoir	825	Aug	Kings Bromley Gravel Pits	641	Aug
Fleet Pond	800	Sep	Tring Reservoirs	618	Nov
Croxall Pits	793	Sep	Abberton Reservoir	608	Aug

† *as site designation does not occur and the 1% criterion is not applied, qualifying levels of 600 and 200 have been chosen to select sites in Great Britain and Northern Ireland, respectively, for presentation in this report*

BARNACLE GOOSE
Branta leucopsis

GREENLAND POPULATION

International threshold:	**320**
Great Britain threshold:	**270**
All-Ireland threshold:	**75**

GB max:	35,123	Nov			
NI max:	0		**% young:**	**6.1**	**brood size:**	**1.95**

The peak British count of just over 35,000 birds comprises counts from the co-ordinated census of main locations in Argyll plus that from South Walls, Orkney, representing the majority of accessible sites used by this population. This figure exceeds the equivalent figure published in the 1996-97 report simply because only counts from Islay were used as a maximum national count previously. Peak counts at the two key sites, as shown in the table below, were in fact lower in 1997-98 than in recent years, with that on Islay representing the first fall in numbers since

1990-91. The very low breeding success in 1997 will have contributed to this decline.

The remainder of the population is thinly spread across a relatively large number of small Hebridean islands. These are censused periodically, usually once every five years, along with sites in the Republic of Ireland, as part of a complete international census. An aerial survey of Barnacle Geese on some of these sites was undertaken in spring 1997 as part of an assessment of potential Special Protection Area for Scottish Natural Heritage (Mitchell *et al.* 1997).

The largest counts were made at the Sound of Harris (1,351), Monach Isles (760), North Uist machair and lochs (600) and the Treshnish Isles (270). The abandonment of these smaller sites, resulting in the population becoming more concentrated at the main haunts, particularly Islay, has been cause for concern in recent decades (e.g. Delany & Ogilvie 1994), so it is encouraging to note that numbers recorded at many of these smaller sites in 1997 were higher than during the previous survey. Using estimates for areas not visited suggests a Scottish total of 37,000 birds in 1996-97, around 22% higher than during the previous full census in 1994.

	93-94	94-95	95-96	96-97	97-98	Mon	Mean
International							
Islay[1]	27,791	28,298	31,099	35,013	32,812	Jan	31,003
Sound of Harris[2]	(474)	1,664	-	1,351	-		1,508
Tiree[3]	684	1,145	1,465	1,479	1,158	Feb	1,186
South Walls (Orkney)[4]	890	1,208	1,138	1,170	1,180	Feb	1,117
Coll[3]	764	991	682	861	715	Jan	803
North Sutherland[2]	630	465	-	792	-		629
North Uist[2]	-	543	-	600	-		572
Monach Isles[2]	485	374	-	760	-		540
Colonsay/Oronsay[3]	500	500	309	429	436	Nov	435
Danna/Keills/Eilan Mor[3]	450	400	120	341	469	Mar	356

1   SNH in litt.
2   Delany & Ogilvie (1994), SNH data and Mitchell et al. (1997)
3   data from the Argyll Bird Report and SNH
4   Orkney Bird Report and J. Plowman in litt.

SVALBARD POPULATION

International threshold: 120
Great Britain threshold: 120

GB max: 23,856   Mar

% young: 16.8   brood size: 1.76

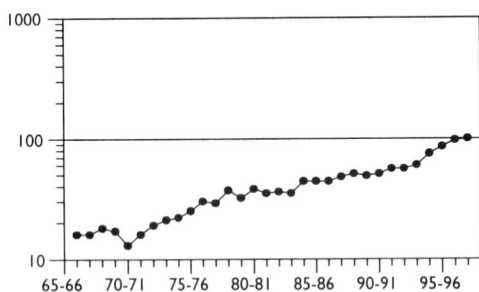

Figure 22. Annual indices for Svalbard Barnacle Goose in GB

Although the peak in 1997-98 was fractionally below that of the previous winter, it confirms the latest jump in population size to around 24,000 birds. The exact pattern of this increase remains unclear, although the simple mathematics of explaining population growth as a function of recruitment and mortality dictate that the rise cannot have been as dramatic as the table below suggests; some birds must have been overlooked in earlier censuses.

Svalbard Barnacle Geese have been the subject of one of the most intensive studies of a population anywhere in the world, with a high proportion of birds marked individually, and a considerable run of data for both counts and life history parameters. It is worth noting that, even with this considerable effort, the natural world continues to confound our best predictions.

	93-94	94-95	95-96	96-97	97-98	Mon	Mean
International							
Solway Estuary[1]	13,700	17,900	17,450	24,360	23,754	Mar	19,433
Lo. of Strathbeg	41	150	533	165	353	Oct	248

1   WWT data

NATURALISED POPULATION

Naturalised establishment[†]

GB max: 619 Oct
NI max: 148 Sep

The peak count of naturalised birds is somewhat below the 925 recorded during the 1991 national survey of introduced geese in Great Britain (Delany 1994), but that survey included 421 free-flying birds in private collections, e.g. at WWT Slimbridge on the Severn Estuary. Given that these sites were not covered in 1997-98 or, as

shown by the table, numbers there are much reduced, the WeBS count of over 600 birds points to an increase during the last decade. A high proportion of this total are located at just a handful of sites, and there has been a marked increase at most in recent years.

	93-94	94-95	95-96	96-97	97-98	Mon	Mean
Great Britain							
Hornsea Mere	169	185	(0)	-	-		177
Eversley Cross/Yateley GP	62	100	218	311	184	Nov	175
Stratfield Saye	75	84	34	141	142	Oct	95
Severn Estuary	27	129	96	46	33	Dec	66
Northern Ireland							
Strangford Lough	83	97	89	129	148	Sep	109

Sites surpassing table qualifying levels in 1997-98

Bramshill Park Lake	132	Sep	Middle Yare Valley	56	Oct
Benacre Broad	80	Nov			

† as site designation does not occur and the 1% criterion is not applied, a qualifying level of 50 has been chosen to select sites for presentation in this report

---

DARK-BELLIED BRENT GOOSE
Branta bernicla bernicla

International threshold: 3,000
Great Britain threshold: 1,000
All-Ireland threshold: +

GB max: 99,045 Dec
NI max: 66 Feb

% young: 9.0   **brood size:** 2.26

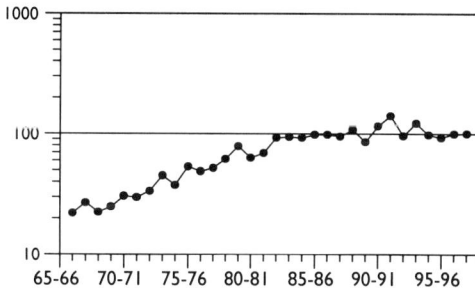

Figure 23. Annual indices for Dark-bellied Brent Goose in GB

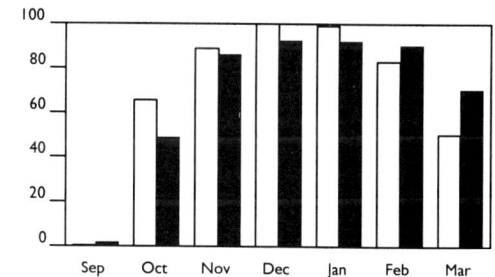

Figure 24. Monthly indices for Dark-bellied Brent Goose in GB (white bars 1997-98; black bars 1992-93 to 1996-97)

With the majority occurring on a small number of well-watched estuarine sites in southern and eastern England, Dark-bellied Brents are probably one of the best monitored species by WeBS. Given suitable weather conditions, national totals are thought to represent 100% of the numbers in Great Britain. The peak of just under 100,000 birds is typical for recent winters, and indices suggest that, with the exception of two or three years in the late 1980s and early 1990s, numbers have remained remarkably steady for the past 15

years. With just 9% young present in scanned flocks, reproductive success was again relatively low.

The table of important sites remains unchanged with no additions or losses following the 1997-98 counts. Numbers on the North Norfolk Marshes returned to high levels following two years of well below average counts, though peaks at several other sites dropped sharply, e.g. Pagham Harbour (62% below the five year mean), Orwell Estuary (59%) and Hamford Water

63

(48%). Small flocks (totalling 66 birds) on three sections of the Outer Ards peninsula in February were unusually large counts for Northern Ireland.

In the summer of 1997, Syroechkovski *et al.* (1998) visited a number of breeding colonies of Brent Geese in northwest Yakutia, East Siberia. At all sites both the nominate *bernicla* and the American wintering *nigricans* were present. Six American colour rings and one Dutch ring collected from local hunters revealed the presence of two flyways and two populations mixing in the Olenyok-western Lena delta region. In addition, there were mixed pairs and individuals of intermediate plumage observed in one of the colonies. Of 22 Geese ringed on the trip, one *bernicla* ringed on an island in the western Lena Delta was controlled on Vlieland in The Netherlands and had therefore travelled more than 5,500 km, the longest distance known for migrating Dark-bellied Brent Geese.

	93-94	94-95	95-96	96-97	97-98	Mon	Mean
International							
Wash	24,446	19,108	21,023	23,001	23,797	Dec	22,275
Thames Estuary	18,733	16,399	10,714	15,393	17,014	Oct	15,651
North Norfolk Marshes	15,061	13,364	8,110	8,793	14,088	Nov	11,883
Blackwater Estuary	12,208	12,763	8,525	10,641	10,290	Dec	10,885
Chichester Harbour	12,647	9,567	10,769	8,997	8,427	Feb	10,081
Hamford Water	8,154	4,395	14,466	9,286	4,194	Dec	8,099
Langstone Harbour	7,776	6,814	6,215	5,520	6,344	Jan	6,534
Crouch/Roach Estuary	5,012	5,022	3,820	4,703	5,644	Jan	4,840
Colne Estuary	4,920	2,929	3,529	3,493	4,263	Dec	3,827
Medway Estuary	5,104	3,121	2,733	2,526	2,725	Feb	3,242
Fleet/Wey	3,983	2,962	2,630	3,529	3,048	Nov	3,230
Great Britain							
Portsmouth Harbour	3,583	2,284	2,773	2,785	2,429	Nov	2,771
Deben Estuary	3,282	2,206	2,536	3,306	2,094	Feb	2,685
Pagham Harbour	2,638	2,611	3,016	2,879	1,071	Feb	2,443
North West Solent	2,650	2,046	2,643	2,279	(440)	Nov	2,405
Swale Estuary	2,823	1,650	1,903	3,141	1,803	Dec	2,264
Dengie Flats	2,780	1,650	2,440	2,000	2,290	Feb	2,232
Humber Estuary	1,795	3,243	2,078	2,366	1,532	Feb	2,203
Southampton Water	2,420	1,475	3,007	1,821	1,191	Mar	1,983
Stour Estuary	1,742	2,293	1,801	1,757	2,173	Mar	1,953
Exe Estuary	2,049	2,056	1,587	1,832	1,768	Oct	1,858
Beaulieu Estuary	1,272	1,417	1,360	2,480	2,283	Feb	1,762
Newtown Estuary	1,708	1,559	1,475	1,676	1,472	Jan	1,578
Poole Harbour	1,486	1,529	1,460	1,644	1,449	Feb	1,514
Orwell Estuary	1,565	1,981	1,290	961	567	Jan	1,273

Other sites surpassing table qualifying levels in 1997-98

Burry Inlet	1,165	Feb

LIGHT-BELLIED BRENT GOOSE
Branta bernicla hrota

CANADIAN POPULATION

International threshold:	200
Great Britain threshold:	+†
All-Ireland threshold:	200

GB max:	26	Feb/Mar			
NI max:	14,910	Oct	% young:	1.5	brood size:	2

The arrival of birds in Northern Ireland from the breeding grounds was early in 1997-98, with monthly indices and counts suggesting that 85% of the peak numbers were already present by the September count date. Many of these birds move on to winter in the Republic of Ireland and small numbers continue on to Wales, the Channel Islands and the north French coast. The peak count was around average for recent years though below that of the previous two winters. Annual indices, however, increased by 14% despite poor reproductive success for the fourth consecutive year: October flocks held just 1.5% young birds.

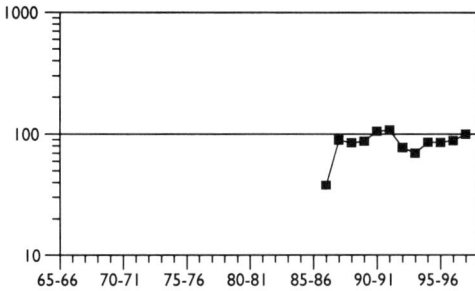

Figure 25. Annual indices for Light-bellied Brent Goose in NI

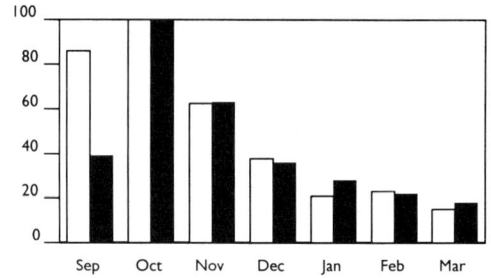

Figure 26. Monthly indices for Light-bellied Brent Goose in NI (white bars 1997-98; black bars 1992-93 to 1996-97)

The second all-Ireland co-ordinated census of the population was carried out in October 1997 and January 1998 (Colhoun *et al.* 1998). The highest total, of 17,180 birds, occurred in October when over 90% of the birds counted were on just two sites, Strangford Lough and Lough Foyle.

The midwinter census, when the population is more widely distributed, recorded less than 60% of this figure (9,921 birds), probably as a result of the dispersal of smaller flocks to more remote western areas and perhaps the use of non-wetland sites also.

	93-94	94-95	95-96	96-97	97-98	Mon	Mean
International							
Strangford Lough	12,795	8,519	11,337	11,614	11,184	Oct	11,090
Lough Foyle	1,934	4,007	5,550	4,757	3,820	Sep	4,014
Carlingford Lough	596	301	189	242	317	Mar	329
Tyrella	-	-	290	-	-		290
Killough Harbour	-	356	122	254	-		244
Larne Lough	290	206	209	177	232	Jan	223
Outer Ards	230	256	196	326	54	Feb	212
Great Britain[†]							
Inland Sea	23	32	36	63	17	Jan	34

Other sites surpassing table qualifying levels in 1997-98[†]
Jersey Shore   29   Dec

† as no British threshold has been set, a qualifying level of 25 has been chosen to select sites for presentation in this report

SVALBARD POPULATION

International threshold: 50
Great Britain threshold: 25*
* 50 is normally used as a minimum threshold

GB max: 2,583   Nov

% young: 9.7   brood size: n/a

Following the exceptional count in January 1997, when virtually the whole of the Svalbard population was present in Great Britain, numbers of Light-bellied Brent returned to more normal levels in 1997-98, with around 2,500 birds present between October and December. Mild weather in January saw this number drop to only 700. Breeding success was reasonable, much below the 1996 level but higher than in the two preceding years.

Although previously included within the Canada population (e.g. Hjort 1995, Scott & Rose 1997), the affiliations of a small population

breeding in northeast Greenland were unclear. Satellite telemetry in 1997 confirmed that these birds wintered within the range of the Svalbard population (Clausen & Bustnes 1998).

Percival *et al.* (1998) showed that recent total winter counts of both Light-bellied Brent Geese and Wigeon at Lindisfarne NNR were both only 40% of the maximum that the food supply at the site could support. This suggests that numbers were held down by some factor other than food supply, possibly hunting disturbance, indicated by the birds preference for feeding as far down the shore as possible.

	93-94	94-95	95-96	96-97	97-98	Mon	Mean
International							
Lindisfarne	1,440	2,150	2,470	4,092	2,567	Nov	2,544

RED-BREASTED GOOSE
Branta ruficollis

Vagrant and escape
Native range: SE Europe and Asia

A single was at Harewood Lake in March.

EGYPTIAN GOOSE
Alopochen aegyptiacus

Naturalised introduction[†]
Native range: Africa

GB max: 373   Aug
NI max: 0

The timing of the peak count was typical for this species, one of the few wildfowl for which summer counts are generally higher than during winter. The 1997-98 peak was over 50% greater than the previous highest recorded by WeBS (in 1991-92), totals boosted by improved summer coverage in the species' East Anglian stronghold. Numbers on the North Norfolk Marshes reached their highest ever levels and represent the largest WeBS count to date at an individual site. Counts at Rutland Water, the only site outside East Anglia to hold significant numbers, continued to rise.

	93-94	94-95	95-96	96-97	97-98	Mon	Mean	
Great Britain								
North Norfolk Marshes	113	179	97	113	198	Aug	140	
St Benet's Levels	28	0	58	85	56	Dec	45	
Lynford Gravel Pit	-	-	-	0	76	Aug	38	▲
Rutland Water	13	18	31	35	46	Sep/Oct	29	
Didlington	-	-	28	4	41	Aug	24	
Nunnery Lakes	15	16	24	11	19	Jul	17	
Blickling Lake	16	-	-	-	-		16	
Middle Yare Valley	6	9	6	4	52	Sep	15	▲
Ranworth/Cockshoot Broads	33	6	6	16	4	Sep/Dec	13	
Gunton Park Lake	12	-	-	-	-		12	
R Wensum: Fakenham - Gt Ryburgh	14	10	-	-	-		12	
Livermere	-	4	12	13	14	Jul	11	▲
Trinity Broads	-	-	8	-	13	Dec	11	▲

Internationally or nationally important sites not counted in last five years
Sennowe Park Lake
Pentney Gravel Pits

Other sites surpassing table qualifying levels in 1997-98

Stanford Training Area	15	Sep	Etherow Country Park	10	Jan
Wash	13	Sep			

† *as site designation does not occur and the 1% criterion is not applied, a qualifying level of 10 has been chosen to select sites for presentation in this report*

RUDDY SHELDUCK
Tadorna ferruginea

Escape
Native range: Asia, N Africa and S Europe

Although the monthly peak was of just 10 birds in September, summed site maxima suggest as many as 40 at the 24 English and one Welsh sites which held this species in 1997-98.

Sites with two or more birds in 1997-98

Bewl Water	6	Sep	Ramsbury Lake	2	several
Severn Estuary	4	Jan	Dee Estuary (Eng/Wales)	2	Oct
Staines Reservoirs	2	Apr	Swale Estuary	2	Nov/Dec
R. Avon: West Amesbury	2	Jun/Jul	North Norfolk Marshes	2	Mar
Ouse Washes	2	Jul			

CAPE SHELDUCK
Tadorna cana

Escape
Native range: S Africa

Four were at Knight & Bessborough Reservoirs in July and one was at Thorpe Water Park in October.

PARADISE SHELDUCK
Tadorna variegata

Escape
Native range: New Zealand

Singles were recorded at Croxall Pits, King George VI Reservoir and Alvecote Pools.

SHELDUCK
Tadorna tadorna

International threshold: 3,000
Great Britain threshold: 750
All-Ireland threshold: 70

GB max: 74,352 Dec
NI max: 4,685 Jan

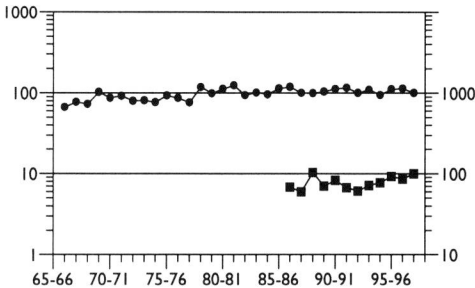

Figure 27. Annual indices for Shelduck in GB (circles, left axis) and NI (squares, right axis)

Figure 28. Monthly indices for Shelduck in GB and NI (white bars 1997-98; black bars 1992-93 to 1996-97)

The peak British count was in the lower range of those recorded in recent winters. Annual indices show wintering numbers have remained relatively stable over the past 35 years, particularly since the mid 1970s, with minor peaks often associated with harsh winters (Ridgill & Fox 1990). An unusually high proportion of birds were present in the early part of the winter, particularly in October, a pattern noted for several other species with high concentrations on estuaries (Dark-bellied Brent Geese, Wigeon). In Northern Ireland, the peak was below that of the previous winter, though above the average of recent years. Counts at Strangford Lough dominate the Northern Irish totals and examination of WeBS Low Tide Count data show that Core Counts may underestimate the numbers of birds on the site by as many as 1,500 birds; this may account for much of the variability in national totals in the province.

Two northwest estuaries, the Mersey and the Dee, recorded exceptional numbers. The count

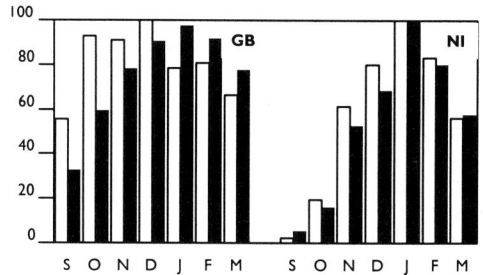

at the former was the highest at any site in the country for five years and, significantly, occurred in August.

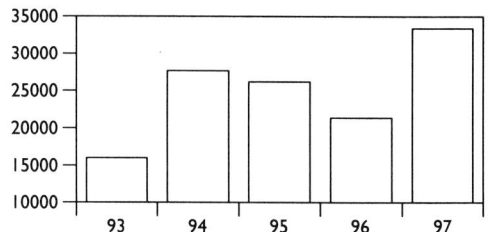

Figure 29. Numbers of Shelduck in late summer in the UK (annual totals calculated by summing the greater of the July and August counts at individual sites).

WeBS counts in recent years have shown a large number of Shelduck to be present in the UK in late summer, and although birds at a number of these, e.g. the Severn Estuary, are known to moult here, most were thought to depart to the

67

Wadden Sea to moult with the majority of the NW European population. The arrival of birds to the Wadden Sea begins in late June; adults and families from Britain arrive slightly later and most birds are in primary moult, and therefore flightless, during July or August (Meltofte *et al.* 1994). Assuming birds remaining in the UK are also flightless at this time, almost 33,400 Shelduck remained to moult here in 1997 (fig. 29).

Using average maxima from just July and August of the last five years, nine estuaries surpass the 1% British threshold at this time: Mersey (7,099), Humber (2,901), the Wash (2,859), Forth Estuary (2,849), Morecambe Bay (1,755), Ribble (1,630), Severn (1,455), Solway (1,164) and Dee (England/Wales) (933). Whilst a number are already known to be moult sites, it would be useful to collect evidence of moult at the others, given their apparently increasing importance at this time of year and, as a result, their heightened conservation significance.

	93-94	94-95	95-96	96-97	97-98	Mon	Mean
International							
Wash	14,242	12,873	14,238	10,352	12,368	Dec	12,815
Dee Estuary (Eng/Wal)	6,229	8,742	5,786	8,047	10,418	Oct	7,844
Morecambe Bay	5,734	8,524	6,098	5,632	8,426	Oct	6,883
Mersey Estuary	3,746	4,584	4,507	7,025	14,516	Aug	6,876
Medway Estuary	6,046	4,463	3,853	5,461	4,160	Jan	4,797
Forth Est.	1,560	5,337	5,077	5,065	5,507	Aug	4,509
Humber Estuary	4,481	3,383	5,240	3,900	4,843	Dec	4,369
Ribble Estuary	5,230	3,278	4,523	3,788	4,106	Dec	4,185
Strangford Lough[1]	4,144	2,189	4,673	3,493	4,142	Dec	3,728
Severn Estuary	2,627	4,466	3,508	4,117	2,371	Nov	3,418
Poole Harbour	2,982	3,177	3,575	4,650	2,662	Feb	3,409
Blackwater Estuary	2,749	2,570	4,356	4,129	2,123	Feb	3,185
Solway Estuary	3,020	2,527	3,293	3,450	3,370	Nov	3,132 ▲
Great Britain							
Swale Estuary	2,587	2,234	2,782	2,760	3,027	Jan	2,678
Thames Estuary	2,923	2,539	2,472	3,094	1,917	Feb	2,589
Stour Estuary	2,967	1,963	2,297	2,044	2,029	Jan	2,260
Hamford Water	1,710	1,508	2,146	3,006	2,781	Jan	2,230
Colne Estuary	1,122	1,533	2,017	1,338	977	Jan	1,397
Chichester Harbour	1,404	1,275	1,980	1,140	1,054	Jan	1,371
Orwell Estuary	1,320	1,221	1,989	1,039	789	Jan	1,272
Alde Complex	1,202	832	1,074	765	1,935	Feb	1,162
Duddon Estuary	1,362	1,567	974	853	900	Nov	1,131
North Norfolk Marshes	1,042	1,185	710	1,335	1,379	Dec	1,130
Tees Estuary	1,496	1,089	1,267	893	837	Jan	1,116
Cleddau Estuary	877	1,178	1,008	1,023	939	Jan	1,005
Eden Estuary	1,031	952	930	942	1,088	Oct	988
Deben Estuary	1,297	925	950	824	875	Mar	974
Lindisfarne	820	930	855	1,295	927	Jan	965
Langstone Harbour	661	698	1,477	889	826	Feb	910
Burry Inlet	1,062	608	695	1,282	883	Mar	906
Montrose Basin	701	818	1039	596	1100	Nov	850
Nationally Important Northern Ireland Sites							
Strangford Lough	2,187	2,189	2,464	3,195	2,978	Jan	2,603
Belfast Lough	509	621	1,062	715	497	Jan	681
Larne Lough	247	373	371	440	505	Feb	387
Lough Foyle	174	215	508	527	439	Jan	373
Carlingford Lough	193	294	172	165	198	Feb	204
Lo. Neagh/Beg	121	236	146	188	240	Mar	186
Dundrum Bay	121	65	76	126	64	Jan	90

Sites no longer meeting table qualifying levels
Crouch/Roach Estuary
Tamar Complex

Other sites surpassing table qualifying levels in 1997-98
Blyth Estuary (Suffolk)   757   Dec

1   *Includes Low Tide Count data*

MUSCOVY DUCK
Cairina moschata

Escape
Native range: South America

GB max:	127	Dec
NI max:	0	

The increase in records of this species continued in 1997-98, with counts from 35 sites (*cf.* a maximum of 20 previously), although the peak count fell slightly. This trend, however, is likely to reflect better reporting by the counter network, rather than a genuine expansion of the species' distribution.

Sites with more than five birds in 1997-98

Lothing Lake & Oulton Broad	52	Dec	Lancaster Canal	8	Jan
Nafferton Mere	30	Dec	Derwent Water	8	Oct/Dec
Rufford Lake	16	Oct-Dec	Gun Knowe Loch	7	Apr
Wilderness Pond	12	Jul/Sep			

WOOD DUCK
Aix sponsa

Escape
Native range: Norther America

Records were received from 15 sites, double the previous number, with summed maxima suggesting up to 19 birds. Most were in south or southeast England, but there was a small concentration in the Manchester area also. All records were of singles except for three at Busbridge Lakes and two each at Strinesdale and Wellington Country Park.

MANDARIN
Aix galericulata

Naturalised introduction[†]
Native range: Eastern Asia

GB max:	315	Nov
NI max:	5	Jun

Despite their gaudy colouration, Mandarin are surprisingly elusive and are probably the most poorly monitored of all inland duck species by WeBS. Numbers in winter 1997-98 exceeded all previous totals, surpassing 300 for the first time. Birds were recorded at 84 sites, the majority in south-eastern England, but also one in Wales and five in Scotland. Dundrum Bay, the only site in Northern Ireland where Mandarin has been recorded during WeBS counts to date, held up to five birds during summer.

Few sites held exceptional numbers, though several new sites joined the table below due to more regular occurrence or detection of the species on count days, with counts at Bramshill Park Lake and Overstone Park Lakes being particularly noteworthy.

	93-94	94-95	95-96	96-97	97-98	Mon	Mean	
Great Britain								
Severn Estuary	79	78	40	113	40	Nov	70	
Cuttmil Ponds	84	32	51	106	44	Jun	63	
Virginia Water	32	74	-	-	-		53	
Arun Valley	27	40	51	48	59	Mar	45	
Panshanger Estate	-	-	18	51	-		35	
Connaught Water	-	-	-	28	39	Jan	34	
Passfield Lake	-	-	-	48	15	Sep	32	
Paultons Bird Park	26	37	-	-	-		32	
Aldford Brook & Eaton Park	-	3	34	34	32	Oct	26	▲
Bramshill Park Lake	-	1	16	7	60	Nov	21	▲
Osterley Park Lakes	-	-	19	24	20	Mar	21	
Overstone Park Lakes	0	10	20	18	32	Sep	16	▲
Thursley Lake	3	17	35	8	6	Aug	14	

	93-94	94-95	95-96	96-97	97-98	Mon	Mean	
Swanbourne Lake	34	15	-	2	2	Jan/Mar	13	
Busbridge Lakes	23	17	0	0	22	Mar	12	
Fonthill Lake	14	5	10	18	12	Feb	12	
Fleet Pond	8	14	1	20	10	Nov	11	▲
R Wensum: Fakenham - Gt Ryburgh	14	7	-	-	-		11	▲
Headley Mill Pond	-	-	-	4	16	Jan	10	▲

Internationally or nationally important sites not counted in last five years
Frenchess Road Pond
Hammer Wood Pond

Sites no longer meeting table qualifying levels
Woburn Park Lakes
Lurgashall Mill Pond

Other sites surpassing table qualifying levels in 1997-98
Chillington Hall Pool	14	Dec/Feb/Jan
Weirwood Reservoir	14	Jun

† as site designation does not occur and the 1% criterion is not applied, a qualifying level of 10 has been chosen to select sites for presentation in this report

---

WIGEON
Anas penelope

International threshold: 12,500
Great Britain threshold: 2,800
All-Ireland threshold: 1,250

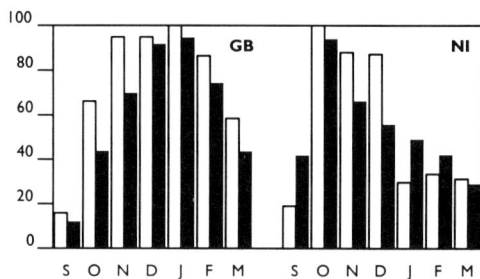

GB max: 327,099 Jan
NI max: 11,278 Oct

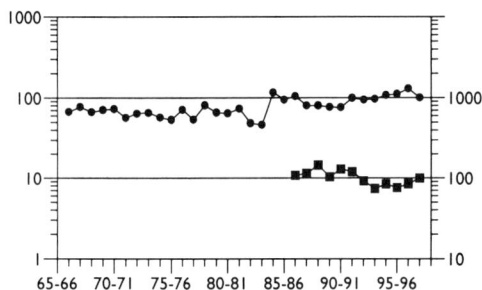

Figure 30. Annual indices for Wigeon in GB (circles, left axis) and squares (right axis)

Figure 31. Monthly indices for Wigeon in GB and NI (white bars 1997-98; black bars 1992-93 to 1996-97)

Numbers wintering in Great Britain in 1997-98 dropped markedly after exceptionally high counts in the previous year, and annual indices fell for the first time, by 24%, since 1992-93. Conversely, the January peak in Northern Ireland was the highest since October 1991, a rise mirrored in the annual indices.

Monthly indices revealed that, whilst peaks occurred, as usual, in January in Britain, and in October in Northern Ireland, much higher numbers were present in both regions in November and December. These patterns are surprising given the generally mild weather, with most winds from the southwest, during these months.

In line with the relatively low national total, numbers at many sites were below average,

notably the Dyfi and Thames Estuaries, Lower Derwent Valley and Upper Lough Erne, whilst the peak on the Ribble Estuary, which has held up to one third of the national total, was almost 20% below the average peak for the previous five year period. High counts were recorded on the Dornoch Firth, consolidating its position as an internationally important site, at Breydon Water & Berney Marshes, more than double the average for this site, and Loch of Harray. The count at Lough Foyle hinted that numbers may return to the levels of the early 1990s, when five figure counts were recorded.

Mayhew & Houston (1999) have shown that grazing by Wigeon, when returning repeatedly throughout the winter to a small number of feeding areas, results in a 52% increase in leaf production over the winter and, at the end of the winter, 4.75% higher protein levels compared to ungrazed plants. The growth of plants is stimulated by defoliation and not caused by a fertiliser effect from the birds' droppings. They suggest that selectively grazing patches of grassland is a deliberate feeding strategy used by Wigeon to obtain improved dietary quality in late winter and early spring, as has previously been shown for some goose species.

Numbers of Wigeon wintering at Strangford Lough have declined drastically from peaks of up to 20,000 in the early 1970s but, as this decline is not mirrored in the UK population, it was thought that the reasons were intrinsic to the lough. Mathers *et al.* (1998) concluded that this may result from an indirect interaction with Pale-bellied Brent Geese through depletion of *Zostera,* a common food source. The geese, which dig for the rhizomes, may alter plant regrowth in future years, thereby affecting Wigeon which feed on the shoots. The earlier arrival of peak numbers of both Wigeon and Brent Geese in recent years may eat out the *Zostera* beds before they reach peak biomass, thereby reducing future growth. Anecdotal data also suggest a change to sandier sediments in the lough, which may allow Brents easier access to *Zostera*, thereby offsetting in the short term the decline in food. The effect of successively earlier and greater exploitation of *Zostera* on its potential for recovery is part of an ongoing study. Human activity, which has also increased through the provision of several car parks and walkways, may also play a part in the changing wildfowl numbers since Brent Geese are much more tolerant of disturbance than Wigeon.

	93-94	94-95	95-96	96-97	97-98	Mon	Mean	
International								
Ribble Estuary	92,465	110,278	83,922	74,068	66,197	Nov	85,386	
Ouse Washes	23,791	28,284	30,545	28,223	26,922	Dec	27,553	
Swale Estuary	10,116	15,039	15,906	40,090	13,292	Feb	18,889	
Somerset Levels	8,880	21,455	24,302	11,000	16,010	Jan	16,329	
North Norfolk Marshes	13,631	16,471	14,377	14,247	12,423	Nov	14,230	
Dornoch Firth	14,501	10,911	12,540	11,615	17,240	Oct	13,361	
Great Britain								
Mersey Estuary	9,121	17,650	11,254	10,885	10,520	Nov	11,886	
Lower Derwent Valley	11,650	14,140	13,060	10,600	7,900	Jan	11,470	
Nene Washes	11,909	11,302	11,526	8,090	12,699	Jan	11,105	
Cromarty Firth	9,603	8,629	11,973	8,516	11,199	Oct	9,984	
Inner Moray Firth	9,417	8,962	8,200	9,305	7,964	Dec	8,770	
Walland Marsh	-	-	-	8,600	5,400	Jan	7,000	
Lindisfarne	6,724	13,476	3,662	4,368	5,600	Oct	6,766	
Morecambe Bay	6,684	7,494	7,045	6,432	6,002	Nov	6,731	
Severn Estuary	3,947	5,689	6,267	11,548	5,304	Dec	6,551	
Middle Yare Valley	7,460	4,335	6,223	7,189	6,306	Dec	6,303	
Breydon Water/Berney Marshes	5,100	4,900	4,300	6,500	10,200	Dec	6,200	
Alde Complex	3,473	6,345	5,827	8,181	6,810	Jan	6,127	
Humber Estuary	5,789	7,502	3,000	5,802	7,668	Nov	5,952	
Martin Mere	2,600	5,580	9,280	2,460	3,620	Mar	4,708	
Rutland Water	4,160	3,859	5,014	4,968	4,669	Feb	4,534	
Arun Valley	2,538	4,804	5,138	4,411	5,155	Jan	4,409	
Dee Estuary (Eng/Wal)	1,866	8,091	2,191	3,682	5,366	Dec	4,239	
Dyfi Estuary	4,831	3,665	4,363	4,681	2,911	Dec	4,090	
Medway Estuary	3,883	4,705	5,131	2,951	3,736	Jan	4,081	
Hamford Water	1,499	2,593	3,785	9,511	2,668	Dec	4,011	
Montrose Basin	3,600	4,233	4,856	2,735	3,170	Nov	3,719	
Fleet/Wey	4,783	5,013	2,957	3,021	2,637	Nov	3,682	
Thames Estuary	2,359	3,537	3,690	5,146	1,223	Oct	3,191	
Loch of Harray	3,105	2,145	3,222	2,384	5,070	Nov	3,185	
Stour Estuary	3,027	3,951	1,958	3,046	3,628	Dec	3,122	
Exe Estuary	3,073	2,173	2,263	3,184	4,344	Oct	3,007	▲
Avon Valley (Lower)	5,000	3,113	2,120	1,570	3,000	Jan	2,961	
Blackwater Estuary	3,999	2,732	2,080	2,534	3,031	Feb	2,875	
Cleddau Estuary	2,088	2,403	3,455	3,351	3,058	Dec	2,871	▲

Nationally Important Northern Ireland Sites

Lough Foyle	3,513	6,094	8,438	6,850	9,440	Oct	6,867
Lo. Neagh/Beg	2,6333,669		3,229	2,398	3,052	Nov	2,996
Strangford Lough	1,870	1,747	2,457	1,900	1,937	Nov	1,982
Upper Lough Erne	1,744	1,707	1,692	1,252	619	Feb	1,403

Sites no longer meeting table qualifying levels
Wash

Other sites surpassing table qualifying levels in 1997-98

Tophill Low Reservoirs	3,200	Feb	Beaulieu Estuary	2,947	Feb
Burry Inlet	3,144	Nov	North Warren/Thorpness Mere	2,800	Feb

AMERICAN WIGEON
Anas americana

Vagrant
Native range: North and Central America

Singles were found on the Fleet/Way, North Norfolk Marshes, and the Dee (Eng/Wales) and Lossie Estuaries.

CHILOE WIGEON
Anas sibilatrix

Escape
Native range: South America

A notable increase, with records from 11 sites and summed site maxima of 14 birds. Swanpool (Falmouth), Ramsbury Lake and Lower Windrush Valley Gravel Pits each held two birds, with singles at the remaining sites. Only one of the four sites with Chiloe Wigeon in the previous winter matched those at which it was noted in 1997-98.

FALCATED DUCK
Anas falcata

Escape
Native range: Eastern Asia

Singles were found on the Dee Estuary (Eng/Wales) and Merryton Ponds, both in March.

GADWALL
Anas strepera

International threshold: 300
Great Britain threshold: 80
All-Ireland threshold: +†

GB max: 13,073 Nov
NI max: 154 Mar

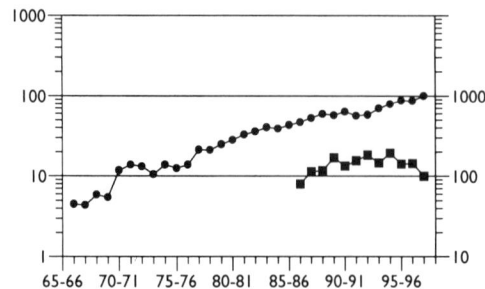

Figure 32. Annual indices for Gadwall in GB (circles, left axis) and NI (squares, right axis)

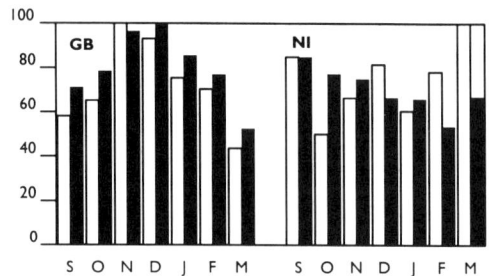

Figure 33. Monthly indices for Gadwall in GB and NI (white bars 1997-98; black bars 1992-93-1996-97)

The seemingly inexorable rise in Gadwall numbers in Great Britain continued in 1997-98: the peak surpassed 13,000 for the first time, having exceeded counts of 5,000 in the mid

1980s and 10,000 only three years ago. Gadwall are the fastest growing 'natural' wildfowl population in Great Britain; only naturalised populations of Greylag Geese and Ruddy Duck are increasing at a greater rate. However, no similar pattern is evident in Northern Ireland where the peak was the lowest this decade.

The peak count at Rutland Water represents by far the largest gathering of Gadwall yet recorded in the UK, and is the only site to have held over 1,000 birds. Sustained high counts saw the Ouse Washes rise to second place in the table and those at Thrapston Gravel Pit pushed the five year average above the threshold for international importance for the first time.

Atkinson-Willes (1963) noted only 12 sites holding 25 or more birds during the 1950s, only two of which surpassed 100 birds. In contrast, 50 sites currently support average peaks in excess of 100 birds. The growth in the national population

is manifested by increases at sites previously holding relatively small numbers of Gadwall: almost one quarter of the nationally important sites held peak counts of less than fifty birds only four or five years ago. Fourteen new sites attained nationally important status following the 1997-98 counts, all in the species' stronghold of central, southern and eastern England.

However, the national population estimate of 8,000 birds, based on data up to 1991-92 (Kirby 1995), is clearly a considerable underestimate of the current picture. The average national total for the last five years is 11,167 birds, and it is likely that the revision of national population estimates in 2000 will result in a figure of at least 12,500 for Gadwall. This would mean only 34 sites qualify as nationally important, compared with the 68 listed below. Consequently, perhaps the most remarkable figure in the table below was the especially low count at Abberton Reservoir.

	93-94	94-95	95-96	96-97	97-98	Mon	Mean	
International								
Rutland Water	933	1,671	1,306	733	2,181	Nov	1,365	
Ouse Washes	455	378	273	942	783	Feb	566	
Avon Valley (Mid)	488	584	491	421	580	Feb	513	
Abberton Reservoir	517	668	829	338	120	Sep	494	
Wraysbury Gravel Pits	426	307	389	528	734	Feb	477	
Lee Valley Gravel Pits	400	393	219	576	609	Nov	439	
Pitsford Reservoir	82	627	471	362	355	Sep	379	
Thrapston Gravel Pit	54	149	139	895	567	Nov	361	▲
Great Britain								
Hornsea Mere	338	300	-	-	(10)	Mar	319	
Severn Estuary	252	270	265	281	250	Dec	264	
Somerset Levels	206	97	293	342	369	Feb	261	
Fen Drayton Gravel Pit	133	276	194	251	388	Dec	248	
Loch Leven	262	252	230	235	248	Oct	245	
Eversley Cross/Yateley Gravel Pits	151	193	184	376	236	Feb	228	
North Norfolk Marshes	260	267	193	163	232	Nov	223	
Colne Valley Gravel Pits	66	173	237	434	141	Nov	210	
Thames Estuary	228	179	252	190	198	Feb	209	
Burghfield Gravel Pits	121	112	393	209	178	Dec	203	
Buckden/Stirtloe Pits	133	147	236	163	277	Nov	191	
Nene Washes	311	168	250	63	151	Feb	189	
Fairburn Ings	128	168	239	202	191	Aug	186	
Gunton Park	186	-	-	-	-		186	
Cotswold WP West	208	162	170	217	147	Jan	181	
Little Paxton Gravel Pits	178	196	69	287	132	Dec	172	
Chew Valley Lake	155	140	180	175	-		163	
Seaton Gravel Pits	293	194	12	201	109	Dec	162	
Chichester Gravel Pits	27	142	161	284	188	Nov	160	
Hampton & Kempton Reservoirs	48	253	145	153	198	Jan	159	
Hollowell Reservoir	212	281	245	45	12	Feb	159	
Hanningfield Reservoir	90	130	157	156	216	Sep	150	
Sutton/Lound Gravel Pits	-	152	191	96	150	Feb	147	
Alton Water	16	127	197	80	312	Dec	146	
Lower Derwent Valley	36	71	67	271	283	Feb	146	
North Warren/Thorpness Mere	103	131	130	141	200	Feb	141	
Stodmarsh	86	142	71	274	122	Sep/Jan	139	
Hardley Flood	360	15	37	-	-		137	
Minsmere Levels	99	141	68	240	130	Oct	136	
Tophill Low Reservoirs	193	90	160	40	190	Sep	135	
Cotswold WP East	101	102	235	125	92	Dec	131	
Dungeness Gravel Pits	58	165	260	76	85	Nov	129	

	93-94	94-95	95-96	96-97	97-98	Mon	Mean	
Bewl Water	60	222	173	120	72	Nov	129	
Earls Barton Gravel Pits	27	71	154	121	264	Dec	127	▲
Ditchford Gravel Pit	58	129	187	115	118	Jan	121	
Langtoft West End Gravel Pits	59	115	152	166	87	Dec	116	
Thorpe Water Park	149	164	60	96	102	Feb	114	
Fleet/Wey	211	171	96	24	70	Jan	114	
Sonning GP	-	71	101	127	143	Jan	111	▲
Middle Tame Valley GP	70	48	135	196	108	Feb	111	
Twyford Gravel Pits	77	94	110	156	89	Jan	105	
Stanford Reservoir	24	10	120	267	100	Dec	104	
Meadow Lane Gravel Pits	22	57	149	157	111	Dec	99	
Wellington Country Park	1	0	174	154	152	Feb	96	▲
Stanford Training Area	119	67	135	32	126	Dec	96	
Swale Estuary	104	167	50	52	106	Feb	96	
Grafham Water	27	27	114	82	223	Mar	95	▲
Fordwich/Westbere GPs	44	163	199	13	57	Nov	95	
Marsh Lane Gravel Pits	165	32	125	100	55	Nov	95	
Rye Harbour/Pett Level	55	82	143	113	69	Dec	92	
Breydon Water & Berney Marshes	67	54	46	129	161	Feb	91	▲
Dinton Pastures	95	102	101	-	64	Dec	91	
Rostherne Mere	22	92	156	134	49	Aug	91	
Swanholme Lake	86	82	79	105	99	Feb	90	
Tattershall Pits	160	96	-	62	35	Oct	88	
Tabley Mere	54	12	150	80	140	Oct	87	▲
Linford Gravel Pits	77	74	102	134	48	Dec	87	
Holme Pierrepont Gravel Pits	13	47	108	152	110	Dec	86	▲
Wash	133	48	94	53	100	Feb	86	▲
Ampton Water	-	-	-	-	86	Mar	86	▲
Middle Yare Valley	61	88	62	85	129	Jan	85	▲
Fort Henry Ponds/Exton Park Lake	44	13	106	179	85	Jan	85	
Blunham Gravel Pit	85	-	-	-	-		85	▲
Lower Windrush Valley Gravel Pits	69	64	63	82	130	Jan	82	▲
Swillington Ings	91	91	54	113	54	Sep	81	
Stoke Newington Reservoirs	90	80	102	-	52	Sep	81	
Tring Reservoirs	59	42	74	79	146	Nov	80	▲
Hilfield Park Reservoir	102	140	80	41	38	Dec	80	▲

Northern Ireland†

	93-94	94-95	95-96	96-97	97-98	Mon	Mean
Lo. Neagh/Beg	144	301	120	124	108	Mar	159
Strangford Lough	112	124	82	118	63	Sep	100
Upper Quoile	10	6	19	58	4	Dec	19

Internationally or nationally important sites not counted in last five years

Lackford Gravel Pits

South Iver Gravel Pits

Sennowe Park Lakes

Clea Lakes

Shrigley Lake

Sites no longer meeting table qualifying levels

Stainhill Reservoirs

Blagdon Lake

Pen Ponds

Hillsborough Main Lake

Upper Lough Erne

Other sites surpassing table qualifying levels in 1997-98

Solway Estuary	220	Dec	Whisby Gravel Pits	98	Sep
Barons Haugh	166	Sep	Leighton Moss	96	Nov
Longside Lake	123	Jan/Feb	Bedfont/Ashford Gravel Pits	94	Dec
Brent Reservoir	121	Oct	Arun Valley	93	Jan
Landbeach Gravel Pits	112	Jan	Kislingbury Gravel Pits	90	Feb
Walland Marsh	111	Feb	Blyth Estuary (Suffolk)	89	Mar
Yarwell Gravel Pits	111	Jan	Besthorpe/Girton Gravel Pits	85	Nov
Leventhorpe Flood Meadows	110	Aug	Ranworth/Cockshoot Broads	83	Jan
Blackwater Estuary	110	Mar	Mere Sands Wood	82	Jan
Crichel Lake	100	Feb	Theale Gravel Pits	81	Dec
Summerleaze Gravel Pits	98	Jan			

† as no all-Ireland threshold has been set, a qualifying level of 10 has been chosen to select sites for presentation in this report

TEAL
Anas crecca

International threshold: 4,000
Great Britain threshold: 1,400
All-Ireland threshold: 650

GB max: 137,754 Dec
NI max: 4,823 Dec

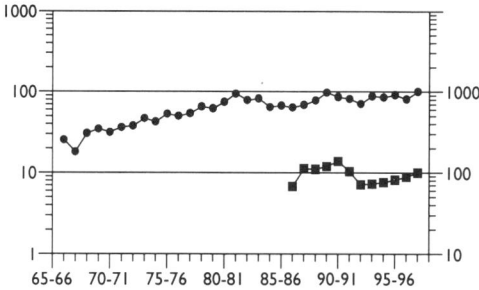

Figure 34. Annual indices for Teal in GB (circles, left axis) and NI (squares, right axis)

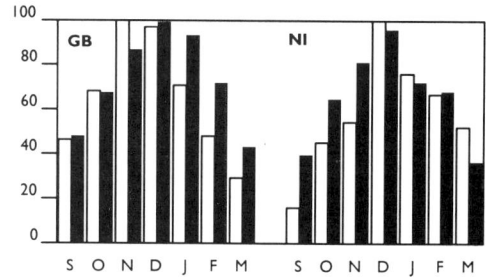

Figure 35. Monthly indices for Teal in GB and NI (white bars 1997-98; black bars 1992-93 to 1996-97)

Peak counts in Great Britain in recent years have fluctuated around the 130,000 mark, and whilst the peak in 1997-98 was therefore seemingly unremarkable, it represents the highest total yet recorded by the scheme. The annual index of the British population also reached its highest ever level, though not notably greater than several peaks during the past 15 years.

In Northern Ireland, counts and annual indices have risen steadily over the past five winters though still remain around 30% below those in the late 1980s and early 1990s. Monthly indices indicate an early peak in November in Great Britain and more typical December peak in

Northern Ireland. Numbers in the second half of the winter were low in Great Britain.

Numbers on the Somerset Levels returned to more typical levels following the unusually low counts in winter 1996-97. Abberton Reservoir joined the six other sites holding internationally important numbers thanks to an exceptional peak count, some 140% above the previous average at the site, and continuing the sustained, large increase over the last five winters. Large counts were also noted at Martin Mere, North Norfolk Marshes, Inner Moray Firth and on Loughs Neagh & Beg.

	93-94	94-95	95-96	96-97	97-98	Mon	Mean	
International								
Somerset Levels	15,256	13,197	24,792	3,305	16,156	Jan	14,541	
Mersey Estuary	13,034	12,098	7,734	14,120	12,065	Nov	11,810	
Ribble Estuary	8,876	5,859	7,343	7,833	6,209	Oct	7,224	
Lower Derwent Valley	8,231	7,050	4,432	3,875	5,900	Feb	5,898	
Abberton Reservoir	1,656	3,022	5,816	6,756	9,381	Nov	5,326	▲
Dee Estuary (Eng/Wal)	3,742	4,085	4,867	6,545	6,254	Nov	5,099	
Hamford Water	3,807	2,746	5,283	6,563	2,633	Dec	4,206	
Great Britain								
Martin Mere	2,760	3,560	4,020	2,560	5,750	Nov	3,730	
Severn Estuary	3,743	4,288	3,806	2,665	2,880	Jan	3,476	
Swale Estuary	5,248	4,278	2,174	2,868	2,457	Jan	3,405	
Ouse Washes	2,349	3,614	2,218	3,661	3,721	Dec	3,113	
North Norfolk Marshes	2,698	2,888	2,665	2,668	3,992	Nov	2,982	
Inner Moray Firth	2,128	2,100	2,873	3,346	3,428	Dec	2,775	
Loch Leven	1,233	2,200	3,884	3,250	3,288	Oct	2,771	
Blackwater Estuary	3,509	3,213	1,825	2,593	2,522	Dec	2,732	
Nene Washes	2,438	3,748	2,602	1,648	2,054	Feb	2,498	
Horsey Mere	-	-	-	-	2,400	Nov	2,400	▲
Thames Estuary	1,654	3,176	2,393	2,575	1,497	Oct	2,259	
Cleddau Estuary	1,787	1,226	2,948	2,220	2,637	Dec	2,164	
Southampton Water	1,437	1,705	2,700	2,356	(1,590)	Dec	2,050	
Alde Complex	2,054	1,497	2,306	1,793	2,078	Jan	1,946	

	93-94	94-95	95-96	96-97	97-98	Mon	Mean	
Dornoch Firth	1,782	2,303	1,759	1,476	2,073	Dec	1,879	
Morecambe Bay	1,334	1,967	2,127	1,439	2,114	Nov	1,796	
Arun Valley	2,615	2,886	1,277	655	1,385	Nov	1,764	
Medway Estuary	1,649	1,767	1,901	1,549	1,466	Jan	1,666	
Mere Sands Wood	2,131	884	1,075	2,525	1,025	Dec	1,528	
Rutland Water	978	805	2,491	1,954	1,402	Dec	1,526	
Chichester Harbour	1,345	1,377	1,172	2,037	1,649	Dec	1,516	
Woolston Eyes	1,500	2,500	1,150	900	1,500	Jan	1,510	▲
Poole Harbour	1,399	1,094	1,661	2,297	972	Feb	1,485	
Pagham Harbour	1,636	1,311	1,870	1,617	969	Dec	1,481	
Tees Estuary	1,558	1,582	1,657	1,059	1,219	Dec	1,415	▲
Northern Ireland								
Strangford Lough	1,363	1,617	1,681	2,302	1,978	Jan	1,788	
Lo. Neagh/Beg	1,481	1,801	1,227	1,076	2,270	Dec	1,571	
Lough Foyle	403	1,007	852	837	575	Oct	735	

Sites no longer meeting table qualifying levels
Loch of Strathbeg

Other sites surpassing table qualifying levels in 1997-98

Minsmere Levels	2,336	Oct	Humber Estuary	1,497	Nov
Pitsford Reservoir	1,538	Dec	Loch Eye	1,465	Feb
Fiddlers Ferry PS Lagoons	1,500	Oct	Forth Estuary	1,411	Jan

SPECKLED TEAL
Anas flavirostris

Escape
Native range: South America

Two were seen at Bramshill Park Lake in March, and singles at Whittlesford Gravel Pits and on the River Derwent at Chatsworth during the summer.

MALLARD
Anas platyrhynchos

International threshold: 20,000**
Great Britain threshold: 5,000†
All-Ireland threshold: 500

GB max: 140,213 Nov
NI max: 8,623 Sep

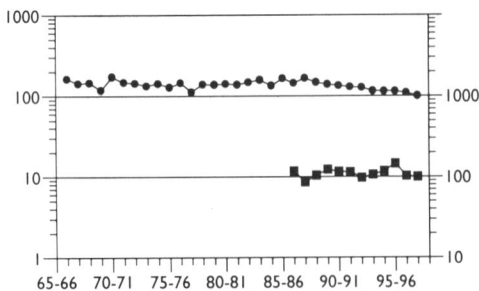

Figure 36. Annual indices for Mallard in GB (circles, left axis) and NI (squares, right axis)

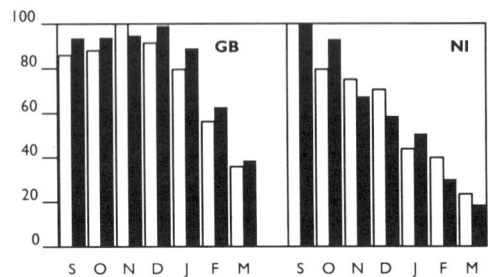

Figure 37. Monthly indices for Mallard in GB and NI (white bars 1997-98; black bars 1992-93 to 1996-97)

The downward trend in numbers of what is traditionally regarded as the UK's commonest species of wildfowl continued in 1997-98 with a further fall of 10% in the British annual index values. Both the peak count and annual index were the lowest recorded since counts began and suggest that the population has fallen by 40% in the last 10 years. Although the absence of national statistics on released stock for hunting and on shooting bags clouds the picture, a decline of this magnitude in several other birds has triggered considerable concern in the conservation community.

The peak count in Northern Ireland was on a

par with those in recent years. No clear trend is apparent in the annual indices, where monitoring of smaller waterbodies away from the major sites is relatively poor and significant numbers of this highly dispersed species undoubtedly go unrecorded.

Patterns in monthly indices for Great Britain and Northern Ireland differ significantly. In the latter, peak numbers occur early in the winter (usually in September) and decline steadily thereafter whilst in Great Britain, values remain high from September through to January, falling only in the last months of winters. The pattern in Great Britain probably reflects a balance between high shooting mortality during autumn and newly arrived immigrants from continental Europe which make up more than one third of the winter population (Owen *et al.* 1986)

Due to the species dispersed distribution, no Great Britain sites support nationally, let alone internationally, important numbers. Indeed, throughout NW Europe, only three sites support more than 1% of the NW European population estimated at five million birds (Scott & Rose 1996).

In line with national trends, most key sites held below average numbers in 1997-98: counts on the Lower Derwent Valley and Solway Estuary were particularly reduced, numbers on the Forth Estuary have declined in each of the last four winters and counts on the Ouse Washes have almost halved in recent years. Amongst key sites, only the count at Strangford Lough exceeded its five year mean.

	93-94	94-95	95-96	96-97	97-98	Mon	Mean
Great Britain[†]							
Morecambe Bay	3,563	4,456	3,798	3,116	3,615	Oct	3,710
Ouse Washes	5,693	4,511	2,868	2,149	2,582	Dec	3,561
Lower Derwent Valley	4,000	3,100	3,200	3,655	2,400	Feb	3,271
Wash	3,518	3,379	3,512	2,636	2,771	Jan	3,163
Tring Reservoirs	2,736	3,250	4,000	2,956	2,200	Aug	3,028
Martin Mere	3,210	3,400	3,100	2,885	2,520	Nov	3,023
Severn Estuary	3,145	2,870	2,383	3,088	2,076	Nov	2,712
Humber Estuary	3,055	3,184	2,621	2,112	2,211	Nov	2,637
Solway Estuary	2,988	2,624	2,637	2,011	1,419	Nov	2,336
Forth Est.	2,717	2,648	2,003	1,672	1,435	Dec	2,095
Northern Ireland							
Lo. Neagh/Beg	3,699	5,713	8,791	5,399	5,463	Aug	5,813
Lough Foyle	2,166	1,699	1,755	1,795	1,592	Dec	1,801
Strangford Lough	1,780	1,886	1,503	1,238	1,753	Sep	1,632

Sites no longer meeting table qualifying levels
Belfast Lough
Upper Lough Erne

Internationally or nationally important sites not counted in last five years
Ballysaggart Lough

Other sites surpassing table qualifying levels in 1997-98
Inner Moray Firth	2,044	Dec
Livermere	2,000	Aug

† as no site in Great Britain is of national importance, a qualifying level of 2,000 has been chosen to select sites for presentation in this report

---

CHESTNUT TEAL
Anas castanea

Escape
Native range: Australia

One was seen on the Lower Windrush Valley Gravel Pits in November.

PINTAIL
Anas acuta

International threshold: 600
Great Britain threshold: 280
All-Ireland threshold: 60

GB max: 24,517 Dec
NI max: 358 Feb

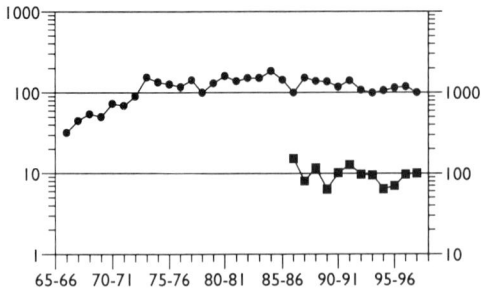

Figure 38. Annual indices for Pinatil in GB (circles, left axis) and NI (squares, right axis)

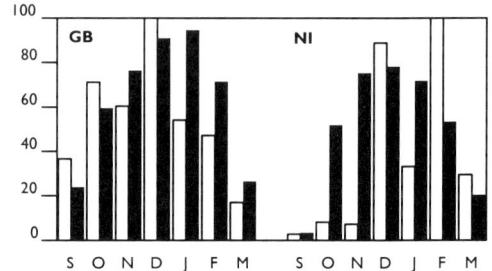

Figure 39. Monthly indices for Pintail in GB and NI (white bars 1997-98; black bars 1992-93 to 1996-97)

In Great Britain, the peak total was around 3,000 below that of the previous winter. Whilst this is still 2-4,000 higher than counts during most of the early 1990s, annual indices suggest the population is near the lower end of recent fluctuations, the figure for 1997-98 being the second lowest for 25 years. Only small numbers of birds occur in Northern Ireland and on very few sites. Consequently, annual indices are quite variable with no clear pattern evident. The peak count, however, was the highest in the province since 1991-92.

Great Britain monthly indices showed a double peak, in October and December, in 1997-98. This unusual phenomenon has occurred on several previous occasions, particularly in the late 1980s and early 1990s, but not since 1992-93. One hypothesis is that the timing of movements of different sub-populations away from the breeding grounds differs. Icelandic birds, for example, may arrive in early winter and rapidly disperse to areas within and beyond the UK, followed by a later arrival of continental birds when suitable weather systems push birds further west. However, insufficient ringing data are currently available to support this suggestion.

Peak 1997-98 counts at many of the 14 sites of international importance were well below their usual levels: those on the Ribble, Medway, Swale and Duddon Estuaries and the Burry Inlet were all 40% or more below their recent averages, whilst continued low counts at Martin Mere saw the site lose its internationally important status for Pintail. Counts were also markedly lower than normal at six nationally important sites, though, somewhat surprisingly, three new sites attained this status in 1997-98.

Pintail is one of only two wildfowl species for which the international 1% threshold for Northwest Europe was reduced in the last review (Rose & Scott 1997), and has an 'unfavourable conservation status' in Europe due to large declines on the breeding grounds in many countries and in wintering areas. This has prompted the drafting of an EU Action Plan to restore the population to a more favourable conservation status, whilst an analysis of wintering trends in Britain has also been undertaken (Kershaw 1998).

Britain supports almost half the Northwest European wintering population, the highest proportion of any duck species. The overall pattern is of an increase up to the early 1980s, with numbers reaching a plateau and possibly declining slightly thereafter, although perceived changes in both Great Britain and Northwest European populations were not statistically significant.

Within Great Britain, there have been significantly different trends between regions, habitats and sites. Northwest England/North Wales is by far the most important region, holding three times the number of birds in the second most important region, East Central England. However, numbers appear to have reached a plateau and are declining in both regions, as well as Southwest England/South Wales. Only in Southeast England and in Cumbria/Southwest Scotland are they still increasing.

Whilst at present, there appears to be no significant decline nationally, there have been large declines on some of the traditionally most important sites, notably the Dee and Mersey

Estuaries. Although the species appears to be highly mobile and able to adapt to changing conditions, given their highly concentrated distribution within Europe, the more serious declines in breeding numbers in Russia/Finland and in wintering numbers in the east Mediterranean, it is important that numbers in Britain are monitored closely.

	93-94	94-95	95-96	96-97	97-98	Mon	Mean	
International								
Dee Estuary (Eng/Wal)	4,566	4,891	5,425	5,749	5,954	Dec	5,317	
Morecambe Bay	2,027	3,427	2,575	3,207	4,411	Oct	3,129	
Solway Estuary	2,356	2,567	4,016	3,852	2,677	Dec	3,094	
Ribble Estuary	1,795	1,587	4,926	4,073	1,271	Nov	2,730	
Ouse Washes	1,082	1,601	2,376	2,055	3,271	Feb	2,077	
Burry Inlet	1,585	942	3,541	2,889	1,093	Dec	2,010	
Nene Washes	2,313	2,569	1,342	264	1,668	Jan	1,631	
Duddon Estuary	2,194	2,261	1,275	1,349	464	Dec	1,509	
North Norfolk Marshes	1,443	923	1,036	1,177	1,668	Dec	1,249	
Mersey Estuary	1,636	1,620	873	904	813	Jan	1,169	
Medway Estuary	399	622	1,214	2,047	489	Jan	954	
Swale Estuary	1,349	1,310	1,029	277	570	Dec	907	
Pagham Harbour	596	604	990	1,210	1,087	Dec	897	
Severn Estuary	664	465	539	698	709	Feb	615	
Great Britain								
Stour Estuary	811	425	397	507	638	Dec	556	
Martin Mere	424	416	499	231	239	Jan	362	▼
Hamford Water	162	116	330	1,117	54	Feb	356	
Tottenhill Gravel Pits	-	108	486	415	397	Oct	352	▲
Orwell Estuary	321	282	821	214	94	Jan	346	
Blackwater Estuary	639	214	362	280	139	Nov	327	
Arun Valley	450	421	211	167	359	Jan	322	
Fleet/Wey	352	270	245	414	276	Feb	311	
Cromarty Firth	228	460	367	370	130	Jan	311	
Poole Harbour	186	231	301	375	451	Feb	309	▲
Somerset Levels	286	611	433	76	118	Feb	305	
Abberton Reservoir	225	242	316	283	430	Nov	299	
Thames Estuary	263	593	398	149	50	Feb	291	
Alde Complex	317	439	203	147	340	Jan	289	▲
Nationally Important Northern Ireland Sites								
Strangford Lough	269	180	159	242	304	Dec	231	

Sites no longer meeting table qualifying levels
Wash

Other sites surpassing table qualifying levels in 1997-98

Lower Derwent Valley	337	Mar
Bayfield Loch	290	Dec
Lough Foyle	67	Feb

BAHAMA PINTAIL
Anas bahamensis

Escape
Native range: South America

One was found on the Fleet/Wey in September.

CAPE TEAL
Anas capensis

Escape
Native range: Africa

A single frequented Beddington Sewage Farm in October and November, and presumably the same bird was at Staines Reservoirs in December.

GARGANEY
Anas querquedula

International threshold: 20,000**
Great Britain threshold: ?[†]
All-Ireland threshold: ?[†]

GB max: 37 Aug
NI max: 0

The typical August peak was well below that of the last two years, though the secondary peak in May was near normal. As expected, most records were from southern and southeastern England, with no records from Scotland or Northern Ireland and only one from Wales. The customary hot-spot in northern England, Fairburn Ings, held no birds in August, and no counts were available for another regular site, Chew Valley Lake. Unusually, small numbers were recorded in most winter months involving birds at ten different sites, stragglers perhaps taking advantage of the mild winter.

Sites with four or more birds in 1997-98

Dungeness Gravel Pits	7	Aug	Nene Washes	4	May
Minsmere Levels	6	May	Walland Marsh	4	Jul
Rye Harbour/Pett Level	5	Aug	Blithfield Reservoir	4	Aug

SHOVELER
Anas clypeata

International threshold: 400
Great Britain threshold: 100
All-Ireland threshold: 65

GB max: 9,268 Oct
NI max: 207 Dec

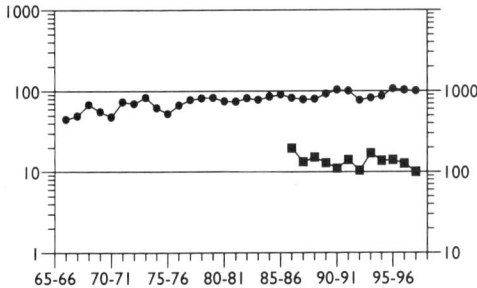

Figure 40. Annual indices for Shoveler in GB (circles, left axis) and NI (squares, right axis)

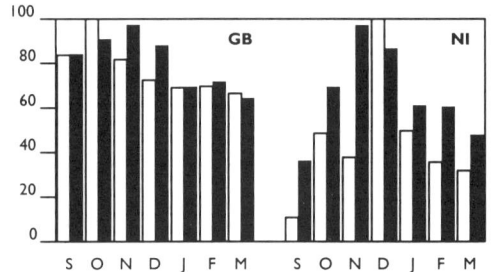

Figure 41. Monthly indices for Shoveler in GB and NI (white bars 1997-98; black bars 1992-93 to 1996-97)

The Great Britain peak was around average for recent winters, though well below the exceptional count of over 12,000 birds in late autumn 1995. However, the total does not include counts from one key site, Chew Valley Lake. Annual indices appear to show a long term increase, though this is less obvious for the last decade. Monthly indices show the normal autumn peak was relatively short lived in 1997-98, with many birds moving on to winter further south in France and Spain.

Annual indices in Northern Ireland reached their lowest level since regular monitoring began in the mid 1980s, although the small numbers involved at just a handful of sites results in large fluctuations for this species. As in Great Britain, numbers in Ireland are usually boosted in November, though monthly indices suggest unusually low numbers in 1997-98.

Peak counts at most sites occur during the autumn period. Due to the large movements of Shoveler through the country at this time, these peaks can fluctuate widely at individual sites and are more variable than for almost any other species of non-maritime wildfowl. High counts on the Nene Washes (104% higher than the previous average) saw the site added to the list of internationally important sites, though all except one other site with this status held below average numbers. Notably high counts were also made at Staines Reservoirs (+395%) and Blithfield Reservoir (+197%), whilst Hollowell Reservoir, Blagdon Lake and Aqualate Mere recorded peak counts well below their respective norms. Peaks at both key sites in Northern Ireland were the lowest of the last five years.

	93-94	94-95	95-96	96-97	97-98	Mon	Mean	
International								
Ouse Washes	1,066	837	212	663	540	Dec	664	
Abberton Reservoir	606	598	937	628	541	Sep	662	
Somerset Levels	373	931	839	435	504	Feb	616	
Rutland Water	701	513	562	704	531	Sep	602	
Loch Leven	458	570	550	541	426	Sep	509	
Swale Estuary	473	530	357	411	551	Dec	464	
Burry Inlet	193	395	745	490	363	Jan	437	
Chew Valley Lake	475	100	875	225	-		419	
King George VI Reservoir	153	246	1,134	310	248	Feb	418	
Nene Washes	367	517	347	143	689	Mar	413	▲
Great Britain								
Walland Marsh	-	-	-	359	325	Feb	342	
Dungeness Gravel Pits	320	163	252	421	260	Oct	283	
Lee Valley Gravel Pits	225	248	178	291	283	Nov	245	
Lower Derwent Valley	251	163 *	257	221	310	Feb	240	
Fairburn Ings	74	134	303	352	272	Oct	227	
Arun Valley	218	319	268	146	176	Nov	225	
Hanningfield Reservoir	110	242	254	211	304	Sep	224	
Knight/Bessborough Reservoirs	90	434	245	185	160	Sep	223	
Stodmarsh	71	199	240	265	328	Nov	221	
Stanford Reservoir	61	117	500	145	276	Nov	220	
Wraysbury Gravel Pits	221	214	341	157	169	Sep	220	
Blithfield Reservoir	329	172	11	77	436	Sep	205	
Thames Estuary	187	264	202	197	173	Feb	205	
Fleet/Wey	223	296	183	133	107	Feb	188	
Blagdon Lake	195	160	115	404	64	Nov	188	
Staines Reservoirs	126	29	74	210	490	Oct	186	▲
Severn Estuary	160	168	270	169	150	Jan	183	
North Norfolk Marshes	162	279	206	135	121	Nov	181	
Tees Estuary	108	122	232	202	201	Nov	173	
Pitsford Reservoir	20	239	196	236	157	Sep	170	
Aqualate Mere	114	97	225	358	50	Sep	169	
Grafham Water	74	73	240	290	160	Jan	167	
Ringstead Gravel Pits	283	262	151	0	86	Nov	156	
North Warren/Thorpness Mere	190	220	110	108	138	Mar	153	
Wraysbury Reservoir	132	138	238	184	69	Oct	152	
Thrapston Gravel Pit	128	258	173	108	88	Sep	151	
Rye Harbour/Pett Level	55	155	164	238	135	Nov	149	
Breydon Water & Berney Marshes	120	213	46	172	183	Feb	147	
Walthamstow Reservoirs	186	116	118	144	143	Jan	141	
Poole Harbour	263	117	156	64	103	Jan	141	
Leighton Moss	92	120	205	188	95	Sep	140	
Swillington Ings	94	204	70	155	141	Sep	133	
Middle Tame Valley GP	168	91	97	186	116	Feb	132	
Alde Complex	40	164	214	120	119	Jan	131	
Hampton & Kempton Reservoirs	49	142	123	234	88	Jan	127	
Fiddlers Ferry Lagoons	180	156	126	50	100	Sep	122	
Woolston Eyes	25	126	152	152	152	Sep	121	▲
Barn Elms Reservoirs	135	197	108	26	-		117	
Rostherne Mere	156	52	103	103	157	Oct	114	▲
Minsmere Levels	108	173	83	69	128	Oct	112	▲
Colne Valley Gravel Pits	93	110	142	123	94	Dec	112	
Swithland Reservoir	65	147	104	98	116	Sep	106	
Blackwater Estuary	100	128	101	122	60	Dec	102	
Hollowell Reservoir	137	106	124	129	16	Nov	102	
Marton Mere	-	142	104	83	77	Nov	102	▲
Nationally Important Northern Ireland Sites								
Lo. Neagh/Beg	319	103	150	89	84	Dec	149	
Strangford Lough	144	104	213	108	101	Dec	134	

Internationally or nationally important sites not counted in last five years
Lackford Gravel Pits
Ashford Common Waterworks

Other sites surpassing table qualifying levels in 1997-98

Mersey Estuary	214	Nov	Little Paxton Gravel Pits	115	Nov
Llyn Penrhyn	191	Dec	Wanstead Park Ponds	106	Dec
Lower Windrush Valley Gravel Pits	170	Feb	Middle Yare Valley	106	Feb
Willen Lake	138	Nov	Brent Reservoir	103	Oct

RINGED TEAL
Callonetta leucophrys

Escape
Native range: South America

One was at Thrapston Gravel Pits in August.

RED-CRESTED POCHARD
Netta rufina

Vagrant and escape[†]
Native range: Europe and Asia

GB max: 94 Nov
NI max: 0

The peak count was similar to that of recent years, given the absence of data for two sites which have formerly held significant numbers of this sedentary species. A total of 36 sites held birds in 1997-98, mainly in southern and eastern England, though most refer to small numbers of individuals and only six sites held five or more birds. The count at Cotswold WP West was the highest recorded at any site to date by WeBS.

	93-94	94-95	95-96	96-97	97-98	Mon	Mean
Great Britain							
Cotswold WP West	50	49	59	54	62	Jan	55
R Wensum: Fakenham - Gt Ryburgh	41	44	-	-	-		43
Paultons Bird Park	36	0	-	-	-		18
Cotswold WP East	8	20	26	15	12	Nov	16

† *as site designation does not occur and the 1% criterion is no applied, a qualifying level of 10 has been chosen to select sites for presentation in this report*

POCHARD
Aythya ferina

International threshold: 3,500
Great Britain threshold: 440
All-Ireland threshold: 400

GB max: 42,091 Jan
NI max: 19,309 Dec

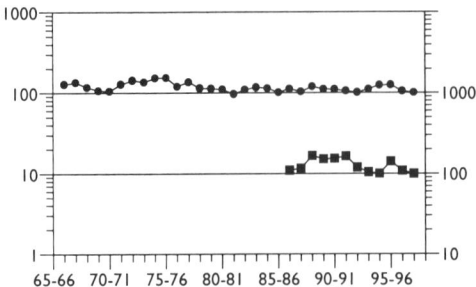

Figure 42. Annual indices for Pochard in GB (circles, left axis) and NI (squares, right axis)

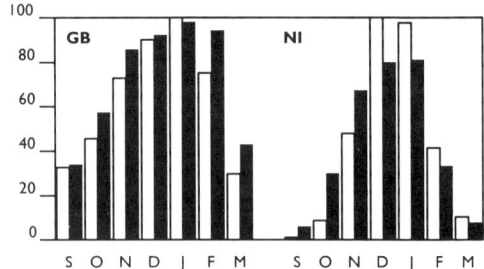

Figure 43. Monthly indices for Pochard in GB and NI (white bars 1997-98; black bars 1992-93 to 1996-97)

The peak count in Great Britain increased markedly from that recorded in the previous winter, and was around average for recent years. However, annual indices, which take account of variations in site coverage between winters, show a different tale, with the 1997-98 figure being the second lowest for over thirty years. Over the last twenty years, the population has shown a degree of fluctuation and the current winter index falls within, albeit at the lower end of, these natural variations.

As for most other species of diving duck, numbers in Northern Ireland are largely equivalent to counts at Loughs Neagh & Beg. The peak for 1997-98 was the lowest since the mid 1980s and just half the numbers occurring in the late 1980s and early 1990s. Consequently annual indices for the province equalled their lowest value ever.

The most noteworthy high count in 1997-98 was that on the Ouse Washes, where record numbers, perhaps attracted by the high water levels at the site, raised the five year average to above the threshold for international importance. Particularly low counts were made on the Humber Estuary and Rostherne Mere.

	93-94	94-95	95-96	96-97	97-98	Mon	Mean	
International								
Lo. Neagh/Beg	21,332	19,908	28,601	25,230	19,205	Dec	22,855	
Ouse Washes	3,087	3,786	3,929	1,413	5,737	Feb	3,590	▲
Great Britain								
Abberton Reservoir	3,240	2,014	3,247	3,079	2,518	Aug	2,820	
Lower Derwent Valley	2,785	5,184	1,020	750	2,350	Jan	2,418	
Severn Estuary	1,470	1,681	1,676	1,576	1,248	Jan	1,530	
Cotswold WP East	1,988	1,690	1,394	1,235	1,151	Jan	1,492	
Salford Docks	-	-	2,042	816	(3)	Oct	1,429	
Walton Lock	-	-	[1]1,400	-	-		1,400	
Rostherne Mere	749	1,186	[1]2,200	2,616	152	Feb	1,381	
Humber Estuary	2,029	700	1,000	2,503	183	Jan	1,283	
Rutland Water	250	2,346	1,776	855	680	Jan	1,181	
Middle Tame Valley GP	898	743	1,036	1,899	1,236	Feb	1,162	
Loch of Boardhouse	2,090	1,375	789	913	613	Oct	1,156	
Loch Leven	536	1,123	1,000	1,692	1,125	Jan	1,095	
Loch of Harray	298	1,081	2,070	1,119	506	Nov	1,015	
Nene Washes	1,675	2,094	528	185	435	Jan	983	
Cotswold WP West	1,151	1,086	1,163	562	922	Jan	977	
Fleet/Wey	949	1,232	913	853	848	Dec	959	
Chew Valley Lake	830	730	1,130	865	-		889	
Lower Windrush Valley Gravel Pits	952	699	1,331	622	780	Jan	877	
Poole Harbour	570	477	946	1,386	298	Jan	735	
Martin Mere	513	508	786	1,111	747	Feb	733	
Dungeness Gravel Pits	713	801	456	633	836	Dec	688	
Chorlton Water Park	1,153	750	417	218	589	Feb	625	
Cheddar Reservoir	1,204	689	632	428	140	Sep\Dec	619	
Kilconquhar Loch	712	412	646	761	468	Nov	600	
Wraysbury Gravel Pits	352	917	488	513	697	Dec	593	
Loch of Hundland	160	1,540	308	193	491	Jan	538	
Hanningfield Reservoir	367	359	1,084	467	377	Dec	531	
Loch Gelly	110	55	475	1,518	490	Oct	530	
Alton Water	236	916	826	370	174	Jan	504	
Baston/Langtoft Gravel Pits	754	396	573	322	456	Dec	500	▲
Staines Reservoirs	70	669	1,285	237	231	Dec	498	
Thames Estuary	576	472	515	539	207	Oct	462	
Fen Drayton Gravel Pit	244	592	975	195	274	Aug	456	
Little Paxton Gravel Pits	240	577	520	330	586	Dec	451	▲
Loch Watten	432	309	296	580	606	Dec	445	▲

Sites no longer meeting table qualifying levels
Eyebrook Reservoir
Avon Valley (Mid)
Arun Valley

Other sites surpassing table qualifying levels in 1997-98

Shustoke Reservoir	685	Jan	Kenfig Pool	498	Dec
Moorgreen Reservoir	527	Jan	Lee Valley Gravel Pits	461	Jan

1   B. Martin (in litt.)

RING-NECKED DUCK
Aythya collaris

Vagrant
Native range: North America

Singles were seen at Rutland Water, Kilconquhar Loch, Timsbury Gravel Pits, Pugney Water, Altofts Ings and Swillington Ings. Birds at Rutland, in August, and Kilconquhar, in September, were unseasonally early.

FERRUGINOUS DUCK
Aythya nyroca

Vagrant and escape
Native range: Europe, N Africa and Asia

One was at Beesands Ley throughout the year, with singles also recorded at Hanningfield Reservoir and on the North Norfolk Marshes.

TUFTED DUCK
Aythya fuligula

International threshold: 10,000
Great Britain threshold: 600
All-Ireland threshold: 400

GB max: 52,004 Dec
NI max: 19,021 Nov

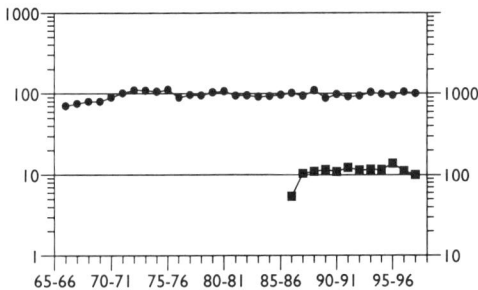

Figure 44. Annual indices for Tufted Duck in GB (circles, left axis) and squares (right axis)

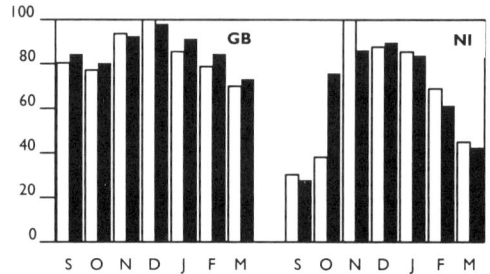

Figure 45. Monthly indices for Tufted Duck in GB and NI (white bars 1997-98; black bars 1992-93 to 1996-97)

Wintering Tufted Duck numbers in Great Britain are the most stable of any wildfowl species: the peak in 1997-98 was, predictably, around average for recent years. By contrast, numbers in Northern Ireland are subject to significant annual variations, influenced strongly by numbers at Loughs Neagh & Beg. Though the relatively low peak count appears to represent a significant fall, it is still well within the bounds of previous fluctuations. Monthly indices suggest that the arrival of birds in Northern Ireland was initially slow, but followed by a large and earlier than usual influx.

The distribution of sites holding important numbers of Tufted Duck is widespread, though concentrated in lowland areas. As with national counts, numbers at individual sites are more consistent than for many other wildfowl species. Large counts at Wraysbury Gravel Pits, Walthamstow Reservoirs and the Ouse Washes, the last coinciding with increased Pochard numbers at this site also, were thus notable, whilst numbers have increased steadily on the Middle Tame Valley Gravel Pits in recent years. It is also worth highlighting that peak counts from several sites in the table below, predominantly in southeast England, occurred in August, presumably gatherings of moulting birds.

	93-94	94-95	95-96	96-97	97-98	Mon	Mean
International							
Lo. Neagh/Beg	22,470	21,101	25,340	27,368	18,697	Nov	22,995
Great Britain							
Loch Leven	3,481	3,800	3,000	4,589	3,310	Sep	3,636
Rutland Water	2,500	2,448	3,775	3,159	3,557	Aug	3,088
Abberton Reservoir	2,126	1,803	1,356	3,218	2,268	Sep	2,154
Middle Tame Valley GP	1,411	1,162	2,018	2,384	2,422	Jan	1,879

	93-94	94-95	95-96	96-97	97-98	Mon	Mean
Wraysbury Gravel Pits	1,140	1,678	844	1,709	2,868	Dec	1,648
Hanningfield Reservoir	1,624	1,213	1,594	1,600	1,747	Aug	1,556
Staines Reservoirs	977	382	3,332	1,405	1,283	Aug	1,476
Pitsford Reservoir	1,428	1,616	854	2,034	1,129	Feb	1,412
Alton Water	463	1,682	1,331	1,536	783	Jan	1,159
Ouse Washes	1,737	761	811	391	1,165	Feb	973
Loch of Harray	774	1,202	1,625	524	713	Oct	968
Walthamstow Reservoirs	781	868	722	1,083	1,368	Aug	964
Lee Valley Gravel Pits	863	1,069	726	1,163	935	Nov	951
William Girling Reservoir	534	1,331	1,300	738	807	Aug	942
Thames Estuary	914	812	1,055	769	434	Feb	797
South Muskham/North Newark GP	760	782	1,075	740	592	Nov	790
Chasewater	792	855	723	-	-		790
Besthorpe/Girton Gravel Pits	452	635	913	1,122	637	Feb	752
King George V Reservoirs	780	500	678	1,020	700	Aug	736
Windermere	802	682	727	565	-		694
Draycote Water	638	630	925	475	645	Nov	663
Severn Estuary	571	662	1,004	610	382	Feb	646
Cotswold WP East	564	559	711	647	707	Nov	638 ▲
Little Paxton Gravel Pits	382	666	696	852	489	Nov	617 ▲
Dungeness Gravel Pits	583	608	493	558	760	Nov	600 ▲

Nationally Important Northern Ireland Sites

Upper Lough Erne	381	293	349	644	509	Feb	435

Sites no longer meeting table qualifying levels
Inner Moray Firth
Cotswold WP West

Other sites surpassing table qualifying levels in 1997-98

Loch Watten	941	Dec	Fen Drayton Gravel Pit	679	Dec
Fairburn Ings	774	Aug	Chichester Gravel Pits	639	Jan
Blithfield Reservoir	769	Jul	Lower Windrush Valley Gravel Pits	624	Feb

---

SCAUP
Aythya marila

International threshold: 3,100
Great Britain threshold: 110
All-Ireland threshold: 30*

* 50 is normally used as a minimum threshold

GB max: 7,529 Dec
NI max: 3,816 Jan

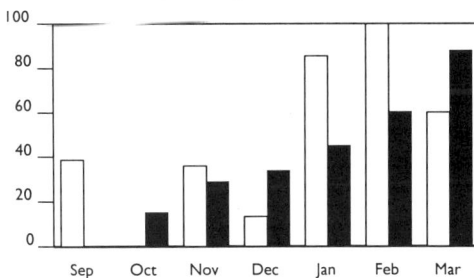

Figure 46. Monly indices for Scaup in NI (white bars 1997-98; black bars 1992-93 to 1996-97)

Even following the incorporation of late data for 1996-97, elevating the maximum for that winter to almost 6,900 birds, the most recent British peak represents a further notable increase in numbers and, with the exception of that in 1988-89, is the highest total since large numbers frequented the Forth Estuary in the mid 1970s. Conversely, the

peak in Northern Ireland fell slightly, and is more than 20% lower than counts in the mid 1990s. Although numbers in the province were much higher than normal in most months, the count in March, when the peak usually occurs, was very low, perhaps simply as result of the premigratory gathering of birds at Loughs Neagh & Beg falling between the WeBS count dates.

UK totals for December, January and February were consistently around 8,000. The particularly high peak on the Solway in December corresponded with low numbers on Loughs Neagh & Beg, strongly suggesting interchange between these two key sites.

The table below illustrates the large fluctuations that occur at some sites. Whilst some of this will be due to the inherent problems involved in counting sea-ducks, Scaup often occur relatively close inshore, enabling counts to be made with some degree of success; that record numbers were recorded at the top three sites in Great Britain during WeBS counts is perhaps a demonstration of this.

Whilst the very low counts on the Humber Estuary and the North Norfolk Marshes might at first suggest that counts of sea-ducks at these sites were hampered in 1997-98, counts of Common Scoter and Long-tailed Ducks at the latter were about average. Further, there is a remarkable similarity in the pattern of numbers at both sites, with large numbers in 1995-96 and 1996-97 but a virtual absence in other years. Given the close proximity of these sites, this suggests that counts indicate a genuine change in numbers of wintering Scaup in the area. It is

possible that this phenomenon also affects use of the Forth, the only other important site on the east coast, given the similar pattern of numbers at this site also. Unlike 1996-97, when many additional east coast sites held large numbers, counts at only one further site exceeded 110 birds in 1997-98, despite the large national total.

No such patterns are obvious at other sites, although there is some suggestion of redistribution between the various firths around the Moray. Around 20 years since detailed studies of birds on the Forth, it appears that much remains to be learned about current movements and site use by Scaup.

	93-94	94-95	95-96	96-97	97-98	Mon	Mean
International							
Lo. Neagh/Beg	2,632	4,934	4,022	4,222	3,671	Feb	3,896
Great Britain							
Solway Estuary	2,084	2,007	(484)	2,341	4,533	Dec	2,741
Loch Indaal	699	969	661	732	1,110	Jan	834
Loch Ryan	350	[1]622	[1]916	600	1,249	Dec	600
Forth Est.	135	77	[2]753	1,031	145	Jan	381
Loch of Stenness	317	267	361	318	258	Feb	304
Cromarty Firth	[3]381	406	279	115	45	Oct	229
North Norfolk Marshes	4	11	517	482	1	Oct	203
Humber Estuary	32	12	353	594	21	Jan	202
Inner Moray Firth	14	90	120	332	[4]416	Jan	180
Loch of Harray	175	194	191	16	208	Oct	157
Irvine to Saltcoats	70	158	160	57	-		111
Dornoch Firth	[3]42	[3]85	219	122	36	Jan	101
Northern Ireland							
Carlingford Lough	877	472	800	404	572	Jan	625
Belfast Lough	103	186	247	243	95	Jan	174

Other sites surpassing table qualifying levels in 1997-98
Clyde Estuary   111   Mar

1   P. Collin (in litt.)
2   SNH funded surveys of SE Scotland (WWT, unpubl. data)
3   RSPB/BP studies (e.g. Stenning 1994)
4   Stenning (1998) RSPB report to Talisman Energy

LESSER SCAUP
Aythya affinis

Vagrant
Native range: North America

A single was at Tophill Low Reservoirs in April.

EIDER
Somateria mollissima

International threshold: 20,000**
Great Britain threshold: 750
All-Ireland threshold: 20*
* 50 is normally used as a minimum threshold

GB max:	24,579	Oct
NI max:	1,091	Dec

Whilst the 1997-98 maximum in Great Britain was the smallest for 10 years, it was only fractionally

below the recent average and was consistent with the relatively high and stable numbers counted by WeBS over the last six years. The peak in Northern Ireland fell in the middle of the normal range of fluctuations for the same period.

The WeBS count on the Tay Estuary was particularly impressive in view of the considerable problems of counting Eider at this site. The number in Morecambe Bay rose markedly but those on the Forth and Clyde Estuaries fell sharply. Counts at most other sites were generally very similar to their five year means. It should be noted, however, that low counts at some sites in recent years might be better classified as undercounts. Consequently, average counts, particularly for sites on the Ayrshire Coast, may in reality be much higher than indicated in the table. The need to determine what areas constitute a 'site' for Eider and to ensure that coverage is co-ordinated will be one aspect considered in the sea-duck monitoring feasibility study being undertaken by WWT for JNCC.

Also of importance for Eider will be coverage of sites in Orkney and Shetland, particularly if the suggestion by Wetlands International that these constitute a separate population (Rose & Scott 1997) is adopted by the UK Government.

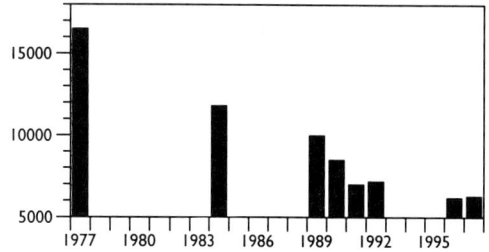

Figure 47. Counts of moulting Eider in Shetland, 1977-1997. Only results from the most thorough surveys are included (from Heubeck 1998)

Biometrics and the apparently sedentary nature of Shetland birds (M. Heubeck 1993, pers. comm.) support this suggestion, whilst the dramatic decline in the last 20 years (Figure 47) from 16,500 to just over 6,000 (Heubeck 1998) gives real cause for concern, irrespective of the population limits.

	93-94	94-95	95-96	96-97	97-98	Mon	Mean
Great Britain							
Tay Estuary[R]	(4)	(4,200)	[1]12,250	(2,500)	9,500	Jan	10,875
Forth Est.	9,698	8,964	[1]9,764	9,166	6,937	Sep	8,460
Morecambe Bay	6,886	6,708	4,882	6,073	8,200	Oct	6,550
Clyde Est.	3,450	3,638	4,238	5,779	3,299	Oct	4,081
Ayr to Troon	775	4,332	8,000	705	3,767	Aug	3,516
Ythan Estuary	3,150	2,420	3,700	3,216	3,366	Jul	3,170
Montrose Basin	1,537	2,120	2,100	2,100	2,163	Nov	2,004
Don Mouth to Ythan Mouth	3,500	3,000	107	1,215	2,159	Sep	1,996
Girvan to Turnberry	250	1,797	1,846	2,835	2,645	Dec	1,875
Lindisfarne	2,055	1,233	2,474	1,255	1,209	Sep	1,645
Lo. Ryan	(93)	[3]893	[3]1,606	(26)	(228)	Jan	1,250
Seahouses to Budle Point	1,520	527	1,221	903	671	Mar	968
Lo. Fleet	[2]1,363	[?]569	(376)	(174)	(33)	Jan	966
Great Cumbrae	679	797	1,077	941	833	Feb	865
Wash	200	231	1,639	1,569	638	Jan	855
Irvine/Garnock Estuary	845	800	1,200	500	-		836
Dee Estuary (Scotland)	620	652	639	1,492	677	Aug	816
Irvine to Saltcoats	1,200	600	600	790	-		798
Northern Ireland							
Belfast Lough	952	695	1,020	448	922	Nov	807
Outer Ards	362	360	255	709	470	Dec	431
Lough Foyle	57	74	83	452	161	Aug	165
Larne Lough	16	61	157	96	39	Sep	74
Strangford Lough	33	26	34	61	52	Mar	41

Internationally or nationally important sites not counted in last five years
Blyth to Newbiggin

Other sites surpassing table qualifying levels in 1997-98
Wemyss Bay to Fairlie   991   Oct

1   SNH funded surveys of SE Scotland (WWT, unpubl. data)
2   RSPB/BP studies (e.g. Stenning 1994)
3   P. Collin (in litt.)

KING EIDER
Somateria spectabilis

Vagrant
Native range: Circumpolar Arctic

A single was on the Forth Estuary in September and October.

LONG-TAILED DUCK
Clangula hyemalis

International threshold: 20,000**
Great Britain threshold: 230[†]
All-Ireland threshold: +[†]

GB max: 1,793 Dec
NI max: 20 Dec

This species is the most pelagic in nature of all UK wildfowl and, consequently, the most difficult to monitor accurately. Thus, WeBS annual maxima, which have continued to fluctuate at around 2,000 birds in recent years, are much lower than the known population.

The resumption of dedicated monitoring of sea-ducks in the Moray Firth enabled a much better assessment of this area in 1997-98 (Stenning 1998). However, this is still likely to be a considerable under-count of Long-tailed Duck for the site, since many birds remain out of sight of land during the day and are best recorded as they flight to roosts. In the absence of data collected using this method, there seems little reason to suppose that numbers have declined since regular records of 10,000 birds in the early

1990s, but proper surveys are now of high priority to confirm the true picture. A feasibility study for national surveys of sea-ducks is currently being undertaken by WWT for JNCC.

Counts at most other sites were relatively low in 1997-98 and only one count exceeded 100 birds away from nationally important sites. The species' occurrence in more remote parts of Scotland is illustrated by non-WeBS counts, e.g. 700 in Orkney in October (Murray 1998) and counts from the two Shetland sites (Yell Sound and Hascosay, Bluemull and Colgrave Sounds) in the table below, obtained during dedicated land and boat-based surveys (Heubeck 1998). Proper survey of these areas is essential to determine the true size of the British wintering population.

	93-94	94-95	95-96	96-97	97-98	Mon	Mean
Great Britain[†]							
Moray Firth	[1]10,115	[1]3,742	(660)	(734)	[2]2,006	Dec	5,288
Forth Est.	942	1,057	461	975	660	Mar	819
Loch Branahuie & Aignish	400	-	-	-	-		400
Hascosay, Bluemull & Colgrave S[3]	483	147	-	421	383	Jan	359
St Andrews Bay	[4]341	(47)	[5]265	106	(29)	Dec	237
South Yell Sound[3]	170	-	157	157	270	Dec	189
Water Sound	222	215	137	88	96	Jan	152
Traigh Luskentyre	-	190	13	146	152	Dec	125
Loch of Stenness	243	118	80	108	48	Jan	119
North Norfolk Marshes	137	57	65	3	34	Jan	59
Lindisfarne	80	82	14	13	55	Dec	49
Seahouses to Budle Point	7	13	8	150	20	Feb	40
Loch of Harray	33	68	26	22	21	Dec	34

Internationally or nationally important sites not counted in last five years
Sound of Taransay

Sites no longer meeting table qualifying levels
Widewall Bay

[†]   as few sites in Great Britain are of national importance, a qualifying level of 30 has been chosen to select sites for presentation in this report
[1]   RSPB/BP studies, e.g. Stenning (1994)
[2]   Stenning (1998)
[3]   Heubeck (1998)
[4]   L Hatton (in litt.)
[5]   SNH funded surveys of SE Scotland (WWT, unpubl. data)

COMMON SCOTER
Melanitta nigra

International threshold: 16,000
Great Britain threshold: 350
All-Ireland threshold: 40*

* 50 is normally used as a minimum threshold

GB max: 8,565   Dec
NI max: 1   Jan/Mar

The British peak in 1997-98 was about average for the last 15 years, but very few have been recorded in Northern Ireland since large numbers in the early 1990s. The table below suggests that this is largely due to a lack of counts of the appropriate sites or in suitable conditions. However, numbers at many UK sites fluctuate considerably both within and between years and it is possible that large numbers of transient birds may simply be missed between WeBS dates, or may no longer use the traditional sites in the province.

Whilst there was a slight increase in numbers at Carmarthen Bay, two years after the

Sea Empress oil spill, aerial surveys revealed numbers remained well below pre-spill levels (Cranswick *et al.* 1998). Interestingly, a large proportion of birds were regularly found in discrete flocks 6-8 km offshore in the east of the bay, in an area little used before the oil spill. A CCW-funded study will conduct benthic surveys of Carmarthen Bay and continue land-based and aerial counts of scoter over the next three years to examine the relationship between bird numbers and distribution and that of their prey.

Apart from a few sites at which counts fluctuate dramatically, most held more or less average numbers of Common Scoter in 1997-98.

	93-94	94-95	95-96	96-97	97-98	Mon	Mean
Great Britain							
Carmarthen Bay	5,012	[1]17,650	[1]10,631	[1]4,323	[2]6,240	Jan	8,771
Cardigan Bay[3]	4,872	4,755	6,720	(636)	5,220	Jan	5,392
North Norfolk Marshes	276	4,750	5,549	2,070	1,860	Dec	2,901
Forth Est.	1128	[4]7,304	[5]2,023	2,320	1,205	Mar	2,796
Moray Firth	[6]2,197	[6]2,988	2,764	(609)	[7]2,061	Dec	2,503
St Andrews Bay	[8]4,420	1,410	1,810	1,704	2,771	Dec	2,139
Rough Firth/Auchencairn Bay[R]	-	-	-	-	2,000	Sep	2,000 ▲
Colwyn Bay	1,300	380	5,480	11	386	Dec	1,511
Clwyd Estuary	56	100	7,000	0	7	Aug	1,433
Don Mouth to Ythan Mouth	300	2,000	[9]1,500	500	525	Sep	965
Earlsferry to Anstruther	-	-	860	-	-		860
Alt Estuary	173	2,000	851	12	811	Jan	769
Wash	70	68	2,002	351	200	Jan	538
Northern Ireland							
Craigalea to Newcastle	-	-	430	-	-		430
Tyrella	-	-	135	-	-		135

Sites no longer meeting table qualifying levels
Dundrum Bay

Other sites surpassing table qualifying levels in 1997-98
Traeth Coch   500   Jan

1   *Stewart et al. (1996)*
2   *Cranswick et al. (1998)*
3   *Data from Friends of Cardigan Bay, e.g. Green & Elliott (1993), and RSPB data (Reg Thorpe, in litt.)*
4   *RSPB/BP studies, e.g. Stenning (1994)*
5   *SNH funded surveys of SE Scotland (WWT, unpubl. data)*
6   *Due to some count sections of the Forth being covered on different dates, it is possible that 2,650 birds may have been counted twice. This would result in a 1994-95 figure of 4,654 and a five year mean of 2,182*
7   *Stenning (1998) RSPB report to Talisman Energy*
8   *L Hatton (in litt.)*
9   *A. Webb (in litt.)*

SURF SCOTER
Melanitta perspicillata

Vagrant
Native range: North America

Three were on the Forth Estuary from mid winter onwards, with one of these or another in St

Andrews Bay in November and December. Two were found at Traigh Luskentyre in February.

VELVET SCOTER
Melanitta fusca

International threshold: 10,000
Great Britain threshold: 30*
All-Ireland threshold: +*

* 50 is normally used as a minimum threshold

GB max: 792 Dec
NI max: 4 Nov

Numbers in the UK in 1997-98 were about normal. Counts since the 1960s are characterised by marked peaks, much closer to the supposed true number of wintering birds in the UK. These peaks do not coincide with increases in other sea-duck in the UK, in particular Common Scoter, which uses similar wintering sites both on the continent and in the UK, suggesting that they

arise due to fortuitous conditions occurring simultaneously at all of the key sites, rather than influxes of birds in those years. Although Velvet Scoter are recorded widely and regularly around much of the UK coastline, the table below demonstrates the marked concentration of the population at the traditional sites and the absence of significant gatherings elsewhere.

	93-94	94-95	95-96	96-97	97-98	Mon	Mean
Great Britain							
St Andrews Bay	[1]1,568	280	1,000	942	520	Nov	862
Moray Firth	[2]1,063	[2]540	(183)	(81)	[3]804	Dec	802
Forth Est.	510	485	[4]1,051	868	528	Mar	688
North Norfolk Marshes	9	101	108	2	18	Oct	48

1	L. Hatton (in litt.)
2	includes data from RSPB/BP studies, e.g. Stenning (1994)

3	Stenning (1998) RSPB report to Talisman Energy
4	SNH funded surveys of SE Scotland (WWT, unpubl. data)

GOLDENEYE
Bucephala clangula

International threshold: 3,000
Great Britain threshold: 170
All-Ireland threshold: 110

GB max: 16,355 Dec
NI max: 6,107 Nov

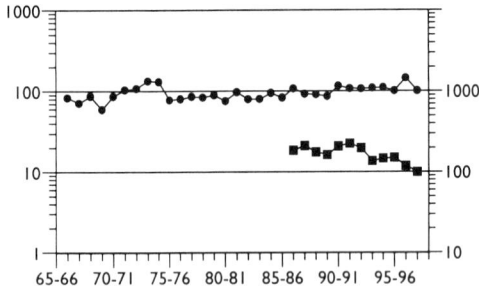

Figure 48. Annual indices for Goldeneye in GB (circles, left axis) and NI (squares, right axis)

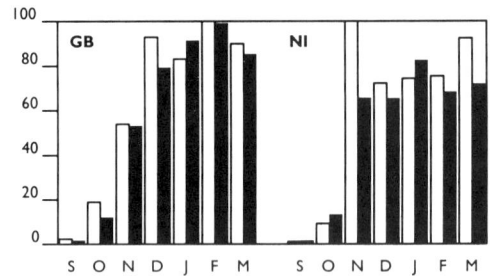

Figure 49. Monthly indices for Goldeneye in GB and NI (white bars 1997-98; black bars 1992-93 to 1996-97)

The peak in Northern Ireland fell to its lowest level since counts began, and was less than half the number regularly recorded between the mid 1980s and early 1990s, a picture reflected by the annual indices. This general decline is not matched by numbers in Great Britain and is contrary to the large increase in the international population during the same period (Rose 1995).

Loughs Neagh & Beg, which holds the majority of these birds, is at the western limit of the wintering range (Scott & Rose 1997), and it might be speculated that birds have been 'short-stopping' further east as a result of milder winters in recent years. Birds presumably pass through southern Scotland from the Scandinavian breeding grounds *en route* to Northern Ireland,

and, although there was an especially large count at the next most important site, the Forth Estuary, in December, this peak was short lived and there has been no discernable change in numbers in previous years.

Unlike Great Britain, where numbers build steadily as winter progresses, monthly indices for Northern Ireland show a relatively early arrival of large numbers, perhaps suggesting a different 'population' moves into the province. Although numbers then declined rapidly in 1997-98, there was no corresponding increase in the Republic of Ireland. Indeed, numbers recorded during the last four winters by I-WeBS have been relatively small and stable, offering no obvious explanation for the 'missing' birds Loughs Neagh & Beg birds.

It is possible that birds move onto rivers, where they remain largely undetected by WeBS and I-WeBS, and a survey of birds on this habitat is required in order to produce an accurate estimate of the wintering population in Britain and Ireland. Alternatively, a change in food availability at Loughs Neagh & Beg, suggested to be affecting some waterfowl species, may have precipitated this switch of habitats, sites or even countries. However, due to the extreme difficulties in catching sea-ducks, no ringing data exist to help determine whether the population as a whole has genuinely declined, or whether birds have simply moved elsewhere.

Peak numbers in Great Britain were about average for recent winters, although there was a marked early arrival in December, due to a remarkably high count on the Forth Estuary, sufficient to elevate its status to internationally

important, although numbers dispersed rapidly thereafter. Counts at most other key sites matched their respective long-term means with the exception of low counts on the Doon Estuary, Lavan Sands and Belfast Lough, and a record high count at Poole Harbour. Lough Money appears in the list of sites important in an all-Ireland context for the first time, although it had been omitted by accident from previous reports.

Comparing counts made on the River Deveron, northeast Scotland, from the 1940s with data from 1996 and 1997, Watson et al. (1998) highlighted the long standing importance of the river for wintering Goldeneye. Numbers were higher during periods of hard frost, indicating the significance of the river when lochs and ponds were frozen. Similar trends were found for Goosander, although the reverse was true for Red-breasted Mergansers and Cormorants.

	93-94	94-95	95-96	96-97	97-98	Mon	Mean	
International								
Lo. Neagh/Beg	10,085	6,712	9,793	8,081	5,587	Nov	8,052	
Forth Est.	2,771	2,369	[1]2,125	2,892	4,864	Dec	3,004	▲
Great Britain								
Inner Moray Firth	[2]552	[2]757	579	903	[3]895	Jan	737	
Tweed Estuary	540	498	617	804	570	Jan	606	
Girvan to Turnberry	100	700	854	672	315	Mar	528	
Clyde Est.	511	439	584	562	509	Dec	521	
Abberton Reservoir	178	431	488	839	426	Feb	472	
Doon Estuary	408	490	607	505	118	Feb	426	
Morecambe Bay	565	379	504	329	310	Dec	417	
Rutland Water	401	448	366	427	424	Feb	413	
Humber Estuary	260	359	331	558	287	Jan	359	
Loch Leven	468	310	300	314	301	Jan	339	
Windermere	409	331	223	296	-		315	
R Tweed: Kelso to Coldstream	225	369	314	268	334	Nov	302	
Blackwater Estuary	301	254	289	353	228	Jan	285	
Hornsea Mere	175	386	-	-	(117)	Mar	281	
Kilconquhar Loch	136	370	167	322	253	Feb	250	
Poole Harbour	132	201	220	232	405	Feb	238	
Lavan Sands	204	465	250	120	86	Mar	225	
Loch of Stenness	247	191	191	259	222	Jan	222	
Fleet/Wey	243	136	175	254	248	Dec	211	
Loch of Skene	250	150	160	202	238	Jan	200	
Irvine to Saltcoats	164	150	340	140	-		199	
North Norfolk Marshes	135	264	182	218	124	Dec	185	
Tay Estuary	121	157	208	251	155	Feb	178	
Northern Ireland								
Belfast Lough	544	977	549	400	259	Nov	546	
Strangford Lough	157	396	216	192	302	Dec	253	
Larne Lough	152	200	297	284	238	Mar	234	
Carlingford Lough	143	154	150	257	200	Jan	181	
Upper Lough Erne	96	104	149	161	91	Feb	120	
Lough Money	101	119	116	136	85	Feb	111	▲

Sites no longer meeting table qualifying levels
Solway Estuary   Cromarty Firth

Other sites surpassing table qualifying levels in 1997-98
Dornoch Firth   216   Feb   Loch Watten   185   Feb
Loch of Strathbeg   204   Oct

1   SNH funded surveys of SE Scotland (WWT, unpubl. data)   3   Stenning (1998) RSPB report to Talisman Energy
2   includes data from RSPB/BP studies, e.g. Stenning (1994)

HOODED MERGANSER
Lophodytes cucullatus

Escape
Native range: North America

One was at Barnstone Pools during late summer.

SMEW
Mergellus albellus

International threshold:			**250**
Great Britain threshold:			**2*[†]**
All-Ireland threshold:			**+***

* 50 is normally used as a minimum threshold

GB max:	300	Jan
NI max:	1	Jan-Mar

As was to be expected during a mild winter, numbers in 1997-98 did not reach the same heights as during the two previous and much colder years. However, it appears that some 'tradition' of wintering in Britain was established in just this short time as numbers in 1997-98 remained higher than normal.

A greater disposition to use British wintering sites is demonstrated in the table below, with counts at both Wraysbury and Lee Valley Gravel Pits particularly remarkable in view of the mild conditions. Indeed, the former represents the highest count at an individual site by WeBS; the only other counts to have exceeded 50 were at

Abberton Reservoir and Besthorpe/Girton Gravel Pits, both during the hard winters of the early 1960s. In addition, three sites feature for the first time in the list of those for which the five year peak mean exceeds the 1% threshold (and Trinity Broads was inadvertently missed from the previous report). The traditional use of sites is perhaps best illustrated by the regular occurrence at Loch of Strathbeg in Northeast Grampian. This site is clearly not simply receiving birds pushed out from the southern Baltic or The Netherlands, the main centres of population during the winter; perhaps these birds arrive using a more direct route from their northern breeding grounds?

	93-94	94-95	95-96	96-97	97-98	Mon	Mean	
Great Britain								
Wraysbury Gravel Pits	29	9	30	43	61	Jan	34	
Dungeness Gravel Pits	33	25	31	16	18	Jan	25	
Lee Valley Gravel Pits	3	3	3	31	23	Jan	13	
Rutland Water	7	15	12	14	18	Jan	13	
Fen Drayton Gravel Pit	1	4	11	22	15	Feb	11	
Thorpe Water Park	3	5	20	11	13	Feb	10	
Earls Barton Gravel Pits	5	5	7	13	15	Feb	9	
Twyford Gravel Pits	3	2	10	13	11	Feb	8	
Eyebrook Reservoir	10	4	3	12	5	Dec	7	
Chew Valley Lake	4	2	7	14	-		7	
Cotswold WP West	1	3	10	10	8	Feb	6	▲
Eglwys Nunydd Reservoir	2	3	6	12	7	Feb	6	▲
Bedfont/Ashford Gravel Pits	0	-	4	16	5	Feb	6	
Loch of Strathbeg	5	4	7	9	5	Feb	6	
Langtoft West End Gravel Pits	6	4	8	6	2	Jan	5	
Trinity Broads	-	-	8	-	2	Dec	5	▲
Pitsford Reservoir	2	1	11	9	2	Dec	5	▲
Croxall Pits	0	0	11	14	0		5	

Internationally or nationally important sites not counted in last five years
Staines Moor Gravel Pits

Other sites surpassing table qualifying levels in 1997-98

Barton Pits	10	Feb	Chichester Gravel Pits	5	Jan/Feb
Seaton Gravel Pits	9	Jan/Feb	Little Paxton Gravel Pits	5	Jan
Fairburn Ings	8	Feb	Middle Tame Valley Gravel Pits	5	Jan
Clifford Hill Gravel Pits	6	Feb			

RED-BREASTED MERGANSER
Mergus serrator

GB max: 4,270 Mar
NI max: 609 Dec

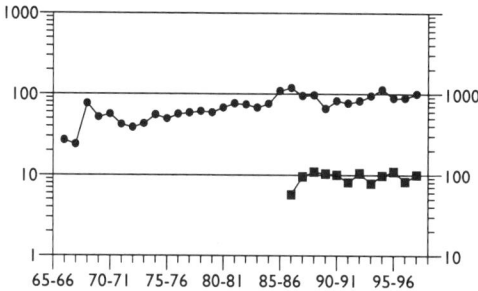

Figure 50. Annual indices for Red-breasted Merganser in GB (circles, left axis) and NI (squares, right axis)

The peak British count was very similar to that in the previous two winters, although annual indices suggest there was a small increase in 1997-98 to near record levels for the population as a whole. A not dissimilar picture is seen in

International threshold: 1,250
Great Britain threshold: 100
All-Ireland threshold: 20*

* 50 is normally used as a minimum threshold

Northern Ireland, with counts around normal but index values suggesting a slight rise.

Such increases are not readily explained by counts at the important sites. On the contrary, the peak count on the Inner Moray Firth was one of the lowest, even with dedicated counts of sea-duck in the Moray area. This species is normally found very close to the shoreline, enabling counts to be made with relative ease, and the genuine decline in numbers may be linked to the suspected decrease in fish stocks at this site. The same surveys recorded exceptionally high numbers on the adjacent Cromarty which may explain at least part of the difference on the Inner Moray. The most notable counts elsewhere were large totals at both Montrose Basin, elevating its status to nationally important, and at Lough Foyle, the highest at that site since WeBS began.

	93-94	94-95	95-96	96-97	97-98	Mon	Mean	
International								
Inner Moray Firth	[1]3,509	[1]1,544	(163)	706	[3]239	Dec	1,500	
Great Britain								
Forth Est.	348	1,053	665	715	675	Sep	691	
Duddon Estuary	365	538	424	382	394	Sep	421	
Poole Harbour	366	375	448	333	502	Mar	405	
Fleet/Wey	417	245	329	344	440	Mar	355	
Morecambe Bay	283	311	297	323	312	Dec	305	
Lavan Sands	[2]380	159	[2]288	[2]330	249	Sep	281	
Langstone Harbour	127	201	419	182	199	Nov	226	
Cromarty Firth	[1]347	179	116	193	[3]508	Dec	287	
Clyde Est.	109	127	292	230	186	Mar	189	
Loch Indaal	259	167	172	159	157	Aug	183	
Chichester Harbour	129	176	120	94	184	Mar	141	
Irvine to Saltcoats	90	131	150	135	-		127	
Irvine/Garnock Estuary	101	87	176	118	-		121	
Montrose Basin	45	79	220	52	204	Jun	120	▲
Exe Estuary	166	95	114	67	133	Mar	115	
North Norfolk Marshes	77	84	141	166	102	Nov	114	
Wash	98	74	104	132	109	Jan	103	▲
Loch Branahuie & Aigrish	100	-	-	-	-		100	
Northern Ireland								
Strangford Lough	217	264	486	276	191	Dec	287	
Larne Lough	174	295	331	201	171	Mar	234	
Belfast Lough	144	265	180	93	270	Nov	190	
Lough Foyle	30	39	197	130	296	Aug	138	
Craigalea to Newcastle	-	-	62	-	-		62	
Outer Ards	54	32	34	65	50	Dec	47	
Lo. Neagh/Beg	101	33	32	27	23	Aug	47	
Carlingford Lough	24	25	29	36	44	Feb	32	
Tyrella	-	-	21	-	-		21	

GOOSANDER
Mergus merganser

International threshold: **2,000**
Great Britain threshold: **90**
All-Ireland threshold: **+***

GB max: 3,628   Feb
NI max: 1   Dec/Jan/Mar

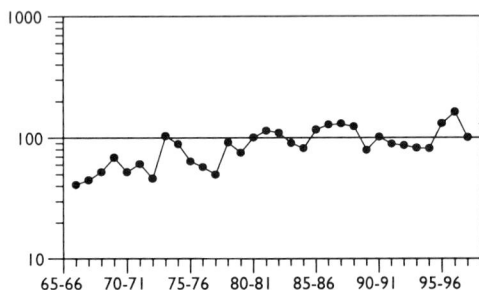

Figure 51. Annual indices for Goosander in GB

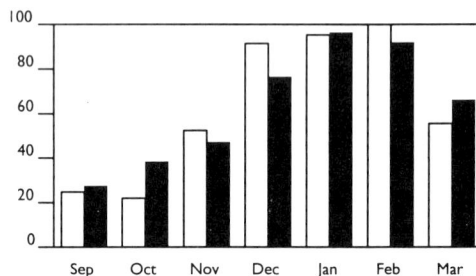

Figure 52. Monthly indices for Goosander in GB (white bars 1997-98; black bars 1992-93 to 1996-97)

Both the peak count and annual index values for Goosander dropped markedly, as expected, as a result of mild weather in 1997-98 following two cold winters. Like Smew, however, there is some suggestion that a number of birds continued to spill into Britain from the continent, with 1997-98 counts remaining higher than normal despite the mild conditions, perhaps having become accustomed to wintering further west.

At only a handful of nationally important sites did 1997-98 peak counts show marked deviation from the five year mean. Most notable are the continuing low numbers on the Inner Moray Firth, coinciding with low numbers of Cormorant and Red-breasted Mergansers also. The Sea Mammal Research Unit have recorded a decline in seal numbers in the Moray since 1993 (P. Thompson pers. comm.), strongly implicating a decline in fish stocks affecting all of these species.

Numbers were also down on the River Tweed and Hirsel Lake. The latter is used primarily as a roost site, and low numbers resulted simply from the absence of roost counts (high numbers in previous years have been recorded during dawn counts). However, Hirsel Lake attracts birds from the Tweed and thus, whilst low counts at both sites may suggest a genuine drop in numbers in the area, comprehensive surveys of the river system and or key roosts is necessary to determine the true picture. Numbers at Loch Lomond continued to increase in 1997-98 to such a level that the site attained nationally important status. The single bird in Northern Ireland frequented the BP Pools of Belfast Lough. Some unusual plumage features initially indicated it may have been of the North American race, although closer examination showed this not to be the case (Garner 1999).

	93-94	94-95	95-96	96-97	97-98	Mon	Mean
Great Britain							
Inner Moray Firth	1341	595	559	65	8	Dec	314
Tay Estuary	186	209	277	225	240	Aug	227
Loch Garten/Mallachie	226	201	-	-	-		214
Lower Derwent Valley	155	127	173	298	182	Feb	187

	93-94	94-95	95-96	96-97	97-98	Mon	Mean	
Hirsel Lake	-	172	180	210	(13)	Nov	187	
Eccup Reservoir	125	141	178	163	163	Feb	154	
R Tweed: Kelso to Coldstream	119	153	149	158	84	Nov	133	
Chew Valley Lake	184	50	208	80	-		131	
Hamilton Low Parks	90	-	-	-	170	Nov	130	
Loch Lomond	7	11	198	184	226	Jul	125	▲
Solway Estuary	200	88	40	134	103	Jan	113	
Tyninghame Estuary	121	143	93	98	107	Jul	112	
Montrose Basin	71	77	156	89	136	Aug	106	
Hay-a-Park Gravel Pit	69	24	120	209	-		106	
Castle Howard Lake	98	-	-	-	-		98	
Eversley Cross/Yateley Gravel Pits	59	38	165	135	74	Jan	94	▲

Internationally or nationally important sites not counted in last five years
Spey Mouth

Sites no longer meeting table qualifying levels
Cults Reservoir

1   *RSPB/BP studies, e.g. Stenning (1994)*

RUDDY DUCK
Oxyura jamaicensis

Naturalised introduction[†]
Native range: North and South America

GB max:	3,585	Dec
NI max:	28	Feb

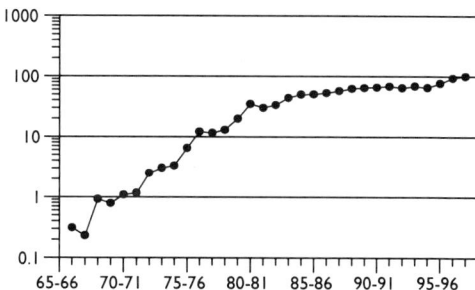

Figure 53. Annual indices for Ruddy Duck in GB

The proposal to investigate wide-scale control of Ruddy Ducks in the UK in order to conserve the globally threatened White-headed Duck has proved controversial in some quarters and a small number of observers opposed to this work have declined to submit counts of this species. Whilst attempts have been made to fill the gaps, this has not always been achieved, and consequently, our understanding of this species at a national level has been confounded to an extent.

Whilst the peak total (in December) fell considerably in 1997-98, it does not include counts from the most important site (two supplementary counts were made in 1997-98, in November and January, enabling a figure to be included in the table below). Thus, it is likely that the real national total was similar to that of the previous winter, but this conclusion remains speculative given the degree to which site totals fluctuate, especially as a result of regional movements in response to changing weather conditions. Annual indices, designed specifically to compensate for missing data, suggest that the population is still increasing, although the inclusion of the missing data would allow greater confidence that these values accurately track the true population.

Continental records of Ruddy Ducks are highly correlated with the increase in the UK population, with over 900 records of 1,500 birds in 19 Western Palearctic countries between 1965 and 1996 (Hughes *et al.* 1999). Birds have been recorded as widely as Iceland, Finland, Morocco and Turkey, the last particularly worryingly given its importance as a stronghold for White-headed Ducks. Records of birds in Spain continue, with at least 30 birds arriving in January 1997 during a cold spell in northwest Europe. Between 1965 and 1996, at least 85(but possibly as many as 175) Ruddy Ducks were seen in Spain, 68 of which (46 males and 22 females) were shot, as well as 51 Ruddy x White-headed Duck hybrids, all of which were shot.

The conservation arguments for the control of Ruddy Ducks are overwhelming and the UK Government recently announced the move to phase 2 of the control programme, with a trial of methods currently proceeding in three regions of the country. Consequently, it is hoped that all observers will submit counts of this species through WeBS so that its efficacy can be monitored properly at a national level, and so that the decision over whether or not to proceed with phase 3 control can be made with a full understanding of its effects. Other countries in Europe have also started to act, including France which has established a working group and has controlled 22 birds to date.

White-headed Duck numbers in Spain now exceed 1,300 and are increasing at a rate of 22% per year. Recent genetic work has shown White-headed Ducks and Ruddy Ducks to be two different species, separated for 2-5 million years.

The characteristic large fluctuations at individual sites render an assessment of the most recent year's counts alone unwise. Site averages and the sites' positions in the table below have changed little since 1993-94, suggesting a degree of stability at the traditional strongholds. Notable exceptions are large increases on the Middle Tame Valley (rising from 9th to 4th) and at Blagdon Lake (16th to 5th). It is surprising that the latter has only recently increased in importance, in view of its close proximity to Chew Valley Lake which has long been one of the most important sites. The majority of changes have been at the lesser sites, with increases at many holding 100 or fewer birds, including five added to the table below in 1997-98, suggesting the continuing growth in the population is accommodated by expansion onto new sites. By comparison, there have been obvious declines at only three sites: Swithland Reservoir (5th to 21st place), Belvide Reservoir (6th to 30th) and Woolston Eyes (15th to 35th). Although the peak count at Loughs Neagh & Beg dropped to the lowest level since 1989-90, this site, too, has shown an overall increase over the period.

	93-94	94-95	95-96	96-97	97-98	Mon	Mean	
Great Britain								
Chew Valley Lake	552	362	851	789	700	Nov	651	
Rutland Water	304	311	231	1,078	727	Feb	530	
Blithfield Reservoir	578	602	380	566	327	Jan	491	
Middle Tame Valley GP	133	192	226	501	457	Dec	302	
Blagdon Lake	155	409	145	296	213	Feb	244	
Eyebrook Reservoir	¹442	73	151	239	275	Dec	236	
Hilfield Park Reservoir	107	96	190	306	186	Dec	177	
Pitsford Reservoir	234	131	154	87	98	Jan	141	
Stanford Reservoir	104	86	107	221	181	Jan	140	
Hanningfield Reservoir	73	207	174	76	162	Nov	138	
Swillington Ings	211	238	130	82	3	Jun	133	
Fairburn Ings	39	144	150	83	243	Aug	132	
Farmwood Pool	94	123	133	99	61	Oct/Nov	102	
Llyn Traffwll	74	50	89	156	122	Sep	98	
Attenborough Gravel Pits	81	70	49	265	18	Feb	97	
Abberton Reservoir	76	77	135	88	99	Sep	95	
Clumber Park Lake	73	118	100	111	60	Feb	92	
Holme Pierrepont GP	36	73	101	134	99	Oct	89	
Llyn Penrhyn	86	103	-	71	92	Dec	88	
Alaw Reservoir	78	91	58	18	133	Dec	76	
Swithland Reservoir	145	46	73	12	67	Mar	69	
Kilconquhar Loch	8	66	62	85	62	Sep	57	
Knight/Bessborough Reservoirs	57	32	28	126	32	Jan	55	
Worsborough Reservoir	37	65	90	29	-		55	
Rostherne Mere	48	6	38	152	24	Sep	54	
Colwick Country Park	34	13	88	68	41	Dec	49	
Hollowell Reservoir	27	169	38	7	3	Jan	49	
Pugney Water	20	12	50	65	94	Nov	48	
Staines Reservoirs	19	38	67	112	3	Feb	48	
Belvide Reservoir	24	110	7	31	(2)	Aug	43	
Aqualate Mere	34	20	60	73	25	Mar	42	▲
Cropston Reservoir	3	3	33	156	16	Mar	42	
Tophill Low Reservoirs	25	27	35	51	56	Dec	39	
Wath Main Ings	30	20	56	25	42	Mar	35	
Woolston Eyes	25	42	44	40	23	Mar	35	▲

	93-94	94-95	95-96	96-97	97-98	Mon	Mean	
Ellesmere Lakes	32	32	77	22	11	Nov	35	
King George VI Reservoir	0	0	92	20	50	Dec	32	▲
Thoresby Lake	25	16	-	20	63	Jan	31	▲
Sutton/Lound Gravel Pits	-	19	45	25	36	Mar	31	▲
Avon Valley (Mid)	41	28	44	23	15	Jan	30	
Northern Ireland								
Lo. Neagh/Beg	59	45	73	89	28	Feb	59	

Sites no longer meeting table qualifying levels
Combermere

Other sites surpassing table qualifying levels in 1997-98

Mersey Estuary	55	Dec	Hogganfield Loch	35	Dec
Cotswold WP West	42	Feb	Moorgreen Reservoir	34	Nov
Houghton Green Pool	42	Oct	Bolton-on-Swale Gravel Pits	34	Nov
Pirton Pool	42	Sep	Dungeness Gravel Pits	33	Nov
Doddington Pool	40	Aug	Cefni Reservoir	30	Feb
Anglers Country Park Lake	39	Nov			

† as site designation does not occur and the 1% criterion is not applied, a qualifying level of 30 has been chosen to select sites for presentation in this report
1 B. Hughes, unpubl. data

WATER RAIL
Rallus aquaticus

International threshold:	?
Great Britain threshold:	?†
All-Ireland threshold:	?†

GB max: 345 Dec
NI max: 2 Feb

The peak count in 1997-98 was similar to that of the previous winter, and fell in the middle of the range of values for recent years. The peak count in December immediately followed the first and only real cold spell of the winter which may simply have resulted in birds being more conspicuous; counts for the remainder of the winter were much lower and relatively constant.

The difficulties in monitoring this species make interpretation of any variation in counts at individual sites problematic. It is interesting, however, to note that the table of key sites remains unchanged since the last report with the sole exception of the loss of Poole Harbour (Rye Harbour & Pett Level was previously omitted by accident).

	93-94	94-95	95-96	96-97	97-98	Mon	Mean	
Great Britain								
Somerset Levels	24	36	46	42	29	Dec	35	
Grouville Marsh	3	50	40	20	20	Oct/Nov/Jan	27	
Pannel Valley	50	-	-	-	-		25	
Lower Derwent Valley	6	12	29	26	27	Dec	20	
Stodmarsh	14	28	30	9	14	Nov	19	
Fleet Pond	6	12	15	20	20	Nov-Jan	15	
Christchurch Harbour	21	8	-	-	-		15	
Longueville Marsh	20	10	20	10	5	Oct/Jan/Feb	13	
Leighton Moss	-	-	15	18	25	Jan	12	
North Norfolk Marshes	7	15	23	6	11	Nov	12	
Rye Harbour/Pett Level	2	13	5	40	2	Nov	12	▲
Knockshinnoch Lagoons	-	4	28	4	9	Dec/Feb	11	
Fleet/Wey	21	8	10	9	4	Dec/Mar	10	

Sites no longer meeting table qualifying levels
Poole Harbour

Other sites surpassing table qualifying levels in 1997-98

Doxey Marshes	11	Jun/Jul	Loe Pool	10	Feb
Slapton Ley	11	Sep	Chichester Harbour	10	Jan
Conwy Estuary	10	Dec			

SPOTTED CRAKE
Porzana porzana

International threshold:	?
Great Britain threshold:	?†
All-Ireland threshold:	?†

GB max: 2   Jun/Sep
NI max: 0

Transient or possibly breeding birds were recorded on the Dee Estuary (England/Wales), Doxey Marshes and North Norfolk Marshes.

MOORHEN
Gallinula chloropus

International threshold:	?
Great Britain threshold:	?†
All-Ireland threshold:	?†

GB max: 11,843   Feb
NI max: 266   Nov

The peak British total had risen in each of the previous four winters, but fell slightly in 1997-98. This pattern is partly explained by the changing winter weather during this period, with birds featuring more prominently in WeBS counts during cold weather, perhaps as a result of congregating at ice-free sites. An explanation for the general increase over the period, however, might be that observers have become more aware of this ubiquitous species which previously attracted little attention, and perhaps are more skilled at locating and counting Moorhen after five years in the WeBS scheme.

There were no big surprises in 1997-98 counts at the key sites, although large increases over the five year period are notable at Martin Mere, Lower Derwent Valley, Burry Inlet and River Wye, and will have contributed the the national picture.

By contrast, there has been a steady fall in numbers in Northern Ireland, strongly influenced by low counts at Loughs Neagh & Beg. Although low numbers of several diving ducks were recorded at the same site in 1997-98, there is no obvious link between the markedly different ecologies of Moorhen and these species to suggest a link. It may be that enthusiasm for counting this species still differs at individual sites, particularly where WeBS counts are that much more demanding due to problems of access, viewing and the sheer numbers of other species present.

	93-94	94-95	95-96	96-97	97-98	Mon	Mean	
Great Britain								
Severn Estuary	705	(13)	(24)	829	(21)	Sep	767	
Martin Mere	369	360	665	739	710	Jan	569	
Lower Derwent Valley	41	220	367	816	680	Mar	425	
Somerset Levels	228	286	356	253	250	Feb	275	
North Norfolk Marshes	281	234	332	179	334	Nov	272	
Ouse Washes	182	269	241	124	201	Mar	203	
Lee Valley Gravel Pits	131	253	194	237	197	Oct	202	
Arun Valley	68	221	161	234	190	Mar	175	
Durham Coast	121	137	176	175	254	Nov	173	
Chew Valley Lake	160	155	180	125	-		155	
Thames Estuary	48	116	234	219	144	Sep	152	
Chichester Gravel Pits	88	129	179	167	176	Nov	148	
Lancaster Canal	143	128	147	106	207	Jan	146	
Leighton Moss	150	-	160	170	95	Nov/Jan	144	
Burry Inlet	24	68	121	220	281	Nov	143	
Blackwater Estuary	102	125	113	181	165	Mar	137	
Bewl Water	80	200	196	119	60	Mar	131	
Rutland Water	97	163	178	71	119	Aug	126	
R. Wye: Bakewell to Haddon	58	50	153	118	160	Feb	108	▲
Sutton/Lound Gravel Pits	-	85	104	120	106	Feb	104	▲
North Warren/Thorpness Mere	147	129	70	80	80	Jan	101	
Pitsford Reservoir	42	159	163	116	21	Sep	100	▲

Northern Ireland

Lo. Neagh/Beg	618	266	265	132	137	Sep	284
Upper Lough Erne	100	174	70	164	39	Feb	109
Broad Water Canal	25	33	-	83	71	Mar	53
Portavo Lake	47	42	43	-	-		44

Internationally or nationally important sites not counted in last five years

Wantsum Marshes   Little Stour Valley
Ash Levels

Other sites surpassing table qualifying levels in 1997-98

Tring Reservoirs	148	Sep		Tullynagee Lough	43	Oct
Fairburn Ings	112	Aug		Upper Quoile	31	Feb

† as no thresholds for national importance have been set, qualifying levels of 100 and 30 have been chosen to select sites in Great Britain and Northern Ireland, respectively, for presentation in this report

---

COOT
Fulica atra

International threshold: 15,000
Great Britain threshold: 1,100
All-Ireland threshold: 250

GB max: 102,507 Nov
NI max: 6,645 Nov

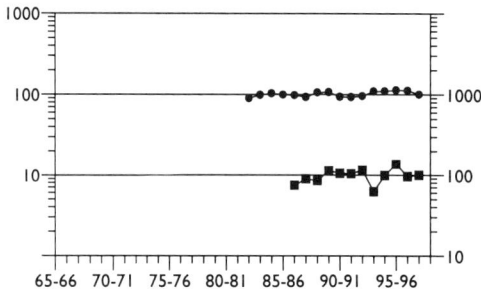

Figure 54. Annual indices of Coot in GB (circles, left axis) and NI (squares, right axis)

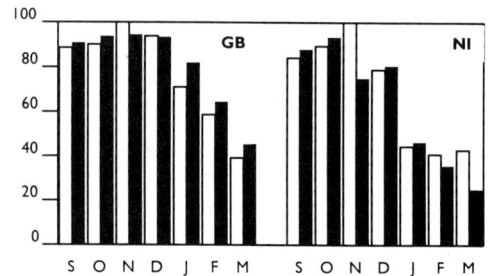

Figure 55. Monthly indices for Coot in GB and NI (white bars 1997-98; black bars 1992-93 to 1996-97)

Annual indices and the national total were the lowest in Great Britain during the last five winters, although both remain around average for the last 15 years. Peak counts in Northern Ireland, in effect, match that of Loughs Neagh & Beg, which has held around 90% of the total in recent winters. As in Great Britain, the 1997-98 peak and annual index value fell in the middle of the normal range of variation.

The 1997-98 maxima at most key sites were around average, and there were no obviously large counts as have been noted at some in recent years. The most significant count was at Abberton Reservoir, where numbers continued the recent decline to the lowest level since counts began in the early 1980s, although numbers at this site are prone to large fluctuations (Figure 55). The magnitude of these variations is greater than for the national total, suggesting that these birds are redistributing within Great Britain. The lack of any obvious correlation with nearby sites, e.g. Alton Water,

suggests any movements are fairly broad scale.

The most notable counts at other sites were continued high totals at Lee Valley Gravel Pits, and consistent increases in recent years at Lower Windrush Valley Gravel Pits and Fen Drayton Gravel Pits, Stanford Reservoir and Loch Leven, the last attaining national importance as a result. The most obvious drop was at Little Paxton Gravel Pits, particularly marked following the recent increase.

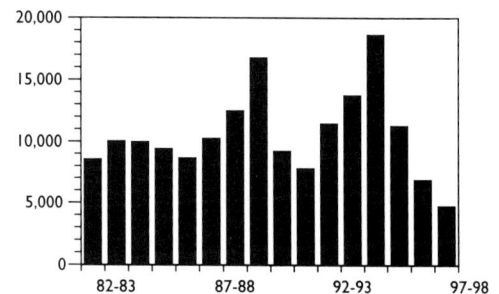

Figure 56. Peak numbers of Coot at Abberton Reservoir

	93-94	94-95	95-96	96-97	97-98	Mon	Mean	
Great Britain								
Abberton Reservoir	13,768	18,632	11,319	6,897	4,784	Sep	11,080	
Rutland Water	3,036	4,165	6,184	3,935	4,663	Dec	4,397	
Cotswold WP West	3,357	4,119	3,946	3,110	3,560	Nov	3,618	
Hanningfield Reservoir	1,941	3,300	4,540	4,986	3,181	Nov	3,590	
Cotswold WP East	1,953	2,185	5,199	2,268	2,094	Nov	2,740	
Cheddar Reservoir	2,066	1,967	2,381	3,100	2,300	Dec	2,363	
Ouse Washes	3,182	2,225	1,755	1,497	3,082	Mar	2,348	
Lee Valley Gravel Pits	1,654	2,012	1,812	3,023	2,926	Nov	2,285	
Fleet/Wey	2,568	1,931	2,417	2,501	1,562	Nov	2,196	
Windermere	1,893	1,660	2,077	2,310	-		1,985	
Chew Valley Lake	2,010	1,135	1,880	2,500	-		1,881	
Lower Windrush Valley GPs	1,447	1,417	1,802	2,068	2,629	Nov	1,873	
Middle Tame Valley GP	1,560	1,497	1,781	2,804	1,196	Jan	1,768	
Hornsea Mere	1,722	1,650	-	-	(60)	Mar	1,686	
Avon Valley (Mid)	1,664	1,828	1,509	1,571	1,821	Nov	1,679	
Alton Water	423	1,427	2,845	1,142	2,135	Dec	1,594	
Pitsford Reservoir	1,373	1,961	1,177	1,222	1,310	Oct	1,409	
Fen Drayton Gravel Pit	804	987	1,528	1,675	1,709	Dec	1,341	
Sutton/Lound Gravel Pits	-	1,145	1,360	1,716	1,072	Nov	1,323	
Wraysbury Gravel Pits	1,345	1,318	1,126	1,382	1,385	Dec	1,311	
Stanford Reservoir	610	1,430	1,200	1,270	1,865	Dec	1,275	
Fairburn Ings	751	1,013	1,784	1,572	959	Jul	1,216	
Loch Leven	995	802	1,020	1,546	1,551	Sep	1,183	▲
Little Paxton Gravel Pits	757	850	1,314	2,173	485	Oct	1,116	
Northern Ireland								
Lo. Neagh/Beg	3,134	7,222	8,788	8,262	5,890	Nov	6,659	
Strangford Lough	515	307	254	378	407	Nov	372	
Upper Lough Erne	228	244	166	441	412	Feb	298	

Internationally or nationally important sites not counted in last five years
Ballysaggart Lough

Other sites surpassing table qualifying levels in 1997-98

Blithfield Reservoir	1,717	Aug	Dungeness Gravel Pits	1,166	Nov
Draycote Water	1,430	Nov	Alaw Reservoir	1,159	Sep
Lower Derwent Valley	1,203	Feb	Thrapston Gravel Pits	1,134	Oct

CRANE

Scarce

Grus grus

A group of up to six birds frequented Heigham
Holmes between November and February.

OYSTERCATCHER

Haematopus ostralegus

International threshold: 9,000
Great Britain threshold: 3,600
All-Ireland threshold: 500

GB max: 251,410 Nov
NI max: 17,799 Jan

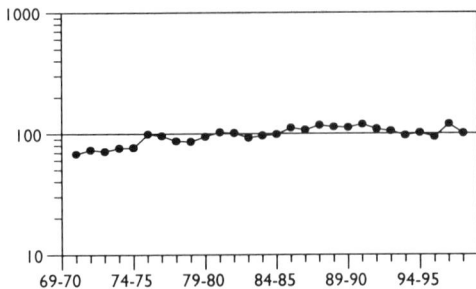

Figure 57. Annual indices for Oystercatcher in UK

In 1997-98 the winter index dropped 17% on the previous year. This may have been, in part, due to the drop in numbers of this species found at several of the internationally and nationally important sites. The Burry Inlet and the Wash, both internationally important sites, saw numbers dip by 38% and 31% respectively on the recent average. Numbers of Oystercatcher at Lavan Sands were also considerably lower than in recent years, with the annual peak being 71% down on the recent average. In contrast peak numbers on the Ribble Estuary were 77% up on the recent average, whilst those on the Inner Moray Firth and the Duddon Estuary were up by 58% and 45% respectively.

Declines in numbers of Oystercatcher have been linked to low food abundance, particularly in areas where cockles, a major prey item, are commercially farmed. A recent study, using WeBS data, examining changes in the number of Oystercatchers wintering in the Burry Inlet, found that their abundance during winter was not significantly related to the biomass of cockles at the start of the winter, or the biomass landed by the fishery (Norris *et al.* 1998). Abundance was, however, correlated with the number of birds wintering in the UK. Conversely, the abundance of Oystercatchers during spring was positively correlated with the biomass of cockles at the start of winter and negatively correlated with the amount of cockles landed by the fishery during the winter. The most likely explanation for this is thought to be that birds disperse from the Burry Inlet earlier in spring in years when the biomass of cockles at the start of the winter is small and/or the amounts landed by the fishery are high.

Studies by Goss-Custard *et al.* (1998) suggested that stable numbers of Oystercatchers on an estuary did not necessarily imply that carrying capacity had been reached. Examining the Exe Estuary, where, since 1976, numbers have fluctuated independently of the increasing numbers of Oystercatchers wintering in Britain as a whole, they found that numbers on the main mussel feeding areas increased even though mussel bed quality remained unchanged, food abundance decreased and disturbance on some important beds increased.

Preliminary estimation of the current East Atlantic Flyway population indicates a significant increase (19%) in numbers since the mid 1980s (Davidson 1998). International thresholds will be revised later in 1999.

	93-94	94-95	95-96	96-97	97-98	Mon	Mean
International							
Morecambe Bay	48,861	53,869	34,811	57,670	56,511	Nov	50,344
Solway Estuary	33,240	41,344	31,031	(47,729)	34,446	Nov	37,558
Dee Estuary (Eng/Wal)	24,809	26,658	21,800	24,897	25,142	Nov	24,661
Wash	25,382	22,300	(14,233)	(16,363)	17,126	Dec	21,603
Ribble Estuary	9,620	(8,294)	12,048	20,846	28,701	Nov	17,804
Thames Estuary	9,741	(23,142)	12,251	14,425	14,615	Feb	14,835
Burry Inlet	9,188	13,958	20,461	19,067	9,423	Dec	14,419
Great Britain							
Forth Est.	8,509	8,851	6,956	6,826	8,045	Jan	7,837
Duddon Estuary	5,645	6,122	6,046	5,630	9,314	Feb	6,551
Inner Moray Firth	4,962	5,419	(5,178)	5,261	8,334	Jan	5,994
Clyde Est.	4,838	4,796	5,303	5,414	(4,781)	Dec	5,088
Lavan Sands	5,055	5,935	(3,621)	5,780	1,611	Dec	4,595
Carmarthen Bay	5,545	4,558	(2,202)	3,474	3,926	Dec	4,376
Exe Estuary	4,502	4,202	4,733	4,215	3,078	Nov	4,146
Medway Estuary	(4,986)	2,732	(3,704)	3,162	(5,521)	Feb	4,021
Swale Estuary	4,230	3,328	3,122	5,780	3,349	Nov	3,962
Humber Estuary	6,140	3,329	3,029	1,360	4,200	Mar	3,612
Northern Ireland							
Strangford Lough	4,554	6,424	6,091	7,276	6,904	Dec	6,250
Belfast Lough	6,349	5,573	4,814	4,091	6,482	Jan	5,462
Lough Foyle	2,334	1,687	2,590	(3,352)	(2,865)	Nov	2,566
Dundrum Bay	1,524	1,150	1,553	1,660	1,763	Nov	1,530
Outer Ards	1,343	1,454	1,390	1,523	(1,385)	Jan	1,428
Carlingford Lough	873	913	938	(812)	860	Jan	896
South Down	-	-	1,138	-	-		

BLACK-WINGED STILT
Himantopus himantopus

Vagrant
Native range: worldwide distribution

The long-staying individual was recorded on the North Norfolk Marshes in most months.

AVOCET
Recurvirostra avosetta

International threshold: 700
Great Britain threshold: 10*
All-Ireland threshold: +*

* 50 is normally used as a minimum threshold

GB max: 3,859 Dec
NI max: 0

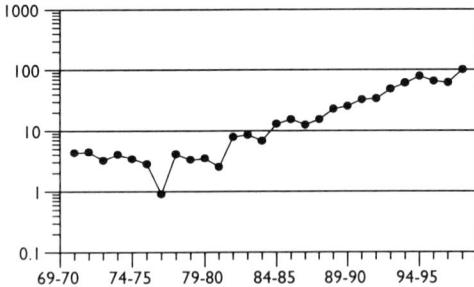

Figure 58. Annual indices for Avocet in UK

Numbers of Avocet continued to rise dramatically, with the 1997-98 winter maximum reaching a new high, and the annual index jumping 68% on the previous year. The peak count is more than three times the population estimate derived from late 1980s/early 1990s data

(Cayford & Waters 1996), whilst three counts of over 3,000 during 1997-98 confirm that the substantial increase is genuine, whilst the international population has also increased (Davidson 1998).

The Alde Complex, the only British site ever to attain internationally important status, held its highest peak in six years, pushing the five year mean still closer to requalifying as internationally important. The majority of nationally important sites also held counts well in excess of their recent averages, most notably Hamford Water (+75%), the Tamar Complex (+177%), North Norfolk Marshes (+132% despite being an incomplete count), Colne Estuary (+104%) and Breydon Water & Berney Marshes, a newly qualified site of national importance (+241%).

	93-94	94-95	95-96	96-97	97-98	Mon	Mean	
Great Britain								
Alde Complex	717	656	744	437	884	Dec	688	
Poole Harbour	396	584	505	520	585	Dec	518	
Thames Estuary	341	647	367	450	488	Feb	459	
Hamford Water	326	418	(249)	299	587	Dec	408	
Medway Estuary	285	(498)	(256)	(368)	(200)	Jan	384	
Blyth Estuary	403	260	489	242	422	Jan	363	
Exe Estuary	331	260	303	339	369	Feb	320	
Tamar Complex	254	81	272	301	595	Dec	301	
Swale Estuary	218	329	285	(208)	340	Mar	293	
North Norfolk Marshes	(160)	194	(41)	(51)	(318)	Mar	256	
Wash	121	188	(106)	(83)	196	Feb	168	
Colne Estuary	15	139	203	150	214	Jan	144	
Deben Estuary	93	79	106	100	102	Jan	96	
Breydon Water & Berney Marshes	13	24	97	77	157	Feb	74	▲
Minsmere	25	115	(2)	(0)	70	Mar	70	▲
Blackwater Estuary	(11)	60	0	14	24	Mar	25	
Abberton Reservoir	0	0	9	0	64	Jan	15	
Pagham Harbour	4	8	14	10	12	Jan	10	

Other sites surpassing table qualifying levels in 1997-98

Abberton Reservoir	64	Nov	Crouch/Roach Estuary	18	Mar
Humber Estuary	28	Mar	Dengie Flats	15	May
Blackwater Estuary	24	Mar	Pagham Harbour	12	Jan
Orwell Estuary	22	Mar			

BLACK-WINGED PRATINCOLE
Glareola nordmanni

Vagrant
Native range: Central Asia and Africa

One was recorded at Martin Mere in August.

LITTLE RINGED PLOVER
Charadrius dubius

International threshold: ?
Great Britain threshold: ?
All-Ireland threshold: ?

GB max: 288   May
NI max: 0

Peak numbers occurred on spring passage, as birds gathered at key arrival sites. Numbers declined as summer progressed, presumably due to poorer coverage during these months. A small

number had already arrived by the count date in March 1998, with two birds at both Portworthy Mica Dam and Stanwick Gravel Pits.

Sites with 10 or more birds in 1997-98

Croxall Pits	21	Jun	Rutland Water	14	May/Jul
Sutton/Lound Gravel Pits	15	May	Holme Pierrepoint GPs	13	May
Upton Warren	14	Jun	Wellington Gravel Pits Gravel Pits	12	Jul
Wath Main Ing	14	Jun			

RINGED PLOVER
Charadrius hiaticula

International threshold: 500
Great Britain winter threshold: 290
Great Britain passage threshold: 300
All-Ireland threshold: 125

GB max: 23,163   Aug
NI max: 659   Dec

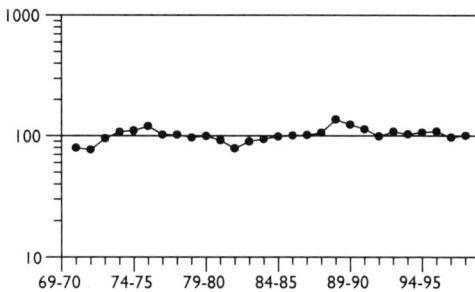

Figure 59. Annual indices for Ringed Plover in UK

In 1997-98, the UK winter index for Ringed Plover rose slightly on the previous year, as did the peak winter count, though it remained below 10,000 birds, representing around one third of the national population. Preliminary results from the national survey of non-estuarine coast suggest that there has been a national decrease in Ringed Plover since the 1984-85 Winter Shorebird Count. Comparison of numbers on coastal sections covered by both surveys revealed a drop from 12,673 to 9,286 (down 27%), with the largest of the statistically significant declines occurring

along the coasts of Strathclyde, Western Isles, Gwynedd, Cornwall, Dumfries & Galloway, Isle of Wight and Fife. However, the peak WeBS Core Count, in autumn, was substantially higher than the 18-19,000 recorded in the previous two years.

Langstone Harbour continued its ascendency of recent years, reaching international importance, though four sites, Hamford Water, Chichester Harbour, North Uist and the Stour Estuary, all dropped to national importance level. Hamford Water, where numbers of Ringed Plover have steadily declined over the last few years, held a peak 50% below the recent average, and whilst those at Chichester recovered after a steady decline in recent years, there were still insufficient to maintain international importance. The Humber Estuary and Blackwater Estuary both regained national importance status after a year when the average dropped below the qualifying level.

It is notable that counts during passage periods exceeded 500 at 17 sites, numbers that if maintained regularly would qualify the sites as internationally important.

	93-94	94-95	95-96	96-97	97-98	Mon	Mean
International							
Tiree	-	1,192	-	-	¹534	Jan	863
Thames Estuary	672	1,127	755	535	792	Nov	776
Medway Estuary	599	1,206	682	(378)	313	Jan	700
South Uist (West Coast)	²506	705	-	-	¹376	Jan	529
Langstone Harbour	388	224	519	660	739	Nov	506 ▲

Great Britain

East Sanday Coast	[3]284	[4]712	-	-	[1](282)	Jan	498	
Hamford Water	(252)	641	(546)	482	281	Mar	488	▼
Colne Estuary	273	382	707	(306)	(568)	Dec	483	
North Norfolk Marshes	329	411	371	775	405	Mar	458	
North Uist (West Coast)	[2]808	333	-	-	[1]244	Jan	462	▼
Chichester Harbour	615	542	435	204	483	Nov	456	▼
Morecambe Bay	342	473	401	528	515	Jan	452	
Stour Estuary	382	502	306	597	(87)	Feb	447	▼
Solway Estuary	489	553	321	223	214	Feb	360	
Wash	330	375	390	(147)	311	Jan	352	
Orwell Estuary	643	349	(411)	226	63	Mar	338	
Forth Est.	400	291	413	259	317	Dec	336	
Jersey Shore	302	212	446	(253)	264	Feb	306	
Humber Estuary	229	336	316	249	382	Nov	302	▲
Blackwater Estuary	220	474	(242)	144	337	Dec	294	▲

Northern Ireland

Outer Ards	562	389	(317)	575	350	Dec	469
South Down	-	-	422	-	-		422
Strangford Lough	218	313	253	324	134	Jan	248
Belfast Lough	214	183	(135)	133	137	Dec	167
Carlingford Lough	(376)	64	85	(131)	86	Dec	148
Kilkeel To Lee Stone Point	-	-	132	-	-		132

Other sites surpassing table qualifying levels in 1997-98

Pagham Harbour	300	Mar	Thanet	297	Nov
Swale Estuary	301	Feb			

Internationally or nationally important sites not counted in last five years
South Ford
Traighear

Sites surpassing passage threshold in Great Britain in 1997-98

Humber Estuary	3,664	Aug	Langstone Harbour	604	Aug
Ribble Estuary	2,350	May	Forth Est.	571	Aug
Solway Estuary	1,876	May	Stour Estuary	502	Sep
Wash	1,570	Aug	Dee Estuary (Eng/Wal)	467	Aug
North Norfolk Marshes	1,460	Aug	Duddon Estuary	430	Sep
Mersey Estuary	1,428	Aug	Blackwater Estuary	422	Aug
Pagham Harbour	1,035	Aug	Taw/Torridge Estuary	416	Aug
Chichester Harbour	1,007	Aug	Medway Estuary	366	Oct
Severn Estuary	933	Aug	Eden Estuary	350	Aug
Swale Estuary	929	Aug	Ayr to Troon	329	Aug
Morecambe Bay	862	Sep	Hamford Water	319	Sep
Tees Estuary	851	May	Swansea Bay	315	Aug
Thames Estuary	774	Aug/Oct	Colne Estuary	312	Oct
Exe Estuary	612	Aug			

1   NEWS data
2   RSPB Report two: Western Isles Winter Shorebird Counts
3   SNH Research Survey and Monitoring Report NE/92/215
4   RSPB Report One: Orkney Winter Shorebird Counts

KENTISH PLOVER   Scarce
Charadrius alexandrinus

A single was noted at Chichester Harbour in April.

DOTTEREL
Charadrius morinellus

Scarce

Three were recorded on the Humber Estuary during May and singles at Dungeness Gravel Pits in April, the Tees Estuary in May and the Hayle Estuary, unusually, in December.

GOLDEN PLOVER
Pluvialis apricaria

International threshold: 18,000
Great Britain threshold: 2,500
All-Ireland threshold: 2,000

GB max: 175,445 Dec
NI max: 14,380 Dec

Numbers in both Great Britain and Northern Ireland peaked in December and declined the following month, considerably more marked in Britain. The winter coastal maximum occurred a month earlier. This may be the result of changeable weather in December, causing more birds to seek refuge from high winds by moving to inland sites.

Comparison of peak figures with those in the previous two winters amply demonstrates the westerly movements of this species in response to cold weather: in the mild winter of 1996-97, the British peak was just 120,000, but with 29,000 in Northern Ireland, whilst these figures were 148,000 and 18,000, respectively, in the previous year when conditions were much colder.

Low counts at Sutton/Lound Gravel Pits and Netherfield Gravel Pits resulted in them ceasing to be of national importance, although peak counts at most sites in 1997-98 exceeded current averages: that on the Wash was 165% above the recent average and was the only count, other than on the Humber, to have exceeded the threshold for international importance in the last five years. High counts at Hamford Water, the Somerset Levels, Colne Estuary, Swale Estuary and North Norfolk Marshes were also particularly notable.

	93-94	94-95	95-96	96-97	97-98	Mon	Mean	
International importance								
Humber Estuary	(29,201)	60,661	(32,532)	8,741	34,444	Dec	34,615	
Great Britain								
Wash	(4,591)	(12,535)	(12,919)	6,879	26,461	Nov	16,670	
Carmarthen Bay	3,000	9,080	11,000	10,003	(3,300)	Feb	8,271	
Breydon Water/Berney Marshes	6,400	6,100	5,300	7,550	7,200	Dec	6,510	
Solway Estuary	2,132	7,464	7,049	4,617	7,572	Feb	5,767	
Lower Derwent Valley	4,000	4,300	8,900	3,000	7,950	Feb	5,630	
Blackwater Estuary	1,752	(9,543)	2,055	(6,631)	6,800	Jan	5,356	
Hamford Water	3,417	4,411	5,073	4,611	8,275	Jan	5,157	
Morecambe Bay	6,211	4,223	3,616	4,310	4,745	Mar	4,621	
Ribble Estuary	3,823	4,017	2,347	6,530	5,325	Nov	4,408	
Thames Estuary	1,449	(8,982)	2,515	3,875	4,925	Dec	4,349	
Somerset Levels	(2,265)	4,104	3,027	683	8,909	Feb	4,181	
Swale Estuary	1,791	3,132	1,393	2,227	9,535	Jan	3,616	
Criddling Stubbs Quarry Pools	-	-	-	2,000	5,000	Nov	3,500	▲
Fairburn Ings	31	3,800	7,000	3,700	2,700	Dec	3,446	
Colne Estuary	1,680	6,302	1,034	(1,500)	4,350	Jan	3,342	
Clifford Hill Gravel Pits	3,000	5,000	2,400	2,000	4,000	Jan	3,280	
Lindisfarne	2,805	(3,180)	4,580	2,604	2,580	Nov	3,150	
St Marys Island	3,700	1,500	600	(3,000)	6,500	Nov	3,075	▲
Mersey Estuary	2,278	2,323	3,850	4,000	2,750	Dec	3,040	
New Road Pits	3,000	-	-	-	-		3,000	
Forth Est.	1,897	3,080	5,260	2,363	2,147	Nov	2,949	
Abberton Reservoir	815	5,778	4,550	233	3,057	Dec	2,887	
North Norfolk Marshes	2,102	2,121	3,258	2,040	4,772	Dec	2,859	
R Idle: Bawtry To Misterton	-	2,400	1,000	5,000	-		2,800	

Northern Ireland

Strangford Lough	3,123	6,420	7,444	14,095	6,221	Jan	7,461
Lough Foyle	(7,006)	4,605	2,050	(5,207)	5,456	Feb	4,865
Lo. Neagh/Beg	1,805	5,758	4,470	3,902	4,300	Jan	4,047
Outer Ards	2,290	1,684	(3,517)	5,869	(735)	Jan	3,340

Sites no longer meeting table qualifying levels
Sutton/Lound GP   Netherfield GP

Other sites surpassing table qualifying levels in 1997-98

Nene Washes	6,109	Feb	Loch of Strathbeg	3,000	Dec
Tamar Complex	5,260	Dec	Port Meadow	2,800	Dec
Dungeness	5,000	Dec	Blyth Estuary	2,760	Feb
Ouse Washes	3,519	Dec	Wigtown Bay	2,671	Feb
Ayr to Troon	3,474	Oct	Taw/Torridge Estuary	2,500	Mar
Cotswold Water Park	3,195	Nov	Walland Marsh	2,500	Dec
Loch of Strathbeg	3,000	Feb	Hayle Estuary	2,500	Dec

GREY PLOVER
Pluvialis squatarola

International threshold: 1,500
Great Britain threshold: 430
All-Ireland threshold: 40*
* 50 is normally used as a minimum threshold

GB max: 46,776 Feb
NI max: 352 Jan

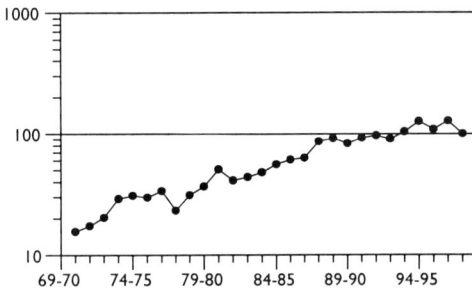

Figure 60. Annual indices for Grey Plover in UK

Great Britain and Northern Ireland winter maxima in 1997-98 were both nearer the bottom of the range for recent years, reflected by a down-turn in the annual index of 22% following the 1996-97 peak.

Despite the apparent declines in 1997-98, a number of main resorts held peak counts well in excess of their recent averages, notably those in the southeast, e.g. Dengie Flats, and the Swale and Medway Estuaries. Large counts at two more east coast sites, the Humber Estuary and North Norfolk Marshes, were sufficient for both to reach international importance, the former recording a peak count of Grey Plover 177% above the recent average. Whilst this may reflect a more easterly centre of distribution during the mild winter, the Thames Estuary and Hamford Water both saw a marked decline in numbers, and local redistribution may have been responsible for the increases nearby.

Initial estimates for the current East Atlantic Flyway population suggest a 46% increase since the 1980s (Davidson 1998). This will certainly lead to revisions of the population estimate and threshold later in 1999.

	93-94	94-95	95-96	96-97	97-98	Mon	Mean
International							
Wash	6,840	17,404	(7,396)	8,952	9,790	Nov	10,747
Ribble Estuary	5,178	(10,802)	(3,211)	12,856	(4,465)	Nov	9,612
Thames Estuary	7,419	(7,820)	(7,515)	9,680	3,708	Feb	7,228
Hamford Water	1,618	2,207	(8,186)	(7,033)	3,270	Jan	4,463
Blackwater Estuary	(6,609)	2,442	(4,230)	(2,383)	(3,549)	Dec	4,208
Stour Estuary	2,424	(4,253)	3,249	3,159	(1,705)	Jan	3,271
Dengie Flats	1,800	3,300	(1,560)	2,160	4,156	Mar	2,854
Medway Estuary	2,605	3,104	1,899	1,979	4,612	Jan	2,840
Swale Estuary	1,337	2,425	1,543	(2,822)	5,313	Feb	2,688
Lindisfarne	(1,545)	(1,810)	(1,728)	2,118	2,950	Feb	2,534
Chichester Harbour	2,862	(3,629)	2,060	2,117	1,434	Dec	2,420
Langstone Harbour	1,821	2,802	1,266	(1,480)	2,157	Nov	2,012
Dee Estuary (Eng/Wal)	1,565	886	(2,567)	2,422	1,186	Nov	1,725
Morecambe Bay	1,859	2,184	1,557	1,695	1,243	Dec	1,708
Humber Estuary	1,662	1,231	1,533	539	(3,368)	Feb	1,667 ▲
North Norfolk Marshes	949	1,766	850	1,867	2,273	Nov	1,541 ▲

	93-94	94-95	95-96	96-97	97-98	Mon	Mean
Great Britain							
Pagham Harbour	1,023	1,624	1,120	1,198	2,452	Nov	1,483
Alt Estuary	896	983	1,456	1,702	2,316	Mar	1,471
Solway Estuary	843	1,215	1,429	1,276	990	Jan	1,151
Colne Estuary	1,016	1,198	1,050	888	1,462	Feb	1,123
Mersey Estuary	2,100	663	417	460	1,410	Nov	1,010
Beaulieu Estuary	830	833	1,021	463	782	Dec	786
Eden Estuary	595	1,403	510	604	491	Feb	721
Forth Est.	605	592	730	658	724	Nov	662
Severn Estuary	647	767	368	519	(436)	Jan	575
Jersey Shore	624	672	452	336	464	Dec	510
Exe Estuary	385	508	513	513	573	Feb	498
Northern Ireland							
Strangford Lough	138	549	170	407	189	Jan	291
Carlingford Lough	63	89	57	(93)	93	Jan	79

Other sites surpassing table qualifying levels in 1997-98
Orwell Estuary   585   Jan

LAPWING
Vanellus vanellus

International threshold: **20,000****
Great Britain threshold: **20,000**†**
All-Ireland threshold: **2,500**

GB max: **464,466**   Dec
NI max: **28,936**   Dec

National totals of Lapwing during 1997-98 remained around the average for recent years and both the Great Britain and Northern Ireland populations peaked in December. Like Golden Plover, whilst coastal maxima occurred in December, those for inland sites in Great Britain and Northern Ireland were both in January.

Sites of international importance remained the same as the previous year, with the Somerset Levels consolidating its position at the top.

Walland Marsh was the only addition to the sites of national importance in Great Britain, provisionally qualifying on the basis of two years counts. Numbers at most sites fluctuate greatly between years, and whilst the 1997-98 peak was markedly different from the five year mean at many, only at a few was it the highest or lowest count in the last five winters. Most notable were high counts on The Wash, Morecambe Bay, the Ouse Washes and the Medway Estuary.

	93-94	94-95	95-96	96-97	97-98	Mon	Mean
International							
Somerset Levels	(26,131)	(78,602)	47,081	16,743	62,886	Jan	51,328
Humber Estuary	22,954	90,288	23,827	9,222	21,884	Dec	33,635
Breydon Water/Berney Marshes	36,000	35,600	24,000	31,000	19,400	Dec	29,200
Wash	13,102	24,225	24,773	9,132	41,538	Nov	22,554
Ribble Estuary	6,495	22,308	(28,270)	18,108	24,932	Nov	20,023
Great Britain							
Morecambe Bay	12,352	16,161	24,293	15,526	26,190	Dec	18,904
Thames Estuary	13,971	(18,365)	8,347	8,391	(17,672)	Jan	13,349
Blackwater Estuary	14,355	(12,186)	(5,280)	4,827	15,897	Jan	11,816
Severn Estuary	6,294	16,251	10,956	10,441	(14,843)	Jan	11,757
Walland Marsh	-	-	-	4,800	18,500	Jan	11,650 ▲
Swale Estuary	9,178	18,424	2,995	(6,271)	(15,430)	Nov	11,507
Mersey Estuary	9,036	16,601	(11,137)	(10,322)	8,658	Jan	11,432
Ouse Washes	2,021	15,591	8,155	(4,675)	15,170	Dec	10,234
Colne Estuary	10,067	9,510	8,222	(1,900)	12,440	Jan	10,060
Lower Derwent Valley	5,200	7,400	14,543	11,941	8,487	Feb	9,514
Arun Valley	1,552	24,457	9,402	692	(7,188)	Jan	9,026
Solway Estuary	3,354	9,067	13,609	6,150	12,004	Nov	8,837
Dee Estuary (Eng/Wal)	5,300	11,514	9,590	6,916	8,828	Jan	8,430
Crouch/Roach Estuary	9,970	(7,480)	5,964	(2,220)	7,440	Dec	7,791
Hamford Water	11,635	7,059	(6,335)	3,220	6,968	Jan	7,221

	93-94	94-95	95-96	96-97	97-98	Mon	Mean
Abberton Reservoir	2,600	6,861	12,425	3,092	10,620	Dec	7,120
Medway Estuary	3,801	5,561	(5,991)	(3,366)	(11,435)	Jan	6,697
Tees Estuary	3,416	7,363	10,505	(5,277)	4,500	Jan	6,446
Poole Harbour	(10,454)	4,583	5,907	4,666	4,912	Feb	6,104
Nene Washes	3,705	7,932	7,190	1,800	8,100	Jan	5,745
Stour Estuary	1,621	8,210	4,228	3,984	7,466	Dec	5,102

Northern Ireland

	93-94	94-95	95-96	96-97	97-98	Mon	Mean
Strangford Lough	3,779	8,266	11,086	13,547	11,989	Dec	9,733
Lo. Neagh/Beg	3,042	5,832	6,758	7,857	6,777	Dec	6,053
Outer Ards	2,592	3,070	3,776	6,104	3,059	Dec	3,720
Lough Foyle	1,078	3,139	7,370	(2,665)	2,315	Jan	3,476

Sites no longer meeting table qualifying levels
Alde complex

Other sites surpassing table qualifying levels in 1997-98

Cotswold Water Park	7,395	Nov	North Norfolk Marshe	6,566	Nov
Forth Est.	6,920	Nov	Alde Complex	6,048	Jan
Dungeness	6,600	Nov	Upper Quoile	3,000	Jan

KNOT

Calidris canutus

International threshold:	3,500
Great Britain threshold:	2,900
All-Ireland threshold:	375

GB max:	294,025	Nov
NI max:	9,655	Jan

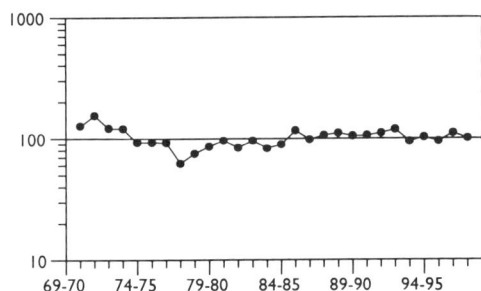

Figure 61. Annual indices for Knot in UK

Whilst the 1997-98 index for Knot dipped 10% to one of the lowest levels during the period of relative stability over the last 15 years, the peak British count was the highest since 1992-93, considerably exceeding the maxima of around quarter of a million since then. The peak in Northern Ireland was in the middle of the widely fluctuating counts of recent years.

The main resorts for this species witnessed increases and decreases in peak counts, as is normal for this highly mobile wader. Whilst Morecambe Bay, the Alt Estuary and Dengie Flats recorded peak counts in excess of the recent average (increases of 139%, 88% and 47%, respectively), counts on the Solway and Cromarty Firths and Dundrum Bay were both much lower than normal, and those at Lindisfarne, Tees Estuary and Duddon Estuary were sufficiently low that all ceased to be of international importance.

The two populations of Knot that pass through the UK have different international thresholds: 5,000 for the sub-species *canutus*, breeding in the Taimyr, and 3,500 for *islandica*, breeding in Greenland and northeast Canada. Separation in the field is, to all intents and purposes, impossible during winter but, since the nominate race winters in Africa and *islandica* in western Europe, it is assumed that all birds in the table below comprise *islandica* and thus the smaller threshold is applied.

The high-arctic breeding Knot has been highlighted as a species likely to provide one of the best early-warning systems of a species' ability to cope with climatic change (Moss 1998). This is because climate change threatens every aspect of the Knot's life cycle, from alterations in the tundra breeding habitat to the threats posed to stop-over and wintering sites vulnerable to sea-level changes.

	93-94	94-95	95-96	96-97	97-98	Mon	Mean
International							
Wash	(90,841)	(77,694)	(47,775)	(72,173)	81,950	Nov	86,396
Ribble Estuary	44,510	61,054	(35,321)	55,752	(36,880)	Nov	53,772
Thames Estuary	50,690	(31,936)	47,191	39,121	55,663	Nov	48,166
Morecambe Bay	22,335	26,711	37,808	44,134	77,344	Feb	41,666
Humber Estuary	24,698	34,663	28,076	22,579	30,283	Dec	28,060
Alt Estuary	17,832	15,020	18,002	25,350	35,881	Nov	22,417
Dee Estuary (Eng/Wal)	11,725	(9,545)	18,520	30,025	14,000	Jan	18,568
Solway Estuary	14,923	16,661	10,516	9,086	5,472	Dec	11,332
Strangford Lough	6,301	7,369	13,444	12,302	9,456	Jan	9,774
Forth Est.	12,688	5,488	8,950	11,299	7,866	Jan	9,258
North Norfolk Marshes	6,143	7,967	5,930	15,236	9,006	Nov	8,856
Dengie Flats	(5,550)	6,820	6,050	6,600	10,490	Feb	7,490
Swale Estuary	6,435	6,406	2,517	7,131	5,420	Feb	5,582
Cromarty Firth	4,782	(2,997)	(6,600)	(6,829)	1,733	Dec	4,986
Montrose Basin	3,500	(1,500)	(6,500)	3,800	3,800	Jan	4,400
Burry Inlet	1,920	2,000	6,353	8,200	2,080	Dec	4,111
Stour Estuary	2,650	3,365	4,748	4,289	3,565	Dec	3,723
Inner Moray Firth	1,491	3,441	2,491	(7,773)	3,097	Jan	3,659
Great Britain							
Lindisfarne	(2,770)	2,022	3,810	4,625	3,218	Jan	3,419 ▼
Tees Estuary	2,050	2,577	5,122	3,520	2,783	Feb	3,210 ▼
Duddon Estuary	3,650	6,520	1,562	2,931	1,286	Mar	3,190 ▼
Northern Ireland							
Belfast Lough	1,001	361	560	580	(430)	Nov	626
Lough Foyle	258	225	1,145	400	270	Mar	460
Dundrum Bay	400	1,200	(546)	123	0	Nov	454

Other sites surpassing table qualifying levels in 1997-98

Hamford Water	4,234	Jan

SANDERLING
Calidris alba

GB max:	11,329	May
NI max:	55	Sep

International threshold:	**1,000**
Great Britain winter threshold:	**230**
Great Britain passage threshold:	**300**
All-Ireland threshold:	**35***

* 50 is normally used as a minimum threshold

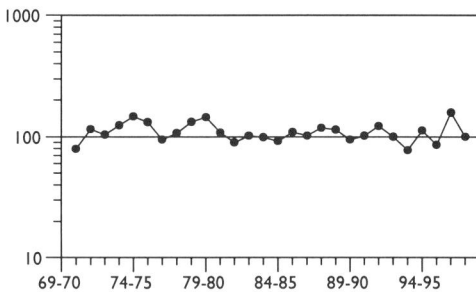

Figure 62. Annual indices for Sanderling in UK

The UK annual index for Sanderling continued to fluctuate dramatically, more so than for any other wader: following the previous year's 63% increase, to the highest level ever reached, it then dropped by 36% in 1997-98. As usual, the highest winter count, of just 7,288, was considerably lower than the peak which occurs on spring passage.

Preliminary analysis of Sanderling numbers from the Non-estuarine Coastal Waterfowl Survey (NEWS) suggests a 33% decline (from 8,637 recorded during the 1984-85 Winter Shorebird Count to 4,803 during NEWS on those coastal sections covered by both surveys). The majority of the NEWS total will be additional to WeBS Core Counts but, even so, it suggests that the number in 1997-98 in Britain peaked at only around half the current population estimate.

During 1997-98, several of the main resorts for this species held peak counts considerably in excess of the recent mean, most notably the Alt Estuary, but also the Tees Estuary Thanet coast, the Wash and Carmarthen Bay. The most notable low counts were the Humber Estuary, South Uist, Tiree and the Ribble Estuary, though this last may be explained by movement to the adjacent Alt.

Four sites held in excess of 1,000 Sanderling during the passage period, all greatly exceeding their respective winter maxima and sufficient to qualify the sites as internationally important if maintained regularly.

	93-94	94-95	95-96	96-97	97-98	Mon	Mean
International							
Ribble Estuary	2,780	3,655	750	3,085	1,134	Mar	2,281
Great Britain							
Alt Estuary	507	500	607	971	1,352	Mar	787
Duddon Estuary	(923)	700	965	404	547	Mar	708
South Uist (West Coast)	[1]289	1,185	-	-	[2]528	Jan	667
Tiree	-	883	-	-	[2]371	Jan	630
Dee Estuary (Eng/Wal)	624	1,180	208	598	429	Jan	608
Thanet Coast	501	654	-	457	776	Dec	597
North Norfolk Marshes	286	681	476	636	594	Nov	535
Humber Estuary	460	665	413	635	345	Dec	504
Wash	250	467	(539)	484	576	Nov	463
Lade Sands	187	-	475	706	460	Mar	457
Tees Estuary	(255)	298	465	331	470	Mar	391
Carmarthen Bay	398	374	323	386	470	Nov	390
Durham Coast	(151)	(176)	255	372	(130)	Jan	314
East Sanday Coast	[3]302	-	-	-	[2](210)	Jan	302
Jersey Shore	449	240	130	(371)	304	Feb	299
Solway Estuary	(195)	316	(220)	384	134	Dec	278
Morecambe Bay	(137)	(138)	312	137	278	Nov	242 ▲
Northern Ireland							
South Down	-	(0)	60	-	-		60

Sites no longer meeting table qualifying levels
North Uist (West Coast)

Internationally or nationally important sites not counted in last five years
South Ford (Outer Hebrides)

Other sites surpassing table qualifying levels in 1997-98

Lindisfarne	292	Jan	Colne Estuary	240	Dec
Tay Estuary	270	Dec	Dundrum Bay	46	Feb
Rye Harbour/Pett Levels	255	Jan			
Forth Est.	252	Dec			

Sites surpassing passage threshold in Great Britain in 1997-98

Ribble Estuary	7,345	Jul	Dee Estuary (Eng/Wal)	612	Oct
Morecambe Bay	2,491	May	Solway Estuary	446	May
Alt Estuary	2,102	May	Thames Estuary	375	Oct
Wash	1,590	May	Thanet Coast	364	Oct
Humber Estuary	885	Aug	Duddon Estuary	360	Aug
North Norfolk Marshes	730	Aug	Chichester Harbour	320	Aug
Tees Estuary	654	May			

1   RSPB Report Two: Western Isles Winter Shorebird Counts
2   NEWS data
3   RSPB Report One: Orkney Winter Shorebird Counts

WESTERN SANDPIPER
Calidris mauri

Vagrant
Native range: Americas

This very rare vagrant to the UK was found during a WeBS count on the Forth Estuary in August.

LITTLE STINT
Calidris minuta

International threshold: 2,100
Great Britain threshold: ?
All-Ireland threshold: ?

GB max: 47 Sep
NI max: 0

Although the annual peak count in the last five years has varied considerably, the 1997-98 figure was particularly small, and considerably lower than the 840 in September 1996, possibly the result of poor breeding success in Western Siberia (Tomkovich & Zharikov 1998). The majority were recorded on autumn passage, and a few overwintered, notably on the Medway and Dee (Eng/Wales) Estuaries.

Sites with five or more birds in 1997-98

Thames Estuary	8	Sep	Chichester Harbour	5	Aug
Wash	6	Aug	Abberton Reservoir	5	Sep
Breydon Water & Berney Marshes	6	May			

PECTORAL SANDPIPER
Calidris melanotos

Vagrant
Native range: Americas, N Siberia and Australia

Three sites recorded single birds, all during September, namely Daventry Reservoir, the Dee Estuary (Eng/Wales) and Hamford Water.

CURLEW SANDPIPER
Calidris ferruginea

International threshold: 4,500
Great Britain threshold: ?
All-Ireland threshold: ?

GB max: 116 Sep
NI max: 6 Sep

As with Little Stint, Curlew Sandpiper proved to be something of a scarcity in 1997 with numbers just a fraction of those the previous autumn.

It is thought likely that both high breeding success and appropriate weather patterns (cyclonic conditions over northeastern Europe) are involved in determining the numbers of Curlew Sandpipers in northwest Europe, the relative importance of the latter factor increasing towards the extremities of the migration range in Britain (Wilson *et al.* 1980). When the two factors are combined, as may have been the case in 1996-97, larger influxes may occur.

As expected, most of the 73 records of Curlew Sandpiper occurred during autumn passage, and just two from November to March, at Blithfield Reservoir and Chichester Harbour.

Sites with more than five birds in 1997-98

Abberton Reservoir	21	Oct	Clwyd Estuary	7	Aug
Breydon Water & Berney Marshes	17	Sep	Dee Estuary (Eng/Wales)	7	Sep
Severn Estuary	11	Sep	Wash	7	Oct
North Norfolk Marshes	9	Sep	Bann Estuary	6	Sep

PURPLE SANDPIPER
Calidris maritima

International threshold: 500
Great Britain threshold: 210[†]
All-Ireland threshold: 10*
* 50 is normally used as a minimum threshold

GB max: 1,061 Jan
NI max: 76 Dec

UK maxima were very similar to those of the previous two winters, though considerably lower than those of the early 1990s. The winter distribution still reflects the pattern previously described by Atkinson *et al.* (1978), with all of the nationally important sites in Britain found on the coasts of northeast England or eastern Scotland. Monthly totals reflect the general pattern by which some birds start arriving on the east coast in July to moult, followed by another wave in October and November of birds arriving on the northern and western coasts of Scotland from different breeding areas. Numbers build to a peak in mid winter before birds start to depart in February and March.

Low counts were recorded at most sites in 1997-98: Core Counts at just one site in each of Great Britain and Northern Ireland exceeded the 1% thresholds, and one site formerly of national importance fell from the table.

Purple Sandpipers are largely confined to rocky shorelines and as such are a species less

well covered by WeBS. Preliminary analysis of data collected by the 1998 European Non-estuarine Waterfowl Survey suggests that, in a comparison of numbers at sites covered both by this survey and the 1984-85 Winter Shorebird Count, Purple Sandpipers have declined by 44% (from 11,045 to 6,161 on sections covered by both surveys), with declines in all east coast Scottish districts being the largest and most significant. Whether these declines are real or a result of redistribution of the population has yet to be determined, although all of the most important British sites have recorded falling numbers since 1994.

	93-94	94-95	95-96	96-97	97-98	Mon	Mean
International							
East Sanday Coast	[1]782	-	-	-	[2](275)	Jan	782
Great Britain[†]							
SE Stronsay	[1]444	-	-	-	-		444
South Uist (West Coast)	[3]304	388	-	-	[2]313	Jan	335
North Ronaldsay	[1]255	-	-	-	[2]400	Jan	328
Tiree	-	314	-	-	[2]262	Jan	288
South Westray	[1]413	-	-	-	[2]159	Jan	286
SE Deerness	[1]274	-	-	-	-		274
North Mainland (Orkney)	[1]413	-	-	-	117	Jan	265
Moray Coast	264	271	268	223	219	Dec	249
Cambois to Newbiggin	296	352	(46)	[R]142	[R]165	Dec	239
Seahouses to Budle Point	250	269	252	144	207	Mar	224
Durham Coast	(179)	(163)	163	153	125	Dec	165 ▼
Fraserburgh to Rosehearty	105	150	-	-	-		128
Northern Ireland							
Outer Ards	(60)	64	(48)	89	(49)	Dec	77

Other sites surpassing table qualifying levels in 1997-98
Belfast Lough   31   Jan

†   as so few British sites are of national importance for Purple Sandpiper, a qualifying level of 100 has been chosen to select sites for presentation in this report.
1   RSPB Report One: Orkney Winter Shorebird Counts
2   NEWS data
3   RSPB Report Two: Western Isles Winter Shorebird Counts

DUNLIN
Calidris alpina

GB max: 462,582   Dec
NI max: 16,803   Dec

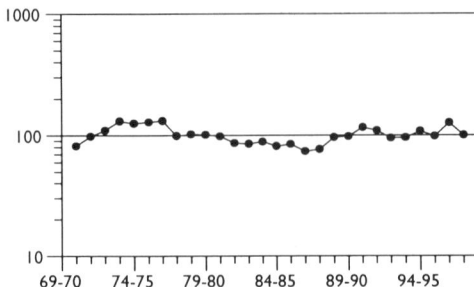

Figure 63. Annual indices for Dunlin inUK

International threshold: **14,000**
Great Britain winter threshold: **5,300**
Great Britain passage threshold: **2,000**
All-Ireland threshold: **1,250**

Following the previous year's peak in index values, there was a 21% decline in 1997-98 to more normal levels. The British winter coastal maximum was also over 100,000 down on the previous year.

Despite these declines, there was a marked increase at Morecambe Bay, and larger than average numbers on the Thames Estuary and Hamford Water, though these were very much the exception in 1997-98. As expected, a greater number of sites saw a downturn in numbers, particularly the Ribble Estuary, Chichester Harbour, Duddon Estuary, Burry Inlet and Orwell Estuary, with counts at the last two representing just one third of the 1996-97 peak. Such fluctuations are not uncommon at a site level as Dunlin is such a highly mobile species. Likewise, annual fluctuations in overall numbers are usual, as movements are greatly influenced by weather conditions. In general, the winter of 1997-98 was mild, particularly in February, and it may be that many Dunlin did not continue their migration once they had reached the Wadden Sea. Provisional analysis of WeBS data suggests that Dunlin distribution in Britain is significantly related to the number of sleet or snow days during the winter, with fewer sleet/snow days, as

is increasingly the case, being reflected by a more easterly distribution of Dunlin populations.

At an international scale, numbers appear to have shown a slight decrease (-14%) since the mid 1980s although the current estimate still lies within the 24-year range of 90,000 to 1,600,000 (Davidson 1998).

During passage periods, six sites held numbers exceeding the international threshold, though all are already identified as internationally important on the basis of winter counts alone. However, numbers an additional four sites exceeded the national threshold for passage that are not listed in the table, namely the Alt Estuary, North Norfolk Marshes, Ythan Estuary and Alde Estuary.

	93-94	94-95	95-96	96-97	97-98	Mon	Mean
International							
Morecambe Bay	41,125	58,914	41,831	57,617	71,731	Dec	54,244
Mersey Estuary	32,000	64,000	40,501	55,430	52,015	Jan	48,789
Severn Estuary	41,209	50,638	(26,150)	29,420	(25,351)	Jan	40,422
Ribble Estuary	51,415	(41,532)	34,215	45,973	18,930	Feb	38,413
Wash	24,930	38,235	(41,487)	38,741	36,054	Nov	35,889
Thames Estuary	16,882	37,368	26,933	34,057	37,979	Dec	30,644
Blackwater Estuary	25,621	(21,960)	27,345	33,512	22,195	Nov	27,168
Dee Estuary (Eng/Wal)	16,378	25,383	(24,695)	31,430	30,318	Nov	25,877
Medway Estuary	23,550	30,540	17,232	33,313	17,200	Jan	24,367
Langstone Harbour	23,294	25,054	21,144	(14,240)	(15,000)	Feb	23,164
Chichester Harbour	19,038	26,087	22,590	19,567	15,629	Nov	20,582
Humber Estuary	22,975	20,145	27,600	10,210	20,695	Dec	20,325
Solway Estuary	18,795	18,498	11,688	20,042	11,982	Jan	16,201
Stour Estuary	18,241	(16,024)	15,343	14,727	14,712	Feb	15,809
Great Britain							
Swale Estuary	(8,278)	12,302	10,971	14,243	15,529	Feb	13,261
Duddon Estuary	8,460	14,790	10,370	14,416	7,232	Jan	11,054
Colne Estuary	10,316	9,810	13,000	8,805	10,510	Jan	10,488
Forth Est.	9,886	7,824	13,830	9,118	9,937	Dec	10,119
Burry Inlet	8,736	8,787	6,966	14,548	4,539	Dec	8,715
Hamford Water	5,918	5,789	(10,113)	9,146	11,970	Dec	8,587
Orwell Estuary	9,900	(6,185)	11,565	9,576	3,210	Nov	8,563
Lindisfarne	8,224	8,027	(7,031)	10,364	6,039	Jan	8,164
Dengie Flats	(6,600)	9,600	4,200	(7,850)	8,100	Dec	7,438
Inner Moray Firth	5,494	(3,805)	7,226	8,567	5,417	Feb	6,676
Southampton Water[1]	(3,654)	5,885	7,796	5,617	7,088	Feb	6,597
Poole Harbour	6,222	5,963	6,424	6,347	6,355	Jan	6,262
Northern Ireland							
Strangford Lough	4,347	6,506	8,317	12,948	7,175	Nov	7,859
Lough Foyle	4,622	(4,417)	7,025	(3,666)	4,106	Feb	5,251
Outer Ards	2,288	1,955	1,709	2,689	1,890	Dec	2,106
Belfast Lough	1,681	1,774	1,811	1,943	1,070	Dec	1,656
Carlingford Lough	(2,410)	650	1,244	(860)	2,002	Jan	1,577
Bann Estuary	1,500	1,260	1,085	2,910	1,075	Jan	1,566
Dundrum Bay	995	1,329	884	1,707	1,893	Feb	1,362 ▲

Other sites surpassing table qualifying levels in 1997-98

Cromarty Firth	5,563	Dec		Cleddau Estuary	5,318	Jan

Sites surpassing passage threshold in Great Britain in 1997-98

Ribble Estuary	33,521	May	Solway Estuary	7,027	May
Wash	27,107	Sep	Dengie Flats	6,200	Apr
Morecambe Bay	26,599	Oct	Alt Estuary	6,027	May
Mersey Estuary	20,000	Oct	Swale Estuary	5,179	Oct
Thames Estuary	18,887	Oct	Dee Estuary (Eng/Wal)	4,918	Oct
Humber Estuary	14,591	Aug	Severn Estuary	4,852	Oct
Blackwater Estuary	11,391	Oct	North Norfolk Marshes	4,423	Oct
Stour Estuary	11,340	Oct	Lindisfarne	3,048	Sep
Chichester Harbour	8,595	Oct	Ythan Estuary	2,400	Sep
Langstone Harbour	8,545	Oct	Alde Complex	2,033	Sep
Colne Estuary	7,200	Oct			

1   *primarily Low Tide Count data*

RUFF

Philomachus pugnax

GB max: 587 Sep
NI max: 2 Sep

International threshold: ?
Great Britain threshold: 7*
All-Ireland threshold: +*

* 50 is normally used as a minimum threshold

Although the maximum British count of Ruff was slightly higher than the previous year, passage numbers in 1997-98 were only around half of those two and three years ago. Numbers dropped sharply in November and, whilst they had increased to around 400 by February, the winter peak was the lowest during the last five winters. Numbers at four of the top five sites were correspondingly well below their five year means and only the count at Martin Mere was

notably higher than the average.

Passage counts usually produce the highest numbers for many sites, e.g. the Humber Estuary held 113 birds in September, 104 were recorded on the North Norfolk Coast in August and 72 at Abberton Reservoir in September. The importance of spring passage sites is undoubtedly under-recorded due the paucity of WeBS counts and the rapid turnover of birds at this time.

	93-94	94-95	95-96	96-97	97-98	Mon	Mean
Great Britain							
Lower Derwent Valley	128	106	189	81	133	Feb	127
Ouse Washes	78	195	139	(113)	60	Mar	118
Nene Washes	138	198	71	19	60	Jan	97
Swale Estuary	186	60	76	4	27	Feb	71
North Norfolk Marshes	24	(43)	(58)	118	21	Jan	55
Martin Mere	18	12	45	67	90	Mar	46
Hamford Water	29	21	81	(23)	32	Feb	41
Abberton Reservoir	5	51	53	18	30	Dec	31
Blackwater Estuary	29	(38)	(12)	19	33	Mar	30
Somerset Levels	12	(33)	37	42	21	Feb	29
Arun Valley	3	34	43	16	34	Jan	26
Walland Marsh	-	-	-	36	9	Nov	23
Thames Estuary	6	21	11	52	15	Jan	21
Sandbach Flashes	-	(30)	(20)	21	10	Nov	20
Ribble Estuary	23	11	17	21	25	Nov	19
Colne Estuary	22	10	30	0	(0)	Nov	16
Humber Estuary	27	20	15	2	14	Nov	16
Rutland Water	0	16	19	21	17	Dec	15
Stour Estuary	4	65	0	0	0	Nov	14
Dungeness	15	1	7	40	4	Nov	13
Breydon Water & Berney Marshes	4	8	1	24	28	Jan	13
Dee Estuary (Eng/Wal)	13	14	17	12	8	Feb	13
Poole Harbour	29	2	16	0	4	Feb	10
Chichester Harbour	19	6	11	(9)	2	Feb	10
Wash	(7)	(0)	17	(4)	3	Dec	10
Buckenham Marshes	(2)	0	0	20	17	Feb	9
Tees Estuary	3	14	16	7	2	Feb	8
Druridge Pool	5	8	7	-	-		7

Loch Leven	37	Sep	Loch of Strathbeg	10	Sep
Ythan Estuary	26	Aug	Tophill Low Reservoirs	9	Oct
Ribble Estuary	25	Nov	Monikie Reservoirs	9	Sep
Middle Yare Valley	22	Feb	Exe Estuary	8	Mar
Forth Est.	20	Aug	Severn Estuary	7	Feb
Mersey Estuary	14	Sep			

JACK SNIPE
Lymnocryptes minimus

International threshold:		?
Great Britain threshold:		?
All-Ireland threshold:		250

GB max: 103   Feb
NI max: 5   Dec

The 1997-98 UK winter peak count was similar to that in previous years. However, the problems entailed in accurately recording such a secretive and well camouflaged species and the fact that its preferred habitat of marsh and wet grassland is poorly covered by WeBS, mean that many records rely purely on the chance of flushing birds by accident. Even summed site maxima for the 107 sites at which Jack Snipe was found results in a total of just 246 birds, certainly just a fraction of the true wintering number. Consequently, little meaningful interpretation can be made of WeBS counts for this species.

As Jack Snipe has unfavourable conservation status in Europe, and is a quarry species in a number of countries, an EU Action Plan is currently being prepared.

Sites with five or more birds in 1997-98

Chichester Harbour	29	Mar	Loch Etive	6	Jan
Inner Moray Firth	16	Dec	Somerset Levels	5	Oct/Feb
Severn Estuary	9	Feb	Langstone Harbour	5	Feb
Lower Derwent Valley	7	Oct	Shipton-on-Cherwell Quarry	5	Feb
Hill Ridware Lake	6	Feb			

SNIPE
Gallinago gallinago

International threshold:		20,000**
Great Britain threshold:		?[†]
All-Ireland threshold:		?[†]

GB max: 7,404   Dec
NI max: 210   Nov

Numbers of Snipe recorded by WeBS in the UK rose in 1997-98, with a peak count in December of 7,404. The same restrictions apply to interpretation of these results as to Jack Snipe. Once again, the Somerset Levels held, by far, the highest numbers Fluctuations at these sites presumably reflect variability in detection of this species or even counter effort as much real differences in numbers: there is no discernable pattern in numbers at key sites that might have been expected as a result of two cold winters followed by a mild one.

	93-94	94-95	95-96	96-97	97-98	Mon	Mean
Great Britain[†]							
Somerset Levels	(2,704)	3,628	(1,929)	1,041	1,975	Dec	2,337
Lower Derwent Valley	1,030	310	472	409	500	Nov	544
Maer Marsh	-	420	650	480	550	Feb	525
Arun Valley	40	832	206	135	272	Nov	297
Morecambe Bay	232	379	198	188	260	Nov	251
Swale Estuary	148	399	335	(46)	38	Mar	230
Newgale Beach	-	-	-	-	230	Jan	230
Exe Estuary	297	145	315	42	(245)	Dec	209
Ouse Washes	204	142	546	71	38	Nov	200

Ribble Estuary   354   Dec
Duddon Estuary   208   Nov

†   *as no British or all-Ireland thresholds have been set for national importance, a qualifying level of 200 has been chosen to selec sites for presentation in this report*

GREAT SNIPE
Gallinago media

Vagrant
Native range: NE Europe, W Asia and Africa

A rare record of this species was made on the Medway Estuary in August.

LONG-BILLED DOWITCHER
Limnodromus scolopaceus

Vagrant
Native range: NE Siberia and N America

Individuals were recorded on the Swale Estuary in several autumn months and at Llangorse Lake in October.

WOODCOCK
Scolopax rusticola

International threshold: **20,000****
Great Britain threshold: **?**
All-Ireland threshold: **?**

This species, being primarily adapted to a woodland rather than wetland existence, combined with its secretive nature and its being active at dawn and dusk, mean that, even if present on a wetland site, it is hard to locate and thus significantly under-recorded by WeBS.

An EU conservation action plan for this species is currently being prepared to address its generally unfavourable conservation status in much of Europe (Tucker & Heath 1994).

Sites with 10 or more birds in 1997-98

Longueville Marsh	12	Dec/Feb
Grouville Marsh	10	Jan
Shell Pond (Carrington)	10	Dec

BLACK-TAILED GODWIT
Limosa limosa

International threshold: **700**
Great Britain threshold: **70**
All-Ireland threshold: **90**

GB max: 16,944   **Aug**
NI max: 404   **Jan**

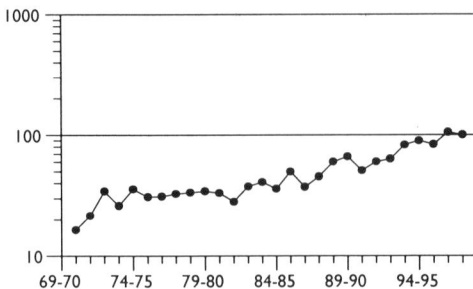

Figure 64. Annual indices for Black-tailed Godwit in UK

Although lower than the previous winters' record values, the annual index and peak winter numbers of Black-tailed Godwit remained very high. Peak numbers, though, occur during autumn passage.

The Black-tailed Godwits which winter in Britain and Ireland are of the Icelandic breeding race *islandica*, and numbers wintering in Britain have historically been related to the prevailing climatic conditions of the time (Prater 1975). During the 1930s, less than 100 Black-tailed Godwit wintered in the UK but this gradually rose and was attributed to higher breeding numbers in Iceland following climatic amelioration. A cooling of the spring climate during the late 1960s may have led to the rapid decline in numbers during the early 1970s. However, numbers have steadily increased since the mid 1970s to an all time high in 1996-97. A re-evaluation of international population sizes suggests a 62% increase in numbers since the mid 1980s (Davidson 1998). The international threshold is certain to be revised upwards in 1999.

As Black-tailed Godwit is a quarry species in some countries, an EU conservation action plan is currently under development.

The table below is characterised by great variation in counts between years at individual sites. Although, as expected, peak counts in 1997-98 were substantially higher or lower than the recent average at a large number of the main resorts, a number of unusual counts are apparent. The high count on the Mersey in the previous winter was sustained in 1997-98 and especially large increases were recorded on the Exe Estuary, Alde Complex, Humber Estuary, Crouch/Roach Estuary and the Burry Inlet. By contrast, there were few significantly low counts, although the decline at Pagham Harbour saw the 1997-98 peak reach just 10% of that in 1993-94. All peak counts of this species in Southampton Water in recent years have been adjudged to be incomplete due to lack of data for a key sector, although the continuing decline is still cause for concern.

The importance of the UK during passage periods is demonstrated by counts at 12 sites exceeding the threshold for international importance. Four of these (Abberton Reservoir, Humber Estuary, Breydon & Berney Marshes and Langstone Harbour) do not regularly hold this number when based on winter counts alone, whilst counts at several of the others greatly exceeded the normal winter counts, with the phenomenal count on the Wash equivalent to half the British winter population.

	93-94	94-95	95-96	96-97	97-98	Mon	Mean
International							
Wash	650	(705)	(1,764)	5,738	3,104	Nov	3,164
Stour Estuary	1,889	(1,882)	3,848	1,848	1,724	Nov	2,327
Dee Estuary (Eng/Wal)	2,033	1,425	1,862	2,203	1,642	Nov	1,833
Poole Harbour	(1,447)	2,046	1,194	1,771	1,895	Feb	1,727
Swale Estuary	1,636	1,910	637	(1,409)	1,010	Feb	1,320
Ribble Estuary	1,690	(845)	180	2,319	911	Mar	1,275
Ouse Washes	1,016	2,068	509	1,019	994	Mar	1,121
Medway Estuary	(380)	1,795	902	(206)	653	Mar	1,117
Mersey Estuary	21	580	494	1,703	2,086	Jan	977
Southampton Water	(801)	(1,450)	(594)	(147)	(31)	Nov	(948)
Blackwater Estuary	630	956	920	(1,088)	(608)	Mar	899
Thames Estuary	(19)	(1,104)	(109)	636	(241)	Jan	870 ▲
Chichester Harbour	664	2,139	551	497	464	Nov	863
Hamford Water	241	1,977	236	732	352	Jan	708
Great Britain							
Exe Estuary	737	479	520	226	1,132	Nov	619
Orwell Estuary	270	728	615	458	253	Mar	465
Nene Washes	472	626	398	80	509	Mar	417
Breydon Water & Berney Marshes	591	437	122	367	503	Jan	404
Alde Complex	55	558	201	254	701	Mar	354
Humber Estuary	80	83	57	544	924	Dec	338
Abberton Reservoir	0	158	724	159	322	Nov	273
Blyth Estuary	430	225	266	200	215	Nov	267
Langstone Harbour	276	284	154	240	327	Jan	256
Newtown Estuary	212	151	(365)	130	148	Feb	201
Crouch/Roach Estuary	163	261	68	87	416	Mar	199
Pagham Harbour	466	260	100	98	46	Nov	194
Deben Estuary	43	111	267	354	154	Feb	186
North West Solent	110	120	200	265	(17)	Mar	174
Colne Estuary	82	227	219	85	214	Feb	165
Eden Estuary	103	(116)	128	176	183	Feb	148
Christchurch Harbour	135	-	-	-	-		135
Beaulieu Estuary	19	0	235	(246)	161	Dec	132

	93-94	94-95	95-96	96-97	97-98	Mon	Mean	
Burry Inlet	76	86	87	114	233	Mar	119	
Tamar Complex	78	100	156	(127)	119	Dec	116	
Fal Complex	71	89	77	131	146	Jan	103	
Solway Estuary	6	6	13	460	9	Jan	99	
North Killingholme Haven Pits	-	-	-	1	190	Mar	96	▲
Severn Estuary	8	32	49	97	230	Dec	83	▲
North Norfolk Marshes	(6)	45	109	(21)	(44)	Mar	77	▲
Portsmouth Harbour	0	0	62	204	100	Jan	73	▲
Caerlaverock WWT	-	4	0	98	176	Nov	70	▲
Northern Ireland								
Belfast Lough	370	359	235	418	178	Jan	312	
Strangford Lough	5	38	43	486	226	Mar	160	

Sites no longer meeting table qualifying levels
Dengie Flats
Morecambe Bay

Other sites surpassing table qualifying levels in 1997-98
Forth Est.   87   Nov

Sites surpassing the international threshold during passing periods in 1997-98

Wash	7,119	Sep	Poole Harbour	1,409	Apr
Ribble Estuary	3,681	Aug	Abberton Reservoir	1,247	Sep
Stour Estuary	2,205	Oct	Humber Estuary	976	Oct
Mersey Estuary	2,043	Aug	Swale Estuary	969	Apr
Ouse Washes	1,721	Apr	Breydon Water & Berney Marshes	881	Aug
Dee Estuary (Eng/Wal)	1,561	Oct	Langstone Harbour	875	Oct

BAR-TAILED GODWIT

Limosa lapponica

International threshold: 1,000
Great Britain threshold: 530
All-Ireland threshold: 175

GB max: 48,313   Jan
NI max: 3,353   Jan

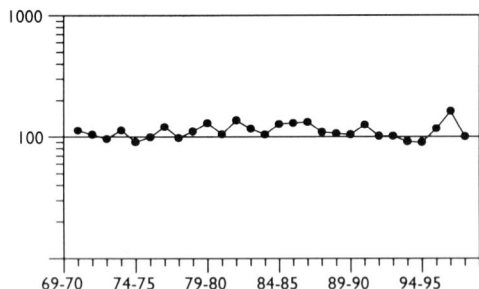

Figure 65. Annual indices for Bar-tailed Godwit in UK

The wintering population has continued to fluctuate considerably between years, with the index value and peak count dropping sharply from the record highs the previous winter, although both values were similar to those recorded in the early to mid 1990s. These variations, with the peak coastal winter count in Great Britain over 30,000 lower than in 1996-97, are thought to result from fluctuations in breeding success and the severity of winter conditions in the Wadden Sea. The majority of Bar-tailed Godwits wintering in the UK are thought to breed in western Siberia and the Russian Arctic. Tomkovich & Zharikov (1998) reported that breeding success of waders had generally been low in the western section of the Russian Arctic and about average for more easterly areas. The low breeding success was thought likely to be a result of a late spring and harsh weather conditions, possibly exacerbated by increased predation resulting from low numbers of alternative prey such as lemmings.

Like so many other species of wader, Bar-tailed Godwits are highly mobile and this is reflected in the large fluctuations in peak counts at some of the major sites. High counts in the previous winter were maintained on the Wash, Alt Estuary and North Norfolk Marshes, and substantially higher than normal counts were recorded on the Humber Estuary and Strangford Lough. Conversely, the Thames Estuary, Morecambe Bay, Dornoch Firth, Chichester Harbour and the Dee Estuary (Eng/Wal) all recorded peak counts 40-50% lower than average.

Despite these fluctuations, the list of sites of international and national importance remained unchanged from the previous year with the exception of the Eden Estuary.

	93-94	94-95	95-96	96-97	97-98	Mon	Mean	
International								
Wash	11,132	8,987	(15,227)	16,246	16,435	Jan	13,605	
Ribble Estuary	16,195	7,100	(7,693)	15,885	(7,063)	Nov	13,060	
Thames Estuary	5,626	(3,547)	11,684	16,164	5,797	Dec	9,818	
Alt Estuary	138	5,511	5,488	9,015	9,424	Nov	5,915	
Lindisfarne	(3,243)	3,324	2,769	2,770	3,225	Feb	3,066	
Solway Estuary	2,407	3,192	2,331	4,273	2,495	Jan	2,940	
Morecambe Bay	890	4,559	(2,985)	3,658	1,818	Jan	2,782	
North Norfolk Marshes	(1,561)	1,205	1,338	3,360	3,108	Nov	2,253	
Dengie Flats	601	1,500	1,300	5,500	1,402	Feb	2,061	
Lough Foyle	1,656	2,428	2,140	(2,120)	1,535	Dec	1,976	
Forth Est.	2,298	1,560	1,988	1,869	2,157	Jan	1,974	
Inner Moray Firth	1,411	1,541	(2,649)	2,792	1,301	Jan	1,939	
Humber Estuary	994	1,233	2,199	1,505	(2,970)	Feb	1,780	
Cromarty Firth	1,055	(2,069)	(1,193)	(1,225)	1,654	Jan	1,593	
Dornoch Firth	1,243	(707)	1,520	(2,125)	847	Jan	1,434	
Tay Estuary	1,387	537	(1,520)	2,305	1,315	Feb	1,413	
Strangford Lough	1,542	843	898	1,269	2,433	Jan	1,397	
Chichester Harbour	1,431	(1,992)	1,250	1,100	820	Jan	1,319	
Dee Estuary (Eng/Wal)	22	(168)	(2,012)	2,167	780	Nov	1,245	
Great Britain								
East Sanday Coast	[1]871	-	-	-	[2](951)	Jan	911	
Eden Estuary	841	1,231	672	603	610	Feb	791	▼
North Uist (West Coast)	[3]641	-	-	-	[2]662	Jan	652	
Hamford Water	284	307	657	1,380	548	Feb	635	
Swale Estuary	350	603	696	824	597	Jan	614	
Northern Ireland								
Belfast Lough	126	572	132	176	225	Jan	246	

Other sites surpassing table qualifying levels in 1997-98

Langstone Harbour	841	Jan	Hamford Water		548	Feb
Swale Estuary	597	Jan				

1   RSPB Report One: Orkney Winter Shorebird Counts
2   NEWS data
3   RSPB Report Two: Western Isles Winter Shorebird Counts

WHIMBREL
Numenius phaeopus

International threshold: **6,500**
Great Britain threshold: **+***
All-Ireland threshold: **+***

* 50 is normally used as a minimum threshold

GB max:	1,735	May
NI max:	331	May

As in previous years, numbers in the UK exhibit two distinct peaks: the first in spring is generally short-lived and typically in May, and a more protracted peak of returning birds in late summer. Tallies during winter months rarely exceed single figures.

Sites with 100 or more birds in 1997-98

Lough Foyle	308	May	Burry Inlet	117	May
Wash	154	Sep	Langstone Harbour	116	May
Morecambe Bay	152	May	Chichester Harbour	172	May
Pagham Harbour	130	May	Exe Estuary	106	May
Severn Estuary	123	May			

CURLEW
Numenius arquata

International threshold: 3,500
Great Britain threshold: 1,200
All-Ireland threshold: 875

GB max: 91,637 Feb
NI max: 7,629 Jan

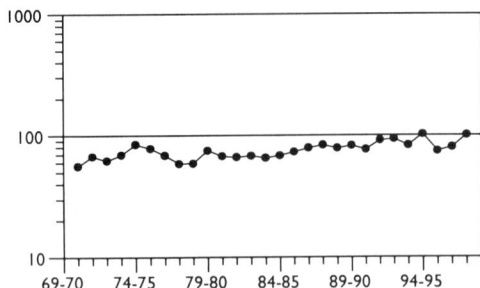

Figure 66. Annual indices for Curlew in UK

Curlew annual indices showed a marked contrast to those of most other waders, not only being one of the few species for which the value increased in 1997-98, but rising 25% over the previous year and almost equalling the peak value in 1994-95. Although maximum counts were about average for recent years, monthly totals, except for a low

in November, remained consistently high throughout the winter in both Britain and Northern Ireland, and will have been the main cause of the high index values. Clearly, the cold weather of the previous two winters caused large-scale Curlew movements, with a significant proportion of birds departing Britain and Ireland altogether.

Consequently, there were no notable low counts at key sites in 1997-98, though higher than normal numbers were recorded on the Dee Estuary (Eng/Wales), Humber Estuary, Forth Estuary and Ribble Estuary. Water Sound regained its status as nationally important, having slipped from the table due to the low count in 1996-97.

The international population appears to have increased by 22% since the late 1980s, an apparently genuine increase (Davidson 1998).

	93-94	94-95	95-96	96-97	97-98	Mon	Mean
International							
Morecambe Bay	10,695	15,654	14,905	12,357	14,858	Dec	13,694
Solway Estuary	5,826	7,562	3,348	4,062	5,716	Feb	5,303
Dee Estuary (Eng/Wal)	3,548	4,127	3,538	4,583	5,370	Jan	4,233
Wash	3,079	3,920	(3,945)	3,241	3,803	Feb	3,598
Great Britain							
Severn Estuary	3,646	5,307	2,682	2,001	(2,903)	Jan	3,409 ▼
Thames Estuary	2,632	4,239	3,006	3,412	2,525	Nov	3,163
Humber Estuary	2,913	2,654	1,973	1,406	(3,282)	Mar	2,446
Duddon Estuary	1,935	2,571	2,019	1,801	2,008	Dec	2,067
Inner Moray Firth	2,200	(2,600)	1,303	1,828	2,334	Feb	2,053
Forth Est.	2,113	1,460	1,607	1,599	2,545	Jan	1,865
Blackwater Estuary	(2,366)	2,271	1,226	(1,533)	1,426	Mar	1,822
Poole Harbour	1,913	1,723	1,428	1,652	1,783	Feb	1,700
Medway Estuary	1,883	2,226	1,474	1,061	1,413	Mar	1,611
Ribble Estuary	931	(1,816)	1,020	1,593	2,507	Feb	1,573
Wigtown Bay	1,183	3,003	656	(1,127)	1,144	Jan	1,497
Swale Estuary	1,992	1,599	832	(1,124)	1,435	Feb	1,465
Cleddau Estuary	1,311	1,732	1,436	1,283	1,330	Jan	1,418
Cromarty Firth	1,462	(986)	1,434	1,092	1,542	Jan	1,383
Chichester Harbour	1,338	1,694	1,296	1,135	1,433	Dec	1,379
Stour Estuary	1,544	(1,560)	912	(1,041)	(1,492)	Dec	1,377
Mersey Estuary	911	1,383	1,439	1,501	1,289	Mar	1,305 ▲
Water Sound	1,600	1,900	(360)	450	1,200	Feb	1,288 ▲
Clyde Est.	1,183	1,201	1,135	(1,088)	1,543	Feb	1,266
Northern Ireland							
Lough Foyle	1,829	1,710	2,231	(2,187)	1,879	Jan	1,967
Strangford Lough	1,483	1,922	2,107	1,344	2,102	Feb	1,792

Sites no longer meeting table qualifying levels
Carmarthen Bay
Lavan Sands

Hamford Water	1,355	Jan	Lindisfarne	1,330	Jan
Alde Complex	1,275	Feb	Dornoch Firth	1,331	Feb
North Norfolk Marshes	1,416	Nov	Newark Bay	1,200	Feb

SPOTTED REDSHANK
Tringa erythropus

International threshold: **1,200**
Great Britain threshold: **+***
All-Ireland threshold: **+***

* 50 is normally used as a minimum threshold

GB max: 221 Aug
NI max: 5 Sep

Although the peak count of Spotted Redshank in Great Britain was noticeably lower than in recent years, autumn passage was rather more protracted than usual, with 179 birds still present in November, largely as a result of 107 birds remaining on the Wash. Numbers in Northern Ireland were about average.

Sites with 10 or more birds in 1997-98

Wash	128	Aug	Humber Estuary	17	Sep
Swale Estuary	71	Jul	Burry Inlet	17	Sep
Abberton Reservoir	45	Sep	Dee Estuary (Eng/Wales)	14	Oct
Medway Estuary	23	Oct	Langstone Harbour	11	Aug

REDSHANK
Tringa totanus

International threshold: **1,500**
Great Britain winter threshold: **1,100**
Great Britain passage threshold: **1,200**
All-Ireland threshold: **245**

GB max: 84,659 Oct
NI max: 7,159 Nov

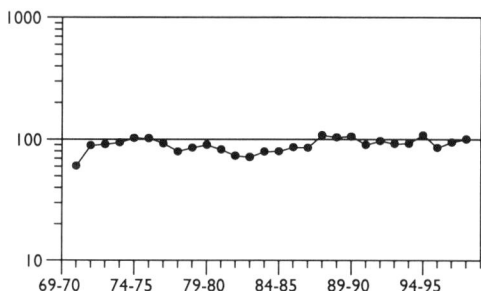

Figure 67. Annual indices for Redshank in UK

Whilst the peak counts in any month and during winter in both Great Britain and Northern Ireland were unremarkable compared to the last five years, the annual index for Redshank rose for the third successive year. This will have resulted from sustained high numbers throughout the winter months, with monthly totals of 75,000 around ten thousand higher than during 1996-97.

Counts of Redshank at key resorts show a reasonable degree of consistency between years, and at only a handful of sites was the 1997-98 peak markedly different from the norm: the Alde Complex (up 144%), Deben Estuary (up 54%), Hamford Water (up 40%), whilst the Blyth Estuary and the North Norfolk Marshes both recorded peak counts well above their averages (62% and 86% up, respectively), sufficient to elevate them to nationally important. The low count on the Stour Estuary was the only notable decline. The list of sites important in an all-Ireland context remained unchanged.

In contrast to a picture of relative stability in the UK, provisional international estimates (Davidson 1998) indicate a 34% increase since the mid 1908s.

The list of British sites surpassing the passage threshold, and the counts obtained during this period, are remarkably similar to the main table, though three sites (Chichester Harbour, Burry Inlet and Tay Estuary) not identified as nationally important on the basis of winter counts alone are included. The tremendous August count on the Humber Estuary is also noteworthy.

	93-94	94-95	95-96	96-97	97-98	Mon	Mean
International							
Morecambe Bay	5,322	7,666	6,847	6,350	6,968	Dec	6,631
Dee Estuary (Eng/Wal)	7,583	5,435	4,651	6,226	7,570	Feb	6,293

	93-94	94-95	95-96	96-97	97-98	Mon	Mean
Mersey Estuary	5,433	4,901	4,710	5,212	4,714	Jan	4,994
Forth Est.	4,190	3,941	5,205	3,602	4,768	Dec	4,341
Humber Estuary	3,437	4,896	4,085	(1,919)	4,575	Mar	4,284
Thames Estuary	2,076	(4,340)	(2,558)	3,295	2,992	Feb	3,176
Wash	2,046	3,814	(2,726)	(3,056)	3,279	Mar	3,049
Solway Estuary	3,012	2,588	(3,746)	2,512	3,196	Dec	3,011
Medway Estuary	3,026	3,264	3,731	1,586	3,020	Feb	2,925
Stour Estuary	1,917	4,178	3,392	2,853	1,908	Feb	2,850
Strangford Lough	2,449	2,817	3,281	2,832	2,713	Dec	2,818
Alde Complex	1,697	2,292	2,233	2,303	5,268	Feb	2,759
Inner Moray Firth	2,657	2,452	2,580	2,177	2,373	Dec	2,448
Montrose Basin	1,780	3,500	1,766	2,508	2,440	Mar	2,399
Duddon Estuary	2,249	2,348	3,888	1,344	1,856	Nov	2,337
Blackwater Estuary	(2,000)	2,653	1,651	(1,930)	2,523	Mar	2,276
Ribble Estuary	2,781	2,238	2,129	2,208	1,901	Mar	2,251
Clyde Est.	1,790	2,829	2,532	(2,092)	1,768	Mar	2,230
Belfast Lough	2,245	2,577	1,634	2,061	2,326	Nov	2,169
Colne Estuary	1,071	5,115	1,537	1,157	1,485	Dec	2,073
Hamford Water	2,011	1,906	1,413	2,322	2,486	Dec	2,028
Deben Estuary	1,590	1,574	1,558	2,632	2,704	Nov	2,012
Severn Estuary	1,328	2,032	2,526	2,072	1,790	Nov	1,950
Orwell Estuary	1,315	2,320	2,485	1,737	1,644	Mar	1,900
Alt Estuary	1,960	1,950	1,600	1,790	(1,000)	Jan	1,825
Chichester Harbour	1,421	1,987	1,287	1,442	1,391	Feb	1,506

Great Britain

	93-94	94-95	95-96	96-97	97-98	Mon	Mean	
Swale Estuary	1,375	1,757	1,325	1,268	1,364	Mar	1,418	
Blyth Estuary	1,246	916	1,000	1,426	1,761	Mar	1,270	▲
Tees Estuary	806	1,087	1,824	1,079	1,408	Nov	1,241	
Poole Harbour	1,172	1,356	1,111	1,028	1,239	Jan	1,181	
North Norfolk Marshes	751	862	(1,088)	1,356	1,729	Nov	1,175	▲
Ythan Estuary	1,448	1,280	660	1,344	1,012	Dec	1,149	

Northern Ireland

	93-94	94-95	95-96	96-97	97-98	Mon	Mean
Outer Ards	766	1,271	773	1,035	957	Jan	960
Carlingford Lough	642	693	789	1,194	1,043	Feb	872
Lough Foyle	(386)	656	1,147	805	720	Mar	832
Dundrum Bay	694	699	608	831	853	Nov	737
South Down	-	-	452	-	-		452
Larne Lough	(172)	377	360	317	362	Dec	354

Sites no longer meeting table qualifying levels

Tay Estuary
Eden Estuary

Other sites surpassing table qualifying levels in 1997-98

Cromarty Firth	1,385	Dec	Tay Estuary	1,156	Mar
Lindisfarne	1,207	Dec	Bann Estuary	420	Mar

Sites surpassing passage threshold in Great Britain in 1997-98

Humber Estuary	10,574	Aug	Ribble Estuary	2,124	Aug
Dee Estuary (Eng/Wal)	7,339	Sep	Clyde Est.	1,996	Oct
Morecambe Bay	6,575	Sep	Swale Estuary	1,993	Oct
Wash	5,958	Sep	Stour Estuary	1,981	Oct
Forth Est.	4,626	Sep	Deben Estuary	1,968	Oct
Mersey Estuary	4,406	Oct	Chichester Harbour	1,761	Oct
Thames Estuary	4,158	Oct	Ythan Estuary	1,660	Oct
Hamford Water	3,143	Sep	Orwell Estuary	1,562	Oct
Montrose Basin	2,896	Oct	Alt Estuary	1,550	Aug
Alde Complex	2,322	Sep	North Norfolk Marshes	1,469	Aug
Severn Estuary	2,227	Oct	Burry Inlet	1,402	Sep
Solway Estuary	2,205	Oct	Duddon Estuary	1,287	Sep
Blackwater Estuary	2,136	Oct	Tay Estuary	1,235	Sep

GREENSHANK
Tringa nebularia

International threshold: ?
Great Britain threshold: +*
All-Ireland threshold: 9*
* 50 is normally used as a minimum threshold

GB max: 1,830 Aug
NI max: 93 Oct/Dec

Peak numbers of Greenshank usually occur on autumn passage but, like several species showing similar phenology, e.g. Little Stint and Curlew Sandpiper, numbers in 1997 were lower than normal. However, the mild weather during 1997-98 resulted in higher than normal numbers

overwintering, particularly in Great Britain. Several of the sites listed below, namely Strangford Lough, Chichester Harbour and Lough Foyle, plus the southwest estuaries Taw/Torridge, Kingsbridge and Tamar, recorded peak counts of over 20 birds during winter months.

Sites in Great Britain with 50 or more birds in 1997-98

Thames Estuary	385	Jan	Hamford Water	62	Sep
Wash	295	Aug	Dee Estuary (Eng/Wales)	61	Aug
North Norfolk Marshes	198	Sep	Morecambe Bay	61	Aug
Chichester Harbour	150	Aug	Stour Estuary	56	Aug
Blackwater Estuary	81	Aug	Exe Estuary	54	Sep
Humber Estuary	79	Aug	Langstone Harbour	50	Sep

Sites in Northern Ireland with nine or more birds in 1997-98

Strangford Lough	67	Sep	Larne Lough	16	Oct
Lough Foyle	30	Dec	Outer Ards	15	Dec
Dundrum Bay	27	Aug	Carlingford Lough	15	Jan

LESSER YELLOWLEGS
Tringa flavipes

Vagrant
Native range: North and South America

Two birds were recorded on the Ribble Estuary in February.

GREEN SANDPIPER
Tringa ochropus

International threshold: ?
Great Britain threshold: ?
All-Ireland threshold: ?

GB max: 452 Aug
NI max: 9 Sep

Like several typical autumn migrants and scarce winterers, numbers of Green Sandpiper were low during passage, but relatively high numbers lingered during winter 1997-98.

Returning Green Sandpipers are one of the first indications that autumn is on the way, and although widely recorded, they are nowhere numerous; no WeBS count at an individual site has recorded more than 50 birds in the last five

years. Although only a handful of sites held 15 or more birds in 1997-98, five or more birds were found at 44 sites during the year. Numbers were even more thinly spread during winter, with only 10 sites holding five or more birds.

Most unusually, nine Green Sandpipers were recorded at Larne Lough during September, the first WeBS record in Northern Ireland since 1994-95.

Sites with 15 or more birds in 1997-98

Blackwater Estuary	36	Aug	Upton Warren	16	Aug
North Norfolk Marshes	29	Aug	Humber Estuary	16	Aug
Swale Estuary	26	Aug	Lower Derwent Valley	16	Nov
Beddington Sewage Farm	18	Jul	Thames Estuary	15	Aug

WOOD SANDPIPER

Tringa glareola

Passage numbers were about average for recent years, with a marked peak of 46 birds in August in Great Britain, but less than 20 in any other month.

None was recorded during winter months or in Northern Ireland.

Sites with five or more birds in 1997-98

Maer Lake	5	Aug	Swale Estuary	5	Aug
North Norfolk Marshes	5	Aug	Rye Harbour/Pett Levels	5	Sep

COMMON SANDPIPER

Actitis hypoleucos

International threshold:	?
Great Britain threshold:	?
All-Ireland threshold:	?

GB max:	1,336	Aug
NI max:	3	Jul

The peak British count was higher in 1997-98 than in any of the last five years, perhaps suggesting high breeding success in Britain, although numbers remained high during winter months,

no doubt encouraged by the mild weather. A count of 31 on the Tamar Complex in February was particularly notable.

Sites with 40 or more bird during 1997-98

Wash	106	Aug	Severn Estuary	46	Jul
Morecambe Bay	74	Jul	Tamar Complex	45	Jul
North Norfolk Marshes	61	Aug	Taw/Torridge Estuary	42	Aug
Abberton Reservoir	53	Aug			

TURNSTONE

Arenaria interpres

International threshold:	700
Great Britain threshold:	640
All-Ireland threshold:	225

GB max:	14,099	Oct
NI max:	1,573	Jan

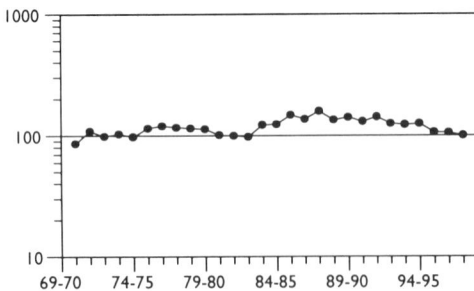

Figure 68. Annual indices for Turnstone in UK

The UK annual index for Turnstone continued its downward trend, reaching its lowest point since 1982-83. This is matched by a slow decline in winter numbers in Great Britain (although the peak in October was slightly higher than in the previous two winters), whilst those in Northern Ireland have shown a more alarming and consistent fall of several hundred birds in each of the last five winters such that the 1997-98 peak represents only half the 1993-94 total.

Preliminary analyses of data collected from the 1998 Non-estuarine Waterfowl Survey suggests a national decrease of 36% (from 26,123 to 16,623 birds on coastal sections counted by the 1984-85 Winter Shorebird Count and the 1998 NEWS survey). The largest of the statistically significant declines were in Cornwall, Grampian, Highland, Lothians and Strathclyde. The only notable increase was on the Orkney Islands.

Peak counts in 1997-98 at all but one of the sites in the table were below their respective five year means, including the Thames Estuary, which dropped below the qualifying level for international importance.

	93-94	94-95	95-96	96-97	97-98	Mon	Mean
International							
Morecambe Bay	(950)	(1,613)	(1,020)	1,248	1,198	Nov	1,353
East Sanday Coast	[1]1,269	-	-	-	[2](734)	Jan	1,269
Outer Ards	1,151	1,074	(750)	1,040	(715)	Jan	1,088
Tiree	-	1,237	-	-	[2]905	Jan	1,071
Dee Estuary (Eng/Wal)	623	383	1,243	1,193	978	Dec	884
Forth Est.	1,091	957	(918)	636	700	Mar	860
Belfast Lough	1,147	984	678	785	612	Nov	841
Alt Estuary	939	850	425	1,092	(461)	Jan	827
Thanet Coast	884	1,048	-	481	784	Nov	799
South Uist (West Coast)	[3]857	913	-	-	[2]554	Jan	775
Great Britain							
Wash	499	1,016	(637)	(766)	444	Jan	681
Thames Estuary	479	(745)	1,034	610	357	Dec	645 ▼
North Mainland (Orkney)	[1]656	-	-	-	[2]624	Jan	640
Northern Ireland							
South Down	-	-	485	-	-		485
Strangford Lough	537	439	369	207	207	Nov	352

Other sites surpassing table qualifying levels in 1997-98

Stour Estuary	710	Oct	Blackwater Estuary	716	Nov

1   RSPB Report One: Orkney Winter Shorebird Counts
2   NEWS data
3   RSPB Report Two: Western Isles Winter Shorebird Counts

GREY PHALAROPE   Scarce
Phalaropus fulicarius

Eight birds were recorded at seven sites over the course of the year: two on the North Norfolk Marshes in October, and singletons at the Hayle Estuary in July, the Camel Estuary in August, Rutland Water in September, Lothing Lake in October, and Chichester Gravel Pits and South Milton Ley in January.

MEDITERRANEAN GULL   Scarce
Larus melanocephalus

GB max: 73 Jan
NI max: 0

Perhaps surprisingly, relatively large numbers were present from mid summer onwards, building to a peak in mid winter, although this was preceded by a count of just 14 in December, presumably reflecting the fact that rarer gulls may be easily overlooked in large mixed species flocks during WeBS counts. Totals were dominated by large counts at several sites on the Isle of Wight. Mediterranean Gulls were noted at 70 sites in total, with summed site maxima suggesting as many as 215 birds.

Sites with five or more birds in 1997-98

Foreland	45	Jan	Newtown Estuary	8	Mar
Brading Harbour	25	Oct	Medway Estuary	8	Feb
Tamar Complex	12	Feb	Alt Estuary	7	Jul
Camel Estuary	10	Aug	Swansea Bay	6	Jun/Feb
Ryde Pier To Puckpool Point	10	Mar	Thames Estuary	5	Apr

LITTLE GULL

Larus minutus   Scarce

GB max: 39   Jul
NI max: 1   Jan

Only small numbers were seen in most months, with the large passage numbers known to occur in the Irish Sea and off the northeast English coast going largely undetected. Peaks were recorded in spring and late summer, with a notable count of nine in January. One was seen on the Outer Ards in Northern Ireland January.

Sites with five or more birds in 1997-98

Alt Estuary	46	Apr	Tophill Low Reservoirs	5	Aug
North Norfolk Marshes	14	Oct/Nov	Humber Estuary	5	Sep
Eden Estuary	13	Jul	Durham Coast	5	Aug
Tees Estuary	6	Jul	Forth Est.	5	Jul

SABINE'S GULL

Larus sabini   Scarce

One was in Filey Bay in August, and three singles were seen in September, at Clifford Hill Gravel Pits, Rutland Water and the Alt Estuary.

BLACK-HEADED GULL

Larus ridibundus

International threshold: 20,000**
Provisional Great Britain threshold: 19,000[†]
All-Ireland threshold: ?[†]

GB max: 253,921   Dec
NI max: 11,837   Jan

Peak totals in Great Britain were the highest yet recorded under WeBS, an increase of 22% on that of the previous year. The timing of the peak was typical, having occurred in January in each of the five years in which gulls have been monitored. However, this total represents just 13% of the provisional British population estimate; the species' use of non-wetland areas and the fact that counts of gulls are optional are the most significant factors in this under-recording. In Northern Ireland, the timing of peak counts has varied, though generally occurring in the autumn period. The 1997-98 maxima was unusual in this respect, with January, as in Britain, being the peak month of occurrence.

Recorded totals in the province were marginally lower than in 1996-97, though higher than in all previous years.

The time of day on which a count was undertaken is perhaps the most influential factor in the number of gulls recorded at individual sites, as birds often disperse from overnight roost sites to feed in non-wetland habitats. Three sites currently support numbers in excess of the threshold for international importance, these also being the only sites to surpass the threshold for national importance. Only two sites recorded peak counts above their respective five year average, though most exceeded the peak of the previous year.

	93-94	94-95	95-96	96-97	97-98	Mon	Mean
International							
Tophill Low Reservoirs	-	34,000	21,710	15,000	43,800	Nov	28,628
Tring Reservoirs	50,000	20,000	(21,000)	(363)	(16,000)	Jan	25,000
Lower Derwent Valley	-	-	32,500	17,500	19,000	Jan	23,000
Other sites in Great Britain supporting more than 10,000 birds[†]							
Morecambe Bay	11,564	15,965	18,998	18,653	25,294	Aug	18,095
Hurleston Reservoir	-	20,000	16,500	-	(3)	Apr	18,250
Poole Harbour	11,283	(10,233)	25,157	(10,732)	15,844	Sep	17,428
Chasewater	15,000	10,000	12,000	-	-		12,333
Wash	8,815	12,355	12,380	13,975	5,780	Nov	10,661

Other sites surpassing table qualifying levels in 1997-98

Portsmouth Harbour	12,642	Dec
Church Wilne Reservoir	15,000	Jan
Eccup Reservoir	10,000	Jan/Feb

† as no British sites are of national importance for Black-headed Gulls and as no all-Ireland threshold has been set, a qualifying level of 10,000 has been chosen to select sites for presentation in this report

---

RING-BILLED GULL
Larus delawarensis

Vagrant
Native range: North America

This species was recorded at 10 UK sites, mostly in the southwest. Singles at Possil Loch, near Glasgow, and at Belfast Lough were therefore noteworthy. Many birds were long stayers and two were found at both Swanpool, Falmouth, and Par Sands Pools.

COMMON GULL
Larus canus

International threshold: 16,000
Provisional Great Britain threshold: 9,000†
All-Ireland threshold:　　　?

| GB max: | 86,528 | Feb |
| NI max: | 3,467 | Feb |

The peak 1997-98 count was 23% up on the previous year's maximum and is the highest from the first five years of gull monitoring under WeBS. Recording of gulls has gradually improved throughout this period and may account for much of this increase. Nevertheless, it represents less than 10% of the Common Gulls wintering in Britain, the lowest percentage for any gull species. As for other gull species, the timing of counts at a few key sites can significantly affect both national and individual site totals. It is probably fair to consider any counts other than of roosting birds, particularly at inland sites, as under-counts.

The northerly distribution of this species is clearly demonstrated by the table below, with just three southern sites creeping onto the bottom of the table. Despite the relatively high peak in Britain, only of the key resorts supported counts above their respective five year averages. The November count at Tophill Low Reservoirs represented 30% of the British total in that month.

	93-94	94-95	95-96	96-97	97-98	Mon	Mean
International							
Tophill Low Reservoirs	-	20,000	18,000	14,000	22,000	Feb	18,500
Great Britain							
Inner Moray Firth	6,100	(40,001)	-	1,850	-		15,984
Lower Derwent Valley	-	-	13,400	6,400	8,000	Jan	9,267
Other sites in Great Britain supporting more than 2,500 birds†							
West Water Reservoir	-	-	-	4,500	12,500	Nov	8,500
Derwent Reservoir	641	5,501	8,769	9,465	9,590	Dec	6,793
Morecambe Bay	3,252	5,274	8,861	4,187	5,536	Aug	5,422
Rutland Water	8,000	1,000	1,000	6,000	8,000	Nov	4,800
Pitsford Reservoir	4,000	6,000	4,000	2,500	4,000	Feb	4,100
Eccup Reservoir	-	3,500	(5,000)	5,000	2,500	Sep	4,000
Doon Estuary	5,500	5,500	6,000	2,700	11	Jul	3,942
Hule Moss	3,900	1	10,730	2,700	1,400	Oct	3,746
Tees Estuary	(2,257)	3,365	2,006	5,014	2,204	Mar	3,147
Thames Estuary	701	3,646	4,146	3,455	(825)	Jan	2,987
Wash	(6,316)	1,188	(2,636)	(1,321)	(887)	Aug	2,865
Tring Reservoirs	2,500	5,000	(1,000)	15	(2,000)	Jan	2,505

Other sites surpassing table qualifying levels in 1997-98

Cameron Reservoir	3,800	Jan
Loch of Harray	2,845	Feb
Wigtown Bay	2,545	Jan

† *as so few British sites are of national importance for Common Gull and as no all-Ireland threshold has been set, a qualifying level of 2,500 has been chosen to select sites for presentation in this report*

LESSER BLACK-BACKED GULL
Larus fuscus

International threshold: **4,500**
Provisional Great Britain threshold: **500**
All-Ireland threshold: **?**

GB max:	59,085	**Jul**
NI max:	1,024	**Sep**

Unusually for most species recorded by WeBS, the peak count of Lesser Black-backed Gulls in Britain occurred during the summer period when site coverage is poorest. The 1997-98 peak far exceeded previous counts, which have barely surpassed 40,000 birds, almost wholly due to an exceptional count at Morecambe Bay, where there is a large colony. Nevertheless, this represents a relatively small proportion of the 83,500 breeding pairs and their young that will be present at that time of year (Lloyd *et al.* 1991). Many birds then depart, and wintering numbers are fewer, though they have increased steadily in recent decades, to around 60,000 in the mid 1980s (Bowes *et al.* 1984). Throughout the winter, monthly totals dropped significantly with fewer than 7,000 birds recorded by January.

	93-94	94-95	95-96	96-97	97-98	Mon	Mean	
International								
Morecambe Bay	10,499	20,479	29,915	30,880	51,829	Jul	28,720	
Great Britain								
Llysyfran Reservoir	-	-	300	8,500	400	Sep	3,067	
Severn Estuary	287	70	57	7,017	6,085	Mar	2,703	
Chasewater	2,000	2,100	3,000	-	-		2,367	
Alde Complex	68	289	162	542	9,633	Mar	2,139	▲
Alt Estuary	1,800	710	886	2,480	1,957	Jul	1,567	
NE Glamorgan Moorland Pools	-	-	-	1,352	1,418	Aug	1,385	
Great Pool Westwood Park	2,000	-	-	1,750	18	Sep	1,256	
Portworthy Mica Dam	-	300	1,000	2,250	1,000	Sep	1,138	
Rutland Water	40	500	1,000	(150)	3,000	Sep	1,135	
Hayle Estuary	1,401	260	1,800	735	1,095	Feb	1,058	
Heaton Park Reservoir	-	20	-	-	2,000	Nov	1,010	▲
Solway Estuary	464	(981)	837	517	1,143	Jun	788	
Camel Estuary	-	741	(1,252)	1,042	48	Aug	771	
Cleddau Estuary	152	(301)	336	2,073	477	Sep	760	
Colliford Reservoir	2,500	121	206	296	600	Aug	745	
Llangorse Lake	500	400	850	820	860	Oct	686	
Caistron Quarry	-	637	730	-	-		684	
Sprotbrough Flash	0	250	350	1,000	1,500	Nov	620	▲
Lower Windrush Valley GP	(4)	16	57	589	1,714	Nov	594	▲
Yarnton Gravel Pits	-	-	-	70	1,050	Dec	560	▲
Hurleston Reservoir	-	476	1,119	-	0	Apr	532	
Blackmoorfoot Reservoir	-	351	656	47	1,037	Oct	523	▲
Bicton Reservoir	-	-	-	850	196	Sep	523	
Wash	145	(234)	331	(1,338)	239	May	513	
Pitsford Reservoir	150	350	300	550	1,200	Oct	510	▲
Fiddlers Ferry Lagoons	-	400	1,000	128	(300)	Oct	509	

Sites no longer meeting table qualifying levels
Frainslake to Freshwater West
R. Arrow/R. Lugg Floodplain
Foremark Reservoir

Wellington Gravel Pits GP	1,400	Nov	Duddon Estuary	606	Aug
Dungeness GP	1,010	Aug	Doddington Pool	600	Aug
Lo. Neagh/Beg	972	Sep	Clarydale Water	600	Nov
Longnewton Reservoir	780	Sep	Kingsbridge Estuary	580	Feb
Somerset Levels	696	Nov	Burry Inlet	549	Aug
Crowdy Reservoir	650	Nov			

HERRING GULL
Larus argentatus

International threshold: **13,000**
Provisional Great Britain threshold: **4,500[†]**
All-Ireland threshold: **?**

GB max: **67,430** **Nov**
NI max: **3,861** **Sep**

British monthly totals of Herring Gull in 1997-98 were about average for recent years, and, throughout the year, slightly higher than normal in Northern Ireland, although numbers were very consistent and the peak was lower than usual.

This species' coastal distribution is reflected in the table below and, as such, it is likely that a higher proportion of counts presented were obtained during Core Counts, rather than at roost, than for other species of gull.

	93-94	94-95	95-96	96-97	97-98	Mon	Mean	
International								
Morecambe Bay	20,840	20,824	19,144	17,260	18,165	Jul	19,247	
Great Britain								
Ribble Estuary	(15)	1,351	27,500	430	1,100	Sep	7,595	
Wash	2,816	6,538	5,142	5,147	12,649	Jan	6,458	
Other sites in Great Britain supporting more than 2,500 birds[†]								
Alt Estuary	3,560	3,021	4,500	5,300	5,500	Feb	4,376	
Solway Estuary	2,986	2,335	(9,397)	4,269	2,884	Aug	4,374	▼
Ayr To Troon	10,000	1,920	5,560	2,460	520	Sep	4,092	▼
Irvine To Saltcoats	5,000	8,000	0	4,500	0	Apr	3,500	
Alde Complex	1,389	734	4,347	312	8,569	Mar	3,070	
Forth Est.	2,640	2,338	3,900	3,747	1,893	Sep	2,904	
Guernsey Shore	(1,120)	(1,187)	2,471	3,073	-		2,772	
Doon Estuary	3,000	3,600	3,570	3,000	246	Jul	2,683	

Other sites surpassing table qualifying levels in 1997-98

Dee Estuary (Sco)	3,085	Feb
Belfast Lough	2,598	Jan
Ythan Estuary	2.600	Jan

[†] as so few British sites are of national importance for Herring Gull, and as no all-Ireland threshold has been set, a qualifying level of 2,500 has been chosen to select sites for presentation in this report

ICELAND GULL
Larus glaucoides

Scarce

Birds were noted at 18 sites, primarily from January to March, but with two in both April and May. Most birds were found in northern England or Scotland, but with several in the southwest also. All records were of singles except for three at Belfast Lough and two at Loch of Hempriggs.

GLAUCOUS GULL
Larus hyperboreus   Scarce

Distribution, both by month and region, was very similar to Iceland Gull, though recorded at 24 sites, with summed site maxima suggesting as many as 30 birds. Only two sites held more than one bird: five at Belfast Lough in January and three at Lower Derwent Valley in February.

GREAT BLACK-BACKED GULL
Larus marinus

GB max:	13,850	Nov
NI max:	596	Sep

International threshold: 4,800
Provisional Great Britain threshold: 400
All-Ireland threshold: ?

Peak counts in 1997-98 were about average for recent years in both Great Britain and Northern Ireland. The pattern of occurrence is rather different compared with the other numerous gull species, which peak either directly after the breeding seasons or in mid-winter; numbers of Great Black-backeds have consistently occurred in late autumn or early winter, then declined steadily to a low in late spring. The table below, containing 20 sites of national importance, suggests that, at least in comparison with tables for other gull species, WeBS is much better at recording numbers of Great Black-backed Gulls, or the provisional population estimate is too low.

	93-94	94-95	95-96	96-97	97-98	Mon	Mean
Great Britain							
Wash	971	2,629	1,150	(1,087)	630	Sep	1,345
Tees Estuary	1,008	1,887	1,325	1,068	1,152	Dec	1,288
Lower Derwent Valley	-	-	617	1,750	1,105	Jan	1,157
Tophill Low Reservoirs	-	981	1,000	835	1,040	Nov	964
Dungeness	0	(6)	2,070	1,600	90	Sep	940
Pegwell Bay	138	2,000	600	750	1,000	Sep	898
Cresswell To Chevington Burn	200	1,000	500	685	2,000	Nov	877
Lossie Estuary	-	629	700	1,053	847	Nov	807
Loch Of Strathbeg	0	545	1,000	(1,200)	670	Dec	683
Morecambe Bay	628	624	554	621	668	Dec	619
Thames Estuary	(175)	271	474	789	505	Jan	510
Fairburn Ings	-	1,000	2	950	12	Apr	491
Portsmouth Harbour	166	772	437	(216)	(420)	Jan	458
Durham Coast	(365)	(229)	(624)	350	(468)	Dec	452
Rutland Water	250	600	300	700	300	Nov	430
Dee Estuary (Eng/Wal)	22	377	1,591	111	25	Jun	425

Other sites surpassing table qualifying levels in 1997-98

Lade Sands	500	Jan	North Norfolk Marshes	46	Oct
Kingsbridge Estuary	473	Nov	Coquet Estuary	414	Nov

KITTIWAKE
Rissa tridactyla

GB max:	4,151	Aug
NI max:	76	Aug

International threshold: ?
Great Britain threshold: ?
All-Ireland threshold: ?

The largest numbers of this species were seen during summer and autumn months, with monthly totals of around 1,000 in Britain, but numbers were considerably higher in both Britain and Northern Ireland in August. In winter, these dropped to around 200 in Britain, and a peak of just three in Northern Ireland, reflecting the pelagic nature of this species at this time of year. Kittiwakes were recorded at 54 sites in 1997-98, with those listed below reflecting post-breeding dispersal from the main breeding areas of eastern Scotland and northeast England.

Sites with 200 or more birds in 1997-98

Tay Estuary	1100	Aug	Dee Estuary (Sco)	430	Aug
Loch Of Strathbeg	1000	Aug	Ythan To Collieston	382	Apr
Tees Estuary	971	Aug	Deveron Est	320	Sep
Beadnell To Seahouses	600	Apr	Forth Est.	280	May
Arran	490	Oct	Wash	240	Oct
Durham Coast	477	May			

SANDWICH TERN
Sterna sandvicensis

International threshold: 1,500
Great Britain threshold: ?[†]
All-Ireland threshold: ?[†]

GB max: 5,718 Aug
NI max: 606 Aug

The peak British total, although similar in timing to the three previous years, was somewhat lower. The peak, as with most species of terns, usually occurs in August when adults are joined by the recently fledged young and birds move beyond the immediate vicinity of the colony.

Although Sandwich Terns were noted at just 101 sites in 1997-98, peak counts by WeBS exceeded those of any other species of tern. All sites of international importance that are regularly monitored by WeBS are listed below, plus sites where the average peak exceeds 200 birds. For the second year in a row, numbers recorded on the Tees Estuary were considerably below average and the average peak fell below the international threshold.

	93-94	94-95	95-96	96-97	97-98	Mon	Mean	
International								
Forth Est.	(877)	(1,708)	1,791	1,352	(1,278)	Aug	1,617	
Great Britain[†]								
Cemlyn Bay	-	-	-	1,450	-		1,450	
Tees Estuary	(1,665)	884	3,774	489	227	Jul	1,408	▼
Dee Estuary (Eng/Wal)	(0)	601	446	2,090	636	Aug	943	
North Norfolk Marshes	(3,000)	395	266	472	311	Aug	889	
Loch Of Strathbeg	0	1,846	220	(750)	710	May	705	
Duddon Estuary	0	487	470	650	808	May	483	
Lindisfarne	(150)	(690)	224	316	(160)	Jul	410	
Wash	-	284	178	186	586	Jul	309	
Ythan Estuary	56	262	300	380	488	Aug	297	
Morecambe Bay	0	865	312	34	143	May	271	
Foryd Bay	-	-	-	390	62	Jul	226	
Tay Estuary	(60)	60	361	401	25	Sep	212	
Eden Estuary	138	(29)	99	(380)	(203)	Aug	205	

Other sites surpassing table qualifying levels in 1997-98

Dundrum Bay	592	Aug	Alt Estuary	249	Aug
Exe Estuary	226	Jul	Don Mouth to Ythan Mouth	212	Jul
Humber Estuary	240	Aug	Medway Estuary	200	May
Filey Bay	200	Aug			

† *as no British or all-Ireland thresholds have been set, a qualifying level of 200 has been chosen to select sites for presentation in this report*

ROSEATE TERN
Sterna dougallii

Scarce

This species is generally only recorded in single figures during the summer months and 1997-98 was no exception. The only record was of two birds at Filey Bay in August. Roseate Terns are often recorded in areas not far from the main breeding areas along the east coast of Britain, e.g. Coquet Island and the Farne Islands, and around the Irish Sea, e.g. Rockabill and Anglesey.

COMMON TERN
Sterna hirundo

International threshold: **6,000**
Great Britain threshold: **?†**
All-Ireland threshold: **?†**

GB max: **4,023**   Jul
NI max: **10**   May

The UK peak total in 1997-98 was lower than during the three preceding years and was slightly earlier in the season. This was by far the most widely recorded tern by WeBS, with records from 265 sites. Once again, the Alt and Tees Estuaries held the greatest numbers.

	93-94	94-95	95-96	96-97	97-98	Mon	Mean
Great Britain†							
Alt Estuary	(1,200)	402	1,500	596	1,038	Aug	947
Tees Estuary	(338)	266	1,575	453	841	Jul	784
Dee Estuary (Eng/Wal)	(0)	429	315	641	225	Jul	403
Wash	-	691	262	310	215	Jun	370
Forth Est.	(144)	276	276	390	(343)	Aug	321
Tay Estuary	(105)	27	450	320	230	Jul	257
North Norfolk Marshes	(200)	326	226	344	107	Aug	251
Loch Of Strathbeg	0	214	325	277	300	Jul	223
Langstone Harbour	-	-	278	138	-		208

Sites no longer meeting table qualifying levels
Thames Estuary

Other sites surpassing table qualifying levels in 1997-98
Ythan Estuary   270   May

† *as no British or all-Ireland thresholds have been set, a qualifying level of 200 has been chosen to select sites for presentation in this report*

ARCTIC TERN
Sterna paradisaea

International threshold: **?**
Great Britain threshold: **?†**
All-Ireland threshold: **?†**

GB max: **1,337**   Aug
NI max: **0**

The peak total in 1997-98 was more than double that of the previous year. Given patchy coverage by WeBS during summer months and the fact that recording of terns is optional, this undoubtedly reflects chance recording of large numbers at just a few sites, rather than provide an indication of population change or breeding success.

Arctic Terns were noted at 46 sites. All with an average of 50 or more, except one, are in Scotland.

	93-94	94-95	95-96	96-97	97-98	Mon	Mean	
Great Britain†								
Tay Estuary	(950)	65	130	40	1,000	Aug	437	
Loch Of Strathbeg	0	10	1,600	3	8	Jul	324	
North Ronaldsay	0	0	563	(210)	0	Sep	155	
Eden Estuary	64	(9)	129	90	190	Jul	118	
St Andrews Bay	0	0	448	121	22	Jul	118	
Ythan Estuary	18	59	80	100	204	Aug	92	
Solway Estuary	(0)	(0)	(0)	82	(0)	Apr	82	
Loch Indaal	44	(82)	(202)	51	29	Jun	82	
Balranald RSPB Reserve	-	-	-	-	80	Jun	80	▲
Morecambe Bay	0	102	44	105	124	Jun	75	
Don Mouth To Ythan Mouth	(0)	(0)	(0)	(0)	66	Jul	66	▲

Sites no longer meeting table qualifying levels
Foryd Bay

Other sites surpassing table qualifying levels in 1997-98
Forth Est.   (58)   Jul

† *as no British or all-Ireland thresholds have been set, a qualifying level of 50 has been chosen to select sites for presentation in this report*

LITTLE TERN
Sterna albifrons

International threshold: **340**
Great Britain threshold: **?[t]**
All-Ireland threshold: **?[t]**

GB max: **457 May**
NI max: **0**

Recorded totals of Little Tern were substantially lower than during the previous three years. Unusually, the maximum count occurred in spring, rather than, as is more normal, in autumn. Such variation between years is likely to reflect differences in site coverage by WeBS during summer months, although low counts at a number of sites may suggest poor breeding success. Little Terns were noted at 34 sites in 1997-98.

	93-94	94-95	95-96	96-97	97-98	Mon	Mean
Great Britain[t]							
Dee Estuary (Eng/Wal)	(0)	156	700	145	160	Jun	290
Langstone Harbour	-	-	210	200	-		205
Thames Estuary	(0)	40	71	467	5	Aug	146
Wash	-	125	43	(330)	30	Jun	132
Tees Estuary	(186)	148	107	47	27	May	103
Pagham Harbour	(0)	16	114	(14)	-		65

Other sites surpassing table qualifying levels in 1997-98

Eden Estuary	52	May
Fleet/Wey	50	May
Solway Estuary	50	Jun

† *as no British or all-Ireland thresholds have been set, a qualifying level of 50 has been chosen to select sites for presentation in this report*

BLACK TERN
Chlidonias niger

International threshold: **2,000**
Great Britain threshold: **?**
All-Ireland threshold: **?**

GB max: **24 Aug**
NI max: **2 Sep**

The maximum count in 1997-98 was down to a third of the previous year's peak, which itself was lower than the two preceding years. Birds were noted at just 22 sites, all in England except for two in Northern Ireland. Singles birds were recorded at 14 of these.

Numbers of Black Tern recorded by WeBS are likely to remain prone to fluctuations, as they are highly dependent on the timing of the spring and autumn passages. In 1997, a huge movement was noted, estimated to include 1,400 birds, but it lasted for just one day (3 May) and was restricted mainly to the midland counties of England (Nightingale & Allsopp 1998). Clearly, WeBS counts stand little chance of recording such transient birds accurately.

Sites with more than three birds in 1997-98

Wash	11	Aug
Severn Estuary	10	May
Church Wilne Reservoir	8	May

WHITE-WINGED BLACK TERN
Chlidonias leucopterus

Vagrant
Native range: E Europe, S Asia and Africa

Two individuals were recorded, both during August, at Stodmarsh and on the Thames Estuary.

KINGFISHER
Alcedo atthis

GB max: 280 Oct
NI max: I several

During 1997-98, Kingfishers were recorded on 567 sites, the majority of which were rivers or gravel pits. The peak UK count was slightly lower than those of the previous two years and occurred a month later than usual. All bar one of the sites with average peak counts of five or more were inland.

	93-94	94-95	95-96	96-97	97-98	Mon	Mean
Sites with average peak counts of five or more birds							
Lee Valley Gravel Pits	7	12	9	26	15	Oct	14
Tamar Complex	6	8	10	9	3	Aug	7
Eversley Cross/Yateley GPs	0	9	4	6	9	Sep	6
Colwick Country Park	(2)	6	-	-	-		6
Somerset Levels	I	4	(7)	(10)	4	Feb	5
Meadow Lane Gravel Pits	3	7	-	-	-		5
Holme Pierrepont GPs	(0)	I	4	8	6	Dec	5

Sites no longer meeting table qualifying levels

Attenborough GPs	Stodmarsh
R. Usk: Pencelli	Thames Estuary
Taw/Torridge Estuary	Deben Estuary
Southampton Water	Middle Tame Valley
Chichester Harbour	Cleddau Estuary

Other sites surpassing table qualifying levels in 1997-98

Cleddau Estuary	6	Oct	R Usk: Pencilli	5	Sep
Fordwich/Westbere GPs	6	Oct	Tyttenhanger Gravel Pits	5	Sep
River Wye: Putson	6	Feb	Arun Valley	5	Sep
Lower Derwent Valley	6	Mar	Dorchester Gravel Pits	5	Sep
Chilham/Chartham GPs	5	Apr	Cotswold WP West	5	Dec
Middle Tame Valley GPs	5	Jun	Tophill Low Reservoirs	5	Mar

PRINCIPAL SITES

In total, 115 sites monitored by WeBS supported at least one species in internationally important numbers and 91 sites currently hold average peak species totals of 10,000 or more waterfowl following 1997-98 counts. The list remains relatively similar each year, with the order of the top ten sites in Table 4 almost unchanged compared with the previous year; the addition of the Swale Estuary in place of the Forth Estuary being the only difference. Of these top ten sites, the Solway Estuary and North Norfolk Coast each added an extra species to their list of waterfowl occurring in internationally important numbers (Shelduck and Grey Plover, respectively).

Of sites with averages of 10,000 or more waterfowl, peak waterfowl totals in 1997-98 fell at approximately forty and rose at just over thirty. Just four sites recorded counts in 1997-98 at least 30% above their five year average, namely Dungeness Gravel Pits (+58%), North Norfolk Coast (+45%), West Water Reservoir (+36%) and Alt Estuary (+35%). At Dungeness, high counts of Golden Plover and Lapwing were predominantly responsible for the large increase in total waterfowl numbers. Both the North Norfolk Coast and the Alt Estuary played host to unusually large numbers of Knot, whilst an exceptional autumn count of Pink-footed Geese at West Water Reservoir was the major influence on its total. Declines of 30% or more in total waterfowl numbers were recorded at Carmarthen Bay (-57%), Lavan Sands (-46%), Ythan Estuary (-41%), Durham Coast (-41%), Orwell Estuary (-40%) and Loch Eye (-33%). Below average counts of Knot and Oystercatcher contributed to low totals at Carmarthen Bay and along the Durham Coast, though patchy overage at both of these sites may have influenced peak counts of some species. Also, no Common Scoter were recorded at the former site during WeBS Core Counts in 1997-98. Low counts of Oystercatchers and Dunlin accounted for the majority of the fall in numbers on Lavan Sands, whilst there were similarly low Dunlin counts on the Orwell Estuary also. Both remaining sites, Loch Eye and the Ythan Estuary, share goose roosts with adjacent sites and the vagaries of the birds' choice of roost on count days probably accounts for the fall in waterfowl numbers at each of these sites.

Table 4. Total number of waterfowl at principal waterfowl sites in the UK, 1993-94 to 1997-98 (includes only Core Count data and roost counts of Pink-footed and Greylag Geese), and species occurring in internationally important numbers at each (based on all survey data). Species codes are listed at the end of the table.

Site	93-94	94-95	95-96	96-97	97-98	Average	Int. imp. species
The Wash	305,645	301,972	331,346	280,811	341,062	312,167	PG DB SU OC GV L. KN DN BW BA CU RK
Ribble Estuary	309,451	340,775	283,739	299,285	266,050	299,860	BS WS SU WN T. PT OC GV L. KN SS DN BW BA RK
Morecambe Bay	213,936	237,247	233,130	252,214	312,709	249,847	PG SU PT OC GV KN DN BA CU RK TT
Humber Estuary	148,185	244,803	150,547	(82,004)	159,536	175,768	SU GP GV L. KN DN BA RK
Thames Estuary	155,743	177,408	156,057	171,722	177,954	167,777	DB OC RP GV KN DN BW BA RK
Solway Estuary	141,629	158,195	143,893	149,948	135,500	145,833	WS PG BY SU PT OC KN DN BA CU RK
Dee (Eng/Wal) Estuary	95,542	124,288	130,833	152,076	129,271	126,402	SU T. PT OC GV KN DN BW BA CU RK TT
North Norfolk Coast	97,694	100,555	100,377	121,979	172,775	118,676	PG DB WN PT GV KN BA
Mersey Estuary	81,628	130,097	91,050	116,641	117,777	107,439	SU T. PT DN BW RK
Swale Estuary	(69,882)	91,034	(64,059)	100,958	89,534	93,842	WN PT SV GV KN BW
Firth of Forth	104,491	88,284	101,747	87,497	85,415	93,487	SZ PG SU GN KN BA RK TT
Loughs Neagh & Beg	78,909	91,654	110,192	101,599	82,373	92,945	GG BS WS PO TU SP GN

Site	93-94	94-95	95-96	96-97	97-98	Average	Int. imp. species
Somerset Levels	64,836	128,169	110,157	38,480	111,755	90,679	BS WN T. SV L.
BlackWater Estuary	(85,257)	80,577	(71,678)	(86,025)	78,493	82,588	DB SU GV DN BW RK
Severn Estuary	79,872	101,701	71,629	81,137	68,881	80,644	BS SU PT DN RK
Strangford Lough	52,773	73,255	81,158	90,697	73,449	74,266	PB KN BA RK
Medway Estuary	67,667	(73,198)	(61,440)	65,118	(60,519)	68,661	DB SU PT RP GV DN BW RK
Ouse Washes	55,050	72,078	63,168	(57,605)	72,870	65,792	BS WS WN GA PT SV PO BW
Hamford Water	(47,793)	45,891	71,043	71,281	56,685	61,225	DB T. SV BW RK
Montrose Basin	61,569	63,278	51,731	38,761	61,874	55,443	PG KN RK
Chichester Harbour	57,133	62,975	58,819	49,471	45,874	54,854	DB GV DN BW BA RK
Breydon & Berney	57,253	53,298	46,760	58,947	55,134	54,278	BS L.
Loch of Sstrathbeg	48,870	67,950	58,424	40,154	50,402	53,160	WS PG BY
Inner Moray Firth	52,428	50,244	46,587	54,707	47,118	50,217	GJ RM KN BA RK
Alt Estuary	38,496	42,468	44,464	54,888	66,628	49,389	KN BA RK TT
Stour Estuary	44,022	54,306	50,034	45,501	48,307	48,434	GV KN DN BW RK
Lower Derwent Ings	44,048	49,074	52,856	43,322	44,371	46,734	T.
Burry Inlet	32,717	39,475	53,270	64,380	35,106	44,990	PT SV OC KN
Lindisfarne	41,351	46,481	40,078	42,868	38,504	41,856	GJ GV BA
Langstone Harbour	45,787	47,844	44,834	32,831	36,785	41,616	DB RP GV DN
Abberton Reservoir	31,204	46,958	49,208	35,707	42,947	41,205	GA T. SV
Duddon Estuary	46,944	48,025	38,631	37,320	34,203	41,025	PT RK
Dupplin Lochs	36,521	62,064	35,030	40,665	29,998	40,856	PG
Colne Estuary	37,233	44,279	38,870	28,032	42,360	38,155	DB RK
Lough Foyle	30,279	35,132	46,959	37,787	37,454	37,522	SZ WS PB BA
Cromarty Firth	38,193	40,189	44,106	34,508	31,681	37,735	PG GJ KN BA
Loch Leven	32,614	31,408	35,852	38,727	32,799	34,280	PG SV
Dengie Flats	32,610	34,780	23,598	39,330	39,363	33,936	GV KN BA
Nene Washes	(29,691)	35,698	28,480	(15,389)	(37,100)	33,759	BS PT SV
Dinnet Lochs	30,817	34,911	41,513	27,095	25,656	31,998	GJ
Poole Harbour	34,796	28,964	31,037	33,283	28,768	31,370	SU BW
Alde Complex	(21,413)	27,912	(32,006)	27,897	35,990	30,951	RK
Martin Mere	25,091	33,093	31,231	27,670	36,903	30,798	BS WS
Carmarthen Bay	34,891	52,367	21,236	(23,430)	13,179	30,418	
Dornoch Firth	31,355	26,219	27,490	29,679	36,982	30,345	GJ WN BA
Rutland Water	20,446	36,959	33,338	25,600	30,803	29,429	GA SV
West Water Reservoir	20,544	26,911	31,500	25,882	39,171	28,820	PG
Ythan Estuary	36,509	31,943	32,373	12,968	15,290	25,817	PG
Tees Estuary	19,490	25,307	33,716	24,465	19,380	24,472	
Orwell Estuary	26,292	24,787	30,712	24,681	14,418	24,178	RK
Walland Marsh	-	-	-	18,357	29,880	24,119	BS
Crouch-roach Estuary	26,794	24,668	(21,560)	19,931	24,640	24,008	DB
Exe Estuary	22,969	25,136	27,177	20,208	24,119	23,922	
Arun Valley	20,444	37,744	21,269	16,013	19,735	23,041	
Inner Firth of Clyde	19,475	20,890	22,634	23,000	21,639	21,528	RK
Belfast Lough	22,042	23,682	21,486	18,370	21,641	21,444	GG RK TT
Tay Estuary	13,672	17,040	20,478	32,646	18,216	20,410	PG BA
Southampton Water	17,239	19,413	23,941	19,473	(8,611)	20,017	
Cleddau Estuary	13,925	20,003	(19,680)	22,077	20,317	19,200	
Pagham Harbour	16,299	19,038	18,452	20,393	18,877	18,612	PT
Deben Estuary	18,152	16,396	16,036	20,069	17,395	17,610	RK
Fleet/Wey	19,776	19,852	16,998	15,961	15,382	17,594	DB
Wigtown Bay	11,450	17,521	20,001	15,934	17,986	16,578	PG
Hule Moss	15,329	9,086	16,498	20,445	20,732	16,418	PG
Beaulieu Estuary	9,926	19,021	15,390	16,623	19,353	16,063	
Cameron Reservoir	29,586	17,381	13,724	5,479	13,060	15,846	PG
Eden Estuary	18,952	18,336	16,273	13,509	15,815	16,577	GJ
North West Solent	15,833	14,748	15,221	(13,715)	(1,669)	15,267	
Taw-torridge Estuary	12,424	14,779	14,959	15,098	17,484	14,949	

Site	93-94	94-95	95-96	96-97	97-98	Average	Int. imp. species
Carsebreck & Rhynd Los	12,421	20,235	18,655	16,091	19,475	17,375	PG
Loch of Skene	19,184	11,222	15,115	14,721	14,818	14,792	GJ
Meikle Loch Slains	12,263	6,094	25,089	17,630	12,226	14,660	PG
Outer Ards Shoreline	13,015	12,681	14,251	21,848	11,090	14,577	PB TT
Dungeness Gravel Pits	8,326	14,901	13,050	12,063	22,349	14,138	
Blyth Estuary	12,918	14,296	12,469	13,745	15,502	13,786	
Durham Coast	14,363	18,666	13,749	10,880	7,660	13,064	
Tamar Complex	13,359	14,774	12,203	9,982	14,958	13,055	
WWT Caerlaverock	-	14,498	10,736	14,681	12,195	13,028	BY
Dyfi Estuary	13,995	12,374	13,699	14,547	9,700	12,863	
Pitsford Reservoir	8,563	11,465	10,247	13,818	14,098	11,638	GA
Hanningfield Reservoir	6,626	9,275	13,699	13,760	14,469	11,566	
Portsmouth Harbour	8,814	9,654	18,597	11,766	8,981	11,562	
Loch of Harray	10,139	12,426	13,080	8,139	13,245	11,406	
Cotswold WP (West)	10,211	10,808	11,325	9,372	13,464	11,036	
Lavan Sands	10,284	12,594	9,834	15,663	5,856	10,846	
Loch Eye	16,244	14,749	10,244	5,830	7,590	10,931	PG GJ
Rye Harbour & Pett Level	8,171	12,017	10,978	10,905	11,879	10,790	
Fairburn Ings	5,471	10,664	14,739	12,186	9,856	10,583	
Middle Tame Valley GPs	7,671	9,274	10,790	12,795	12,275	10,561	
Middle Yare Marshes	11,490	7,302	11,466	10,999	10,513	10,354	
Thanet Coast	11,881	15,193	-	6,279	7,108	10,115	TT
Chew Valley Lake	9,161	8,171	11,637	9,756	-	9,681	SV
Caithness Lochs	5,433	8,503	16,292	5,489	7,551	8,654	GJ
Upper Lough Erne	7,648	9,886	9,553	9,659	4,411	8,231	WS
Haddo House Lakes	5,820	17,735	6,329	9,066	1,250	8,040	GJ
St Benet's Levels	12,124	6,929	11,139	4,614	4,517	7,865	BS
Loch Fleet Complex	10,825	8,678	7,040	7,302	5,259	7,821	GJ
Mid Avon Valley	6,912	7,883	9,230	7,204	7,730	7,792	GA
Carlingford Lough	8,505	5,751	6,845	7,453	8,530	7,417	PB
Lee Valley Gravel Pits	5,333	8,903	5,453	8,422	7,712	7,165	GA
Loch of Lintrathen	6,496	3,444	5,558	4,722	13,360	6,716	GJ
Drummond Pond	7,736	6,892	2,938	8,706	5,610	6,376	PG GJ
Gladhouse Reservoir	3,472	6,095	5,164	8,591	7,385	6,141	PG
Wraysbury Gravel Pits	5,101	6,174	5,010	6,324	7,850	6,092	GA
Tyrella	-	-	5,727	-	-	5,727	PB
Thrapston Gravel Pits	3,712	4,341	4,900	7,735	5,922	5,322	GA
Alloa Inch	-	2,300	8,235	-	-	5,268	PG
Upper cowgill Reservoir	5,400	3,820	4,560	6,060	6,000	5,168	PG
Fala Flow (confidential)	6,450	3,500	2,437	5,000	7,500	4,977	PG
Forth and Teith Valleys	1,180	8,630	-	-	-	4,905	PG
Kilconquhar Loch	4,947	4,249	4,166	4,463	4,326	4,430	GJ
R Clyde: Carstairs Junction	-	1,656	-	28	9,546	4,286	PG
Kinnordy Loch	9,931	4,331	1,437	3,649	1,751	4,220	PG
Loch Tullybelton	4,100	1,800	1,395	4,658	8,000	3,991	PG
Larne Lough	3,339	4,090	4,111	4,220	3,723	3,897	PB
Holburn Moss	2,772	4,327	3,192	3,536	5,005	3,766	GJ
Loch Spynie	6,268	9,406	9,685	6,498	5,686	7,509	GJ
Crombie Reservoir	3,282	-	-	-	-	3,282	PG
Lower Bogrotten	5,626	6,780	3,000	850	0	3,251	GJ
Bridge of Earn	-	-	3,014	-	-	3,014	GJ
King george VI Reservoir	1,026	2,032	7,340	1,864	2,529	2,958	SV
Strathearn (west sites)	-	-	2,665	2,730	-	2,698	GJ
Glenfarg Reservoir	3,920	9,080	0	0	4	2,601	PG
Killimster Loch	-	-	-	2,518	-	2,518	GJ
Loch Mahaick Doune	600	1,064	797	2,789	6,811	2,412	PG
Killough Harbour	-	2,831	3,597	320	-	2,249	PB
Whitton Loch	1,500	3,709	2,200	1,457	-	2,217	PG
R. Spey: TulLochgorum to Boat of BallieFirth	2,216	-	-	-	-	2,216	GJ

Site	93-94	94-95	95-96	96-97	97-98	Average	Int. imp. species
Loch Ken	2,434	2,045	1,813	1,861	2,691	2,169	NW
R Eamont: Watersmeet to Pooley Bridge	2,154	-	-	-	-	2154	GJ
Loch Garten and Mallachie	2,121	2,044	-	-	-	2,083	GJ
Stranraer Lochs	3,131	2,610	794	-	755	1,823	NW GJ
Corby Loch	1,424	-	-	-	-	1,424	GJ
R Tay: Dunkeld	1,400	-	-	-	-	1,400	GJ
Upper Tay sites	765	2,047	750	971	1,334	1,173	GJ
Black Cart Water	457	-	312	238	518	381	WS
Machrihanish	424	294	737	41	172	334	NW
R. Foyle: Grange	304	-	287	387	150	282	WS
South Walls (Hoy)	-	-	-	-	97	97	BY
Appin/Eriska/BenderLoch							NW
Bute							GJ
Coll							NW BY
Colonsay/Oransay							BY
Danna/Keills							NW BY
East Sanday Coast							TT
Islay							NW BY
Moray Firth							SZ
Orkney							GJ
Rhunahaorine							NW
South Uist (west coast)							RP TT
Sw lancashire							PG
Tay/Isla Valley							PG GJ
Tiree							NW BY RP TT

Note that no count data are presented for the last 14 sites in Table 4. These areas are important for geese or swans, or are non-estuarine coastal sites surveyed by local surveys for which WeBS data are not regularly received. Data for any important WeBS sites within these areas, e.g. Lochs Gruinart and Indaal on the islsand of Islay, are presented separately within the Table. Other internationally important sites within these areas are not routinely monitored.

Note that not all species have thresholds for international importance, hence, they do not feature in this table. Gulls and terns have been omitted. Note also that numbers of naturalised and escaped species were included in the above totals due to an oversight. Numbers of these species should be excluded when assessing a site's importance against Ramsar criterion 3(a) (if it regularly supports 20,000 or more waterfowl). In some cases, this will have resulted in site totals being artificially high by several hundred birds (where the site supports large numbers of Canada Geese)

Species codes

AV	Avocet		MA	Mallard
BA	Bar-tailed Godwit		MS	Mute Swan
BS	Bewick's Swan		NW	Greenland White-fronted Goose
BW	Black-tailed Godwit		OC	Oystercatcher
BY	Barnacle Goose		PB	Light-bellied Brent Goose
CA	Cormorant		PG	Pink-footed Goose
CO	Coot		PO	Pochard
CU	Curlew		PT	Pintail
DB	Dark-bellied Brent Goose		RK	Redshank
DN	Dunlin		RM	Red-breasted Merganser
E.	Eider		RP	Ringed Plover
EW	European White-fronted Goose		SP	Scaup
GA	Gadwall		SS	Sanderling
GD	Goosander		SU	Shelduck
GJ	Greylag Goose		SV	Shoveler
GN	Goldeneye		T.	Teal
GP	Golden Plover		TT	Turnstone
GV	Grey Plover		TU	Tufted Duck
KN	Knot		WM	Whimbrel
L.	Lapwing		WN	Wigeon
LN	Long-tailed Duck		WS	Whooper Swan

WeBS Low Tide Counts

AIMS

Despite involving only a relatively small number of sites, estuaries collectively represent the most important habitat for wintering waterfowl in the UK. They are also inherently different from the thousands of inland sites counted for WeBS. The influence of the tide means that the birds have to be much more mobile, both within and between sites. WeBS Core Counts on estuaries have, in general, been based around high tide roosts. Although important in themselves, roost sites are usually secondary in importance to the manner in which waterfowl make use of a site for feeding. Therefore, information gathered about these sites at high tide will only provide part of the picture. The WeBS Low Tide Counts scheme, which was initiated in the winter of 1992-93, aims to monitor, assess and regularly update information on the relative importance of intertidal feeding areas of UK estuaries for wintering waterfowl and thus to complement the information gathered by WeBS Core Counts on estuaries.

WeBS Low Tide Counts provide the crucial information needed to assess the potential effects on waterfowl populations of a variety of human activities which affect the extent or value of intertidal habitats, such as dock developments, proposals for recreational activities, tidal power barrages, marinas and housing schemes. The data gathered contribute greatly to the conservation of waterfowl by providing supporting information for the establishment and management of the UK network of Ramsar sites and Special Protection Areas (SPAs), other site designations and estuary conservation plans. In addition, WeBS Low Tide Counts enhance our knowledge of the low water distribution of waterfowl and provide the data that highlight regional variations in habitat use. In particular, low tide counts should help us to understand, predict and possibly plan for compensation for the effects of sea-level rise on the UK's internationally important waterfowl populations.

METHODS

The scheme provides information on the numbers of waterfowl feeding on subdivisions of the intertidal habitat within estuaries. Given the extra work that Low Tide Counts entail, often to the same counters that carry out the Core Counts, WeBS aims to cover most individual estuaries about once every six years, although on some sites more frequent counts are made. Co-ordinated counts of feeding and roosting waterfowl are made by volunteers each month between November and February on pre-established subdivisions of the intertidal habitat in the period two hours either side of low tide.

DATA PRESENTATION

Tabulated statistics

Table 5 presents three statistics for 18 of the more numerous waterfowl species present on the estuaries covered during the 1997-98 winter: the peak number of a species over the whole site counted in any one month; an estimate of the mean number present over the winter for the whole site (obtained by summing the mean counts of each species for each count section) and the mean density over the site (in birds per hectare), which is the mean number divided by the total area surveyed (in hectares). The statistics differ from those presented in previous years' reports, with the aim of making them more useful and more comparable with the WeBS Core Counts.

Dot density maps

In a change to previous years' reports, WeBS Low Tide Count data are now presented as dot density maps. The maps display the mean number of birds in each count section as dots spread randomly across that count section, thus providing an indication of both numbers and density. It is important to note that individual dots do not represent the precise position of individual birds; dots are randomly placed within the count section. No information about the distribution of birds at a finer scale than the count sector level should be inferred from the dot density maps. In most cases, one dot is equivalent to one bird but for some very dense populations, a scale of one dot per ten birds has been employed. The size of individual dots has no relevance other than for clarity. Additionally, any count sections which were not counted during the 1997-98 winter are marked with an asterisk. The dot density maps enable us to depict a clearer picture of actual bird density, instead of the arbitrary grouping into bands of densities that was presented in previous years. It is hoped that this style of map presentation will lead to an easier and fuller appreciation of low tide estuarine waterfowl distribution. More detailed information concerning WeBS Low Tide Counts can be obtained from the National Organiser (WeBS Low Tide Counts) at the BTO.

ACKNOWLEDGEMENTS

The following counters took part in WeBS Low Tide Counts during the winter of 1997-98; apologies for anyone who may inadvertently have been missed.

Ken Abram, Phil Abram, Martin Alsop, David Andrews, Mike Armitage, John Badley, Peter Bainbridge, Richard Banks, Duncan Bell, Dennis Bill, Sally Brakes, Dave Burges, Keith Burn, Mike Burns, Chris Butterworth, Peter Carr, Alex Carroll, Bob Chapman, Carl Clee, Barry Collins, Jonathan Copp, Trevor Crabb, Mike Creighton, Steve Cross, Oliver Dansie, Janet Dedman, John Dedman, Anne de Potier, Stuart Devonald, Frances Donnan, Jack Donovan, Jean Donovan, Alistair Duncan, Pete Durnell, Bill Emerson, Ian Enlander, Wilton Farrelly, Brian Fellows, Dave Fletcher, Andy Foster, Simon Geary, Stephen Gilbert, Rick Goater, Tom Green, Allan Hansen, Roger Harris, Bob Haycock, Grant Hazlehurst, Tony Heath, Neale Hider, Jane Hodges, Ralph Hollins, Brian Hunt, Philip Johnston, Roy Jones, Jennifer Keddie, Geoff Kelso, Simon King, Dick Lambert, Neil Lawton, Jim Lee, Roy Leeming, Roger Lewis, Charlie Liggett, Alan Livingstone, Paddy Livingstone, Martin Love, Kerry Mackie, Paddy Mackie, Niall McCutcheon, Neil McDowall, Ken McGregor, Jim McNair, Ivor McPherson, Peter Morrison, Sue Morrison, Chris Nickson, Julian Novorol, James Orr, Jess Pain, Mark Painter, Mark Palmer, Andy Parfitt, Tony Parker, Tony Parnell, Terry Paton, Annie Poole, Pete Potts, Trevor Price, Graham Rees, Graham Roberts, Michael Rooney, Duncan Rothwell, Robin Saunders, Brian Savage, Jan Schubert, James Scott, Dave Simpson, Brian Smith, Celia Spouncer, Alan Stewart, Len Stewart, PJ Strangeman, Brian Thomas, David Thompson, Jack Torney, Dave Unsworth, Cliff Waller, Colin Wells, Jo Whatmough, Jerry Williams, Richard Williamson, Jim Wilson, HS Wingfield-Hayes, Eddie Wiseman, Susan Wood, Bill Woodburn, Ingrid Woodburn, Ken Wright and Joan Wright.

ESTUARY ACCOUNTS

The following accounts describe the results of the WeBS Low Tide Counts carried out on 20 estuaries during the 1997-98 winter, namely the Alde Complex, the Alt Estuary, Belfast Lough, the Blyth Estuary, the Cleddau Estuary, Chichester Harbour, the Dee Estuary (North Wirral shore only), Hamford Water, the Mersey Estuary, Montrose Basin, the North Norfolk Coast, the North-west Solent, the Orwell Estuary, Pagham Harbour, Portsmouth Harbour, the Ribble Estuary, Southampton Water, Strangford Lough, the Tamar Complex and the Ythan Estuary were covered. Unfortunately, data from the Alde Estuary and the Orwell Estuary were not received in time for incorporation into this report. Data for each of the estuaries covers the period November to February inclusive. In each case, a list of species present in nationally and internationally important numbers, based on Core Counts, and a description of the estuary are given. This is followed by an outline of the key results. For most of the estuaries, distribution maps are presented for two species for which that site is of particular importance or interest. In the case of Strangford Lough, which is a particularly large and complex site, only one map is presented for clarity.

Species	Alt Estuary			Belfast Lough			Blyth Estuary		
	Peak No.	Mean No.	Mean Dns.	Peak No.	Mean No.	Mean Dns.	Peak No.	Mean No.	Mean Dns.
Brent Goose	0	0	0	1	1	+	1	0	+
Shelduck	322	175	0.11	232	163	0.28	757	605	2.08
Wigeon	0	0	0	133	107	0.19	2,034	1,201	4.13
Teal	0	0	0	414	285	0.49	1,159	490	1.68
Mallard	97	54	0.03	314	272	0.47	113	80	0.27
Pintail	0	0	0	0	0	0	177	111	0.38
Oystercatcher	478	419	0.25	6,974	6,014	10.4	53	34	0.12
Ringed Plover	3	2	+	48	46	0.08	13	6	0.02
Golden Plover	700	454	0.28	450	294	0.51	2,760	1,213	4.17
Grey Plover	1,130	588	0.36	0	0	0	30	21	0.07
Lapwing	729	411	0.25	1,224	1,167	2.02	2,270	958	3.29
Knot	713	298	0.18	300	216	0.37	38	17	0.06
Dunlin	2,650	2,344	1.42	1,906	1,469	2.54	2,893	1,738	5.97
Black-tailed Godwit	0	0	0	163	162	0.28	215	175	0.6
Bar-tailed Godwit	1,800	1,157	0.7	101	56	0.1	22	10	0.03
Curlew	1,202	772	0.47	950	861	1.44	126	76	0.26
Redshank	379	231	0.14	2,148	2,083	3.6	775	665	2.29
Turnstone	13	5	+	250	235	0.41	1	0	+

Species	Chichester Harbour			Cleddau Estuary			Dee Estuary (N. Wirral)		
	Peak No.	Mean No.	Mean Dns.	Peak No.	Mean No.	Mean Dns.	Peak No.	Mean No.	Mean Dns.
Brent Goose	6,530	4,192	1.66	1	1	+	0	0	0
Shelduck	1,063	678	0.27	732	543	0.4	20	18	0.01
Wigeon	896	546	0.22	2,993	1,933	1.43	0	0	0
Teal	763	505	0.2	1,355	1,130	0.84	0	0	0
Mallard	261	225	0.09	295	178	0.13	2	1	+
Pintail	136	53	0.02	0	0	0	0	0	0
Oystercatcher	831	682	0.27	346	328	0.24	3,098	2,354	1.01
Ringed Plover	138	63	0.02	53	29	0.02	98	52	0.02
Golden Plover	1,931	991	0.39	1,272	359	0.27	39	10	+
Grey Plover	720	490	0.19	47	46	0.03	3,143	965	0.41
Lapwing	3,268	1,651	0.65	2,204	1,348	1	800	299	0.13
Knot	1,040	458	0.18	12	5	+	12,132	5,075	2.18
Dunlin	12,555	9,655	3.81	2,516	2,057	1.52	17,738	8,702	3.73
Black-tailed Godwit	413	351	0.14	0	0	0	8	2	+
Bar-tailed Godwit	693	367	0.15	21	14	0.01	5,464	2,643	1.13
Curlew	590	566	0.22	1,071	867	0.64	231	181	0.08
Redshank	726	577	0.23	413	388	0.29	1,848	1,079	0.46
Turnstone	41	27	0.01	52	34	0.03	223	135	0.06

Table 5i. Peak and mean counts, and mean density (birds per hectare), of 18 waterfowl species present on estuaries covered by the 1997/98 WeBS Low Tide Counts. "+" indicates densities of less than 0.01 birds per hectare.

Species	Hamford Water			Mersey Estuary			Montrose Basin		
	Peak No.	Mean No.	Mean Dns.	Peak No.	Mean No.	Mean Dns.	Peak No.	Mean No.	Mean Dns.
Brent Goose	1,831	937	3.75	0	0	0	0	0	0
Shelduck	1,862	822	3.29	4,346	3,235	0.96	1,174	1,287	1.73
Wigeon	1,424	524	2.1	3,428	2,169	0.65	1,797	1,984	2.67
Teal	1,842	552	2.21	8,060	5,534	1.65	95	65	0.09
Mallard	71	26	0.1	1,288	855	0.25	272	222	0.3
Pintail	37	16	0.06	537	375	0.11	161	121	0.16
Oystercatcher	948	399	1.6	2,096	1,700	0.51	1,225	1,132	1.52
Ringed Plover	58	18	0.07	110	80	0.02	0	0	0
Golden Plover	936	172	0.69	1,145	965	0.29	2,306	1,499	2.02
Grey Plover	1,965	477	1.91	4,330	3,058	0.91	0	0	0
Lapwing	5,777	1,924	7.7	10,599	8,834	2.63	688	881	1.19
Knot	957	252	1.01	1,500	571	0.17	2,200	1,550	2.09
Dunlin	5,850	2,141	8.56	48,476	41,218	12.26	1,530	1,463	1.97
Black-tailed Godwit	206	63	0.25	2,655	2,128	0.63	83	34	0.05
Bar-tailed Godwit	424	110	0.44	6	4	+	75	72	0.1
Curlew	222	96	0.38	2,117	1,599	0.48	176	179	0.24
Redshank	769	446	1.78	6,973	4,116	1.22	1,166	1,067	1.44
Turnstone	49	21	0.08	1,188	883	0.26	11	8	0.01

Species	North Norfolk Coast			North-west Solent			Pagham Harbour		
	Peak No.	Mean No.	Mean Dns.	Peak No.	Mean No.	Mean Dns.	Peak No.	Mean No.	Mean Dns.
Brent Goose	8,939	7,036	1.15	1,852	1,325	1.8	538	177	0.46
Shelduck	1,876	1,327	0.22	243	188	0.26	177	131	0.34
Wigeon	6,785	2,337	0.38	511	399	0.54	450	348	0.89
Teal	1,988	1,248	0.2	284	208	0.28	525	273	0.7
Mallard	1,098	888	0.15	81	40	0.05	172	144	0.37
Pintail	142	95	0.02	33	20	0.03	391	292	0.75
Oystercatcher	2,972	2,341	0.38	233	163	0.22	141	78	0.2
Ringed Plover	503	347	0.06	46	21	0.03	46	23	0.06
Golden Plover	2,810	1,992	0.33	148	63	0.09	171	91	0.23
Grey Plover	1,620	1,398	0.23	226	206	0.28	381	248	0.64
Lapwing	9,400	5,401	0.88	895	354	0.48	527	267	0.69
Knot	3,949	1,729	0.28	24	8	0.01	147	52	0.13
Dunlin	6,074	4,458	0.73	5,645	4,024	5.46	2,383	1,658	4.26
Black-tailed Godwit	60	38	0.01	378	193	0.26	124	42	0.11
Bar-tailed Godwit	1,858	1,373	0.22	38	23	0.03	6	3	0.01
Curlew	1,467	1,307	0.21	204	179	0.24	465	295	0.76
Redshank	3,542	3,048	0.5	94	90	0.12	150	123	0.32
Turnstone	587	512	0.08	51	33	0.04	91	62	0.16

Table 5ii. Peak and mean counts, and mean density (birds per hectare), of 18 waterfowl species present on estuaries covered by the 1997/98 WeBS Low Tide Counts. "+" indicates densities of less than 0.01 birds per hectare.

Species	Portsmouth Harbour			Ribble Estuary			Southampton Water		
	Peak No.	Mean No.	Mean Dns.	Peak No.	Mean No.	Mean Dns.	Peak No.	Mean No.	Mean Dns.
Brent Goose	2,505	2,126	2	0	0	0	1,165	1,080	1
Shelduck	120	76	0	2,861	1,822	0	252	151	0
Wigeon	131	100	0	46,465	20,278	3	1,696	1,359	1
Teal	62	29	0	1,375	642	0	1,645	1,145	1
Mallard	96	50	0	219	118	0	209	155	0
Pintail	0	0	0	955	843	0	67	31	0
Oystercatcher	551	519	1	22,860	13,625	2	1,212	1,138	1
Ringed Plover	47	28	0	60	18	+	159	130	0
Golden Plover	44	13	0	2,902	1,051	0	791	386	0
Grey Plover	173	166	0	5,408	2,891	0	408	228	0
Lapwing	465	369	0	8,110	6,162	1	1,227	980	1
Knot	11	3	+	8,905	5,958	1	1	0	+
Dunlin	8,889	7,707	8	45,039	34,863	5	7,088	5,283	3
Black-tailed Godwit	358	217	0	154	45	0	132	69	0
Bar-tailed Godwit	1	0	+	10,431	5,161	1	3	2	+
Curlew	410	390	0	905	754	0	639	519	0
Redshank	372	344	0	1,651	956	0	519	418	0
Turnstone	70	58	0	7	3	+	237	207	0

Species	Strangford Lough			Tamar Complex			Ythan Estuary		
	Peak No.	Mean No.	Mean Dns.	Peak No.	Mean No.	Mean Dns.	Peak No.	Mean No.	Mean Dns.
Brent Goose	7,813	4,438	2	0	0	0	0	0	0
Shelduck	4,111	3,025	1	72	79	0	74	45	0
Wigeon	1,122	727	0	91	64	0	284	222	1
Teal	624	463	0	2	1	+	0	0	0
Mallard	209	167	0	99	73	0	4	3	0
Pintail	242	182	0	0	0	0	0	0	0
Oystercatcher	6,560	5,893	2	97	96	0	305	244	1
Ringed Plover	171	121	0	0	0	0	13	8	0
Golden Plover	12,000	8,945	4	500	294	0	600	201	1
Grey Plover	185	145	0	1	1	+	3	1	0
Lapwing	11,951	6,121	2	221	145	0	4,099	1,837	9
Knot	7,997	4,630	2	1	1	+	190	141	1
Dunlin	8,025	6,434	3	92	92	0	255	191	1
Black-tailed Godwit	436	222	0	80	32	0	0	0	0
Bar-tailed Godwit	1,632	1,203	0	0	0	0	45	26	0
Curlew	1,560	1,237	0	182	199	0	1,571	703	4
Redshank	2,191	2,008	1	180	196	0	475	366	2
Turnstone	152	89	0	11	8	0	44	19	0

Table 5iii. Peak and mean counts, and mean density (birds per hectare), of 18 waterfowl species present on estuaries covered by the 1997/98 WeBS Low Tide Counts. "+" indicates densities of less than 0.01 birds per hectare.

ALT ESTUARY
Merseyside

Internationally important: Knot, Bar-tailed Godwit, Redshank, Turnstone
Nationally important: Cormorant, Common Scoter, Grey Plover, Sanderling

Site description

The River Alt emerges as a creek on the shoreline of Liverpool Bay, between the Ribble and the Mersey Estuaries. The majority of the site is sandy in character, although somewhat muddier around the river mouth where there are also some rocky areas. A large area of saltmarsh used to be present at the mouth of the Alt but has mostly been lost to land claim, principally in the early 19th century. The whole site is backed by one of the most important dune systems in the country, although much of the southern part of this has been lost to housing and dock development at Crosby (Pritchard *et al.* 1992, Davidson 1996a).

Bird distribution

The distributions of Knot and Bar-tailed Godwit were depicted in *Wildfowl and Wader Counts 1996-97* and were essentially the same during the 1997-98 winter, with both species widespread but Knot favouring the Formby Channel (FC) area and Bar-tailed Godwits preferring the flats to the north of here, west of Formby, as well as immediately south of the Alt mouth. Numbers of both of these species were much lower at low tide than high tide, as was noted last winter, and birds presumably continue to roost at the Alt but cross over to feed on the North Wirral shore. Most Redshank were concentrated along the line of the River Alt at low tide, as well as along the shore immediately north and south of the river mouth, with fewer birds elsewhere (Figure 69). Numbers noted at low tide were considerably fewer than those roosting here. Unlike the previous two species, a comparison of Low Tide Count data from adjoining sites suggests that many Redshank may have been moving to Egremont on the western shore of the mouth of the Mersey to feed. The same movement of birds is even more pronounced for Turnstone; despite a peak of over 450 birds roosting at the Alt Estuary during the winter of 1997-98, the maximum number noted at low tide was only 13 with the shortfall apparently made up by the large numbers of Turnstones found feeding at Egremont at low tide.

Figure 69 depicts the distribution of Sanderling along the Alt shoreline and shows how the birds are present along all of the outer parts if the site, with the notable exception of Taylor's Bank (TB) (perhaps due to the distance creating difficulties for shore-base observation?) The distribution on the northern sand flats continues onto the southern parts of the adjoining Ribble Estuary site, although the highest densities on either of these two sites do appear to be just south of the mouth of the River Alt. Unlike many other species, Sanderling do not occur in any great numbers on the adjoining North Wirral or Egremont shores. Both Dunlin and Oystercatcher were widespread although both showed a preference for the outer parts of the site. Curlew were also widespread but had their main concentrations immediately north and south of the Alt mouth. The low tide distribution of Grey Plover was similar to that for Bar-tailed Godwit. Both Lapwings and Golden Plovers favoured the Formby Bank (FB) area, with the latter species also occurring along the inner parts of the Alt. Only a few Ringed Plovers were noted.

The Alt is not particularly important for estuarine wildfowl. The nationally important numbers of Common Scoters frequenting the area were incompletely covered by the WeBS Low Tide Counts, although 182 were noted during the January count (compared with 811 on the January Core Count). As was noted during the previous winter, Cormorants favoured the long spit of Taylor's Bank. Both Mallard and Shelduck occurred along the channel of the River Alt, with the latter species also to be found on Formby Bank. A small number of Great Crested Grebes and Pink-footed Geese were also noted.

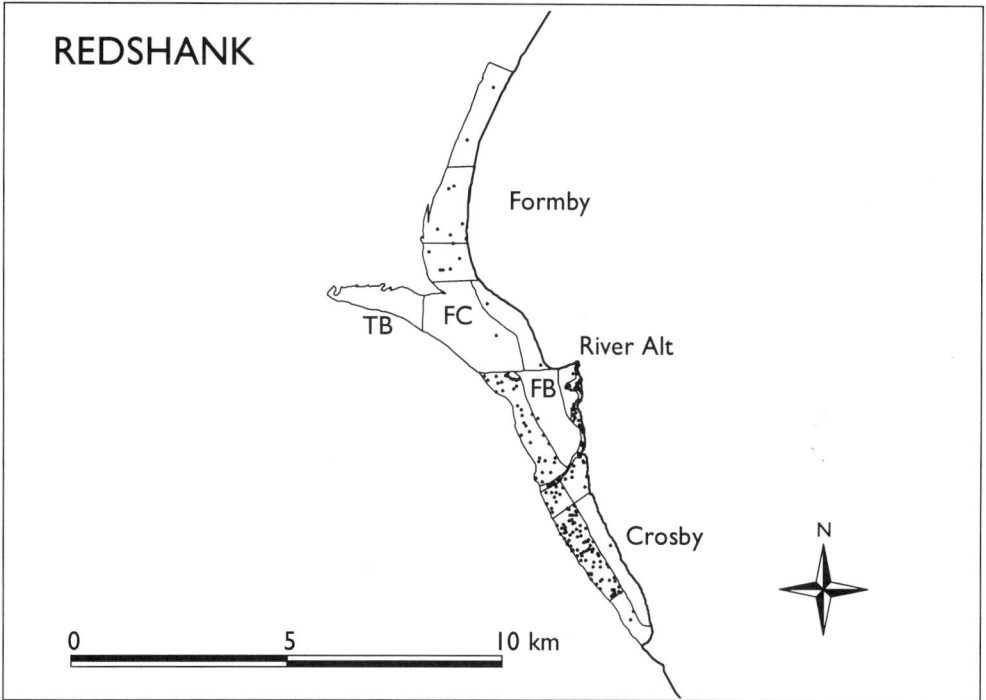

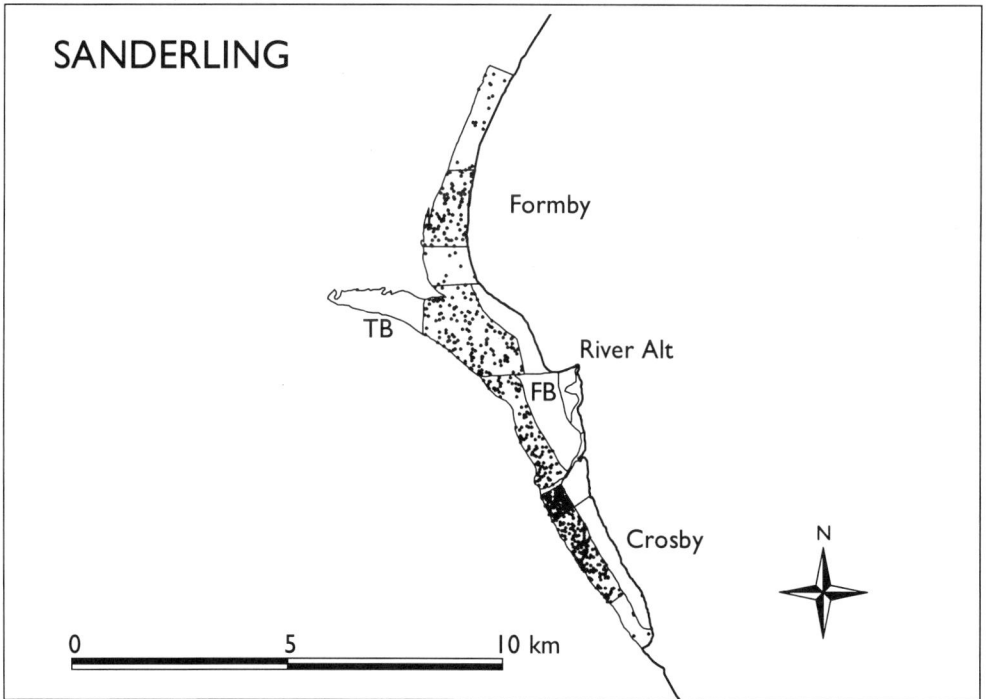

Figure 69. WeBS Low Tide Counts of Redshank and Sanderling on the Alt Estuary, winter 1997-98. (TB=Taylor's Bank, FC=Formby Channel, FB=Formby Bank)

BELFAST LOUGH
Co. Antrim / Co. Down

Internationally important: Great Crested Grebe, Redshank, Turnstone
Nationally important: Shelduck, Scaup, Eider, Goldeneye, Red-breasted Merganser, Oystercatcher, Ringed Plover, Knot, Dunlin, Black-tailed Godwit, Bar-tailed Godwit

Site description

Belfast Lough is a large sea lough in the north-east of Ireland, with the city of Belfast at its head. The area surveyed for the 1997-98 Low Tide Counts comprised the coast from Carrickfergus on the north shore around to the eastern end of Bangor on the south shore. The outer parts of the lough's shore are generally rocky with some sandy bays although more extensive areas of intertidal mud are found toward Belfast. Industrial land claim has, however, reduced the area of the mudflats over the last 150 years, and Belfast has become the main port in Northern Ireland for heavy cargo. More recently, some of the area, including the important Belfast Harbour Pools, has been given a degree of protection. There are also problems of refuse disposal, pollution and general disturbance (Pritchard *et al.* 1992, Buck & Donaghy 1996, Buck 1997b).

Bird distribution

A peak of almost 7,000 Oystercatchers was recorded at low tide during the 1997-98 winter, corresponding well with the Core Counts for the same period. The species was very widespread but the highest concentrations were on the mudflats in the south-west corner of the lough. The same area was also favoured by a number of other wader species, such as Knot (which was found here exclusively), Dunlin, Curlew, Redshank, both godwits (although Black-tailed Godwits favoured Victoria Park and the BP pools also) and a Greenshank. The December count of 1,906 Dunlin was much higher than the peak winter Core Count of 1,070. The BP pools were used by Golden Plover for roosting but although Lapwing favoured this area too, they occurred somewhat more widely with Whitehouse Lake and Victoria Park also favoured. Small numbers of Ruff and Snipe were also found in the inner estuary. Three species of wader, predictably, showed more of a preference for the outer parts of the site. Turnstone were widespread but favoured the rocky shores of the outer lough, with highest densities to the west of Carrickfergus marina. Purple Sandpipers were more localised

between Helen's Bay and Ballymacormick Point. Ringed Plovers were also quite localised, favouring Holywood Flats and the Bangor shore.

An outer estuary distribution was also exhibited by Eider, widespread from the north end of Holywood (on the south shore) and from Green Island (on the north shore) outwards. The highest densities were found between Bangor and Grey Point and at Carrickfergus. Many other sea-duck, however, preferred the inner estuary, including Goldeneye, Red-breasted Merganser and Scaup. All of these ducks were counted in lower numbers at low tide than during Core Counts, possibly due to birds being further away and feeding more actively at low tide. The internationally important numbers of Great Crested Grebes, which at low tide peaked at 1,662 in November, were widespread but also concentrated in the inner estuary, principally along the north shore from Green Island southwards. This count was also substantially lower than the corresponding exceptional November Core Count of 2,403. Little Grebes were restricted to the BP pools. Cormorants were numerous (with 162 counted in December) and widespread but preferred the outer estuary, particularly the Grey Point to Swineley Point stretch, but Victoria Park was also favoured.

Belfast Lough is not of great importance to dabbling ducks and most Wigeon, Gadwall, Teal, Mallard and Shoveler made use of the BP pools, along with small numbers of Pochards and a Common Scoter. Shelduck were restricted to the inner mudflats, along with a single Light-bellied Brent Goose. Victoria Park was favoured by less coastal waterfowl species such as Mute Swan and Coot. Most gulls favoured the inner parts of the lough with Black-headed Gulls very widespread but Common Gulls relatively localised. Interestingly, there was a difference in distribution between the Herring and Great Black-backed Gulls, which favoured the western part of the inner lough (including Whitehouse Lake), and the (admittedly rather small numbers of) Lesser Black-backed Gulls to be found around the BP pools and on the mudflats at Holywood. Single Glaucous and Ring-billed Gulls were also noted in December at Whitehouse Lake.

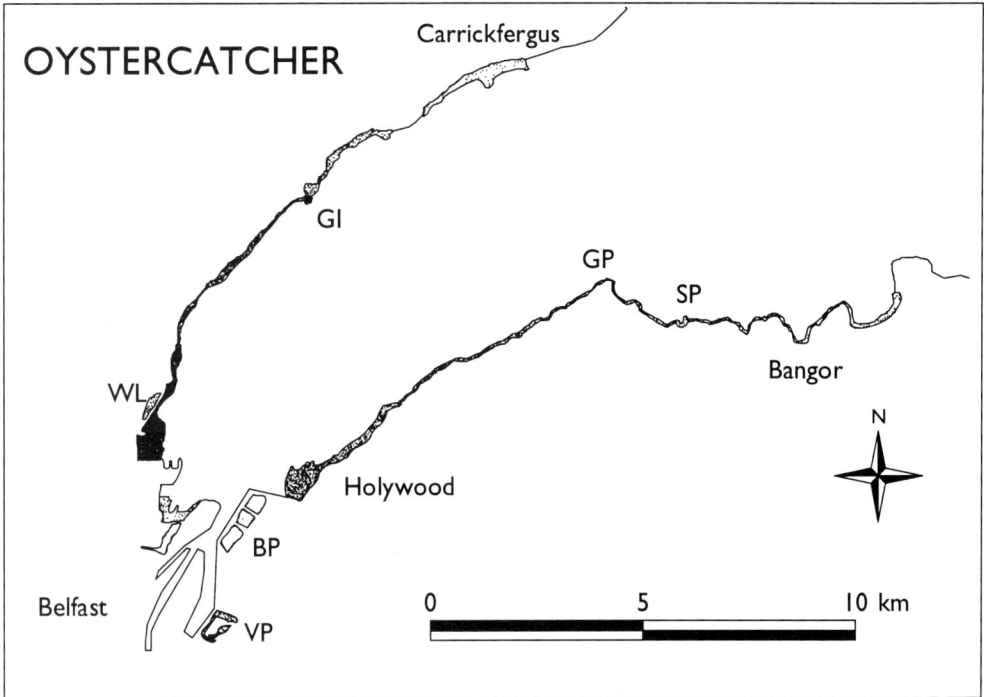

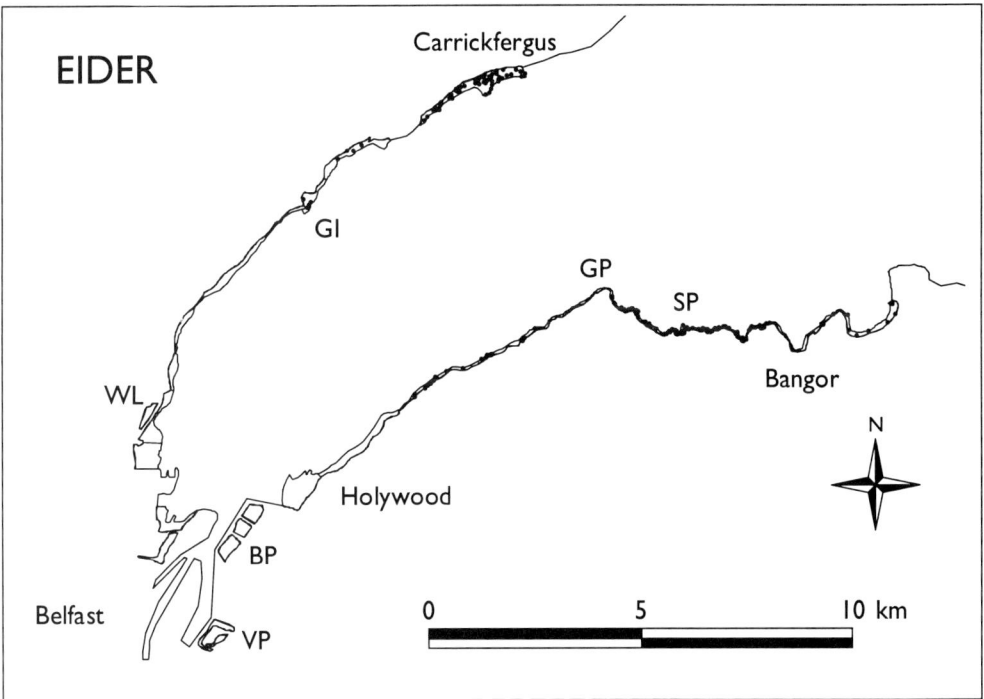

Figure 70. WeBS Low Tide Counts of Oystercatcher and Eider on Belfast Lough, winter 1997-98. (GI=Green Island, WL=Whitehouse Lake, VP=Victoria Park, BP=Belfast Harbour Pools, GP=Grey Point, SP=Swineley Point)

BLYTH ESTUARY
Suffolk

Internationally important: None
Nationally important: Avocet, Black-tailed Godwit, Redshank

Site description

The Blyth is a small estuary, situated immediately inland from Southwold in Suffolk. The inner estuary is a small, muddy basin but the lower reaches have been canalised since the early 19th century and have little in the way of intertidal substrate. The River Blyth finally enters the sea between Southwold and Walberswick. Flanking the narrow lower river channel are a sizeable area of marshes: Tinker's Marshes to the south of the channel and Reydon Marshes and Town Marshes to the north. There have been numerous attempts in the past to 'reclaim' the mudflats and saltmarshes but all have failed which has left a mosaic of breached bunds. There are few threats currently facing the estuary, although the beach at the mouth of the Blyth is frequented by sailors and holiday-makers. The Blyth Estuary forms the northern part of the Walberswick National Nature Reserve (Prater 1981, Pritchard *et al.* 1992, Davidson 1998b).

Bird distribution

The Blyth Estuary supports approximately 11,000 wintering waterfowl and since it is a relatively small site, no species occurs in internationally important numbers. Figure 71 illustrates the low tide distribution of Avocets at the Blyth Estuary and shows that they tend to avoid the eastern end of the site. A peak of 422 birds was recorded in January which was somewhat higher than the peak winter Core Count of 350 Avocets. The distribution of another species occurring in nationally important numbers, the Redshank, is shown in Figure 71. This species does occur in all count sections but it also shows a relative avoidance of the eastern end of the site. The same distribution pattern was shown by Dunlin, the most numerous wader species present. Peak numbers of Redshanks were lower for the Low Tide Counts (775) than for the Core Counts (1,300). The third waterfowl species wintering on the Blyth in nationally important numbers, the Black-tailed Godwit, peaked at 215 birds during the Low Tide Counts (compared with a winter peak of 250 for the Core Counts) and was distributed fairly evenly around the site. Peaks of over 2,000 of both Lapwing and Golden Plover were recorded at low tide with both species widespread although the latter species showed an avoidance of the north-west quarter of the estuary. Other waders were relatively few in number with Curlew, Bar-tailed Godwit and Grey Plover distributed quite evenly and Knot and Oystercatcher occurring throughout but favouring the north-east corner of the site. Small numbers of Ringed Plover, Snipe, Spotted Redshank and Turnstone were also noted.

The most numerous wildfowl species present was Wigeon, with a peak of over 2,000 birds noted throughout the estuary although the north-west corner was less favoured. Teal were also numerous and favoured the eastern end of the estuary. A peak of over 750 Shelduck was noted at low tide; this species was distributed fairly evenly around the estuary although occurred slightly more densely in the north-east corner of the site. The peak of 177 Pintail was quite notable for the size of the site; this species was also evenly distributed, as was Mallard. Peak counts of 61 Gadwall and 41 Shoveler occurred, mostly at the eastern end of the site (i.e. adjacent to Tinker's Marsh). Small numbers of other wildfowl species were also recorded, with Egyptian Goose and Water Rail of most note. A Mediterranean Gull was noted in January, along with the five common species of gull.

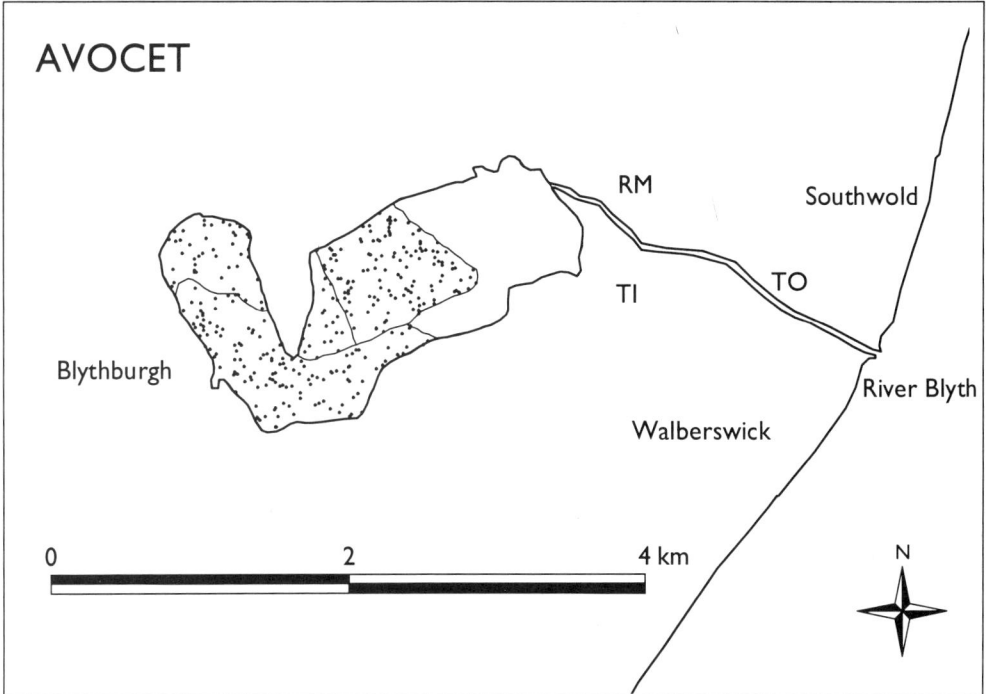

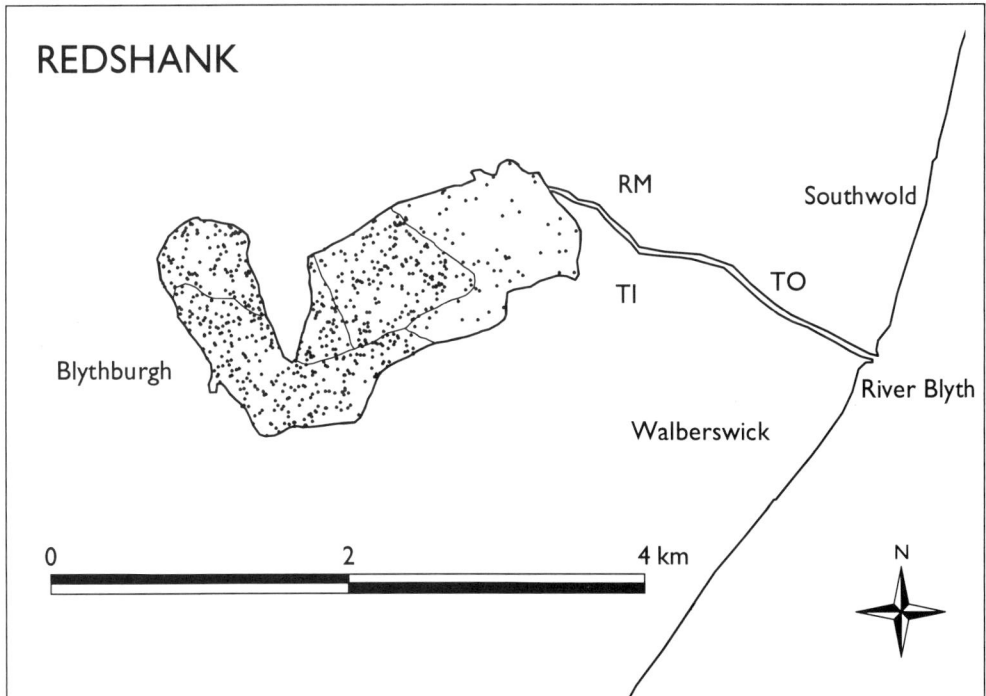

Figure 71. WeBS Low Tide Counts of Avocet and Redshank on the Blyth Estuary, winter 1997-98. (RM=Reydon Marshes, TI=Tinker's Marshes, TO=Town Marshes)

CHICHESTER HARBOUR
West Sussex / Hampshire

Internationally important: Dark-bellied Brent Goose, Grey Plover, Dunlin, Black-tailed Godwit, Bar-tailed Godwit, Redshank

Nationally important: Little Grebe, Shelduck, Teal, Red-breasted Merganser, Ringed Plover, Curlew

Site description

Chichester Harbour is a large and complex site situated between Chichester and Havant and is linked to Langstone Harbour to the west by a channel along the north side of Hayling Island. There are four major arms, Chichester Channel, Bosham Channel, Thorney Channel and Emsworth Channel, originally formed by land sinking along four small river valleys. These run into a wider area near the mouth of the estuary and there is a fairly wide opening to the eastern Solent. The river channels are muddy whereas the intertidal areas south of Thorney Island are much sandier, and also support extensive areas of eelgrass and algae. The estuary is extremely popular with watersports enthusiasts so, although the majority of the shoreline is undeveloped with restricted access, those areas with public access are heavily used. There is always the potential for pressure for further marinas and slipways. Wildfowling also occurs, as does commercial dredging for oysters, hand-gathering of cockles and winkles and bait digging (Prater 1981, Buck 1997a, Pritchard *et al.* 1992, Davidson 1997a, A. de Potier pers. comm.)

Bird distribution

Chichester Harbour is the most important single site on the south coast of England for wintering waterfowl, supporting an average of about 55,000 birds, including six species in internationally important numbers. As has been discussed in previous reports, Bar-tailed Godwits show an obvious preference for outer parts of the estuary, particularly the wide sandy flats south of Thorney Island, but also the adjacent eastern and western shores. The peak Low Tide Count of 693 birds equated relatively well to the peak winter Core Count of 820. Black-tailed Godwit numbers also corresponded well between the two sets of counts, with peaks of 464 at high tide and 413 at low tide. However, the Black-tailed Godwit distribution map shows this species' preference for the muddier tops of creeks and it is possible that some birds may have been missed whilst feeding in creeks or saltmarsh; birds could also have been feeding on nearby non-tidal habitats.

Redshank were very widespread although Chichester Channel held slightly higher densities than average. Dunlin were numerous, although in slightly smaller numbers than during Core Counts, and widespread with Thorney Channel holding many birds. Grey Plovers were somewhat more concentrated around the estuary mouth. Ringed Plovers were particularly concentrated in the southwest corner of the harbour east of Selsmore. Oystercatcher and Curlew were widespread but the latter was especially concentrated at the top of Chichester and Bosham Channels. Most Knot were found in the western half of the harbour, with the intertidal flats southwest of Thorney Island holding the majority of the birds. Lapwing were widespread but Golden Plovers more localised at various points around the harbour. Up to 200 Sanderling were noted at low tide, almost all of which were found on Pilsey Sands south of Thorney Island. Small numbers of Turnstones were widespread and other wader species noted were Avocet, Snipe, Spotted Redshank and Greenshank.

Dark-bellied Brent Geese were found almost everywhere in the harbour, with the highest concentration along the Rookwood shore. A maximum count of 6,530 was noted at low tide. Shelduck were also widespread although occurring slightly more densely in Chichester Channel. Teal had their main concentrations on the western shore as well as at Bosham Hoe. Wigeon and Pintail both favoured the north ends of Bosham and Chichester Channels and Thorney Great Deep. Mallard preferred the Emsworth and Langstone mill ponds and around Bosham, and Mute Swans were also found at high densities at Emsworth mill pond as well as along Chichester Channel. Goldeneyes and Red-breasted Mergansers were both widespread along the channels. Both Little and Great Crested Grebes were widespread but favoured Chichester Channel and up to nine Slavonian Grebes were also noted. Small numbers of Eider were found at Pilsey Sands and other ducks noted were Mandarin, Common Scoter and Smew. Cormorants and Grey Herons were widespread in small numbers. The latter species was outnumbered every month by Little Egret, with up to 23 noted widely around the harbour.

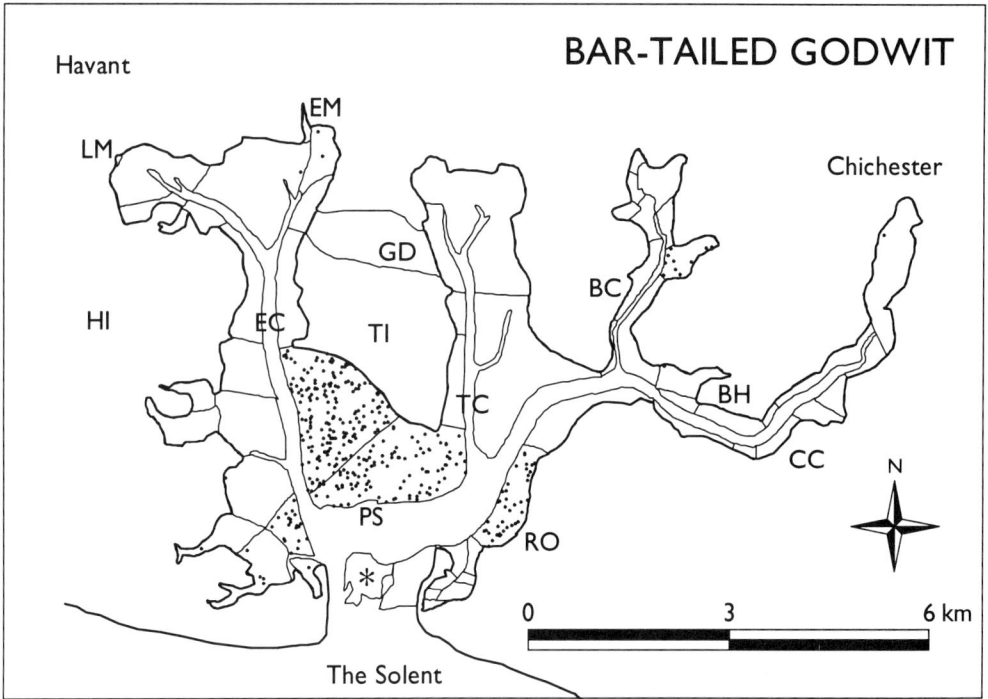

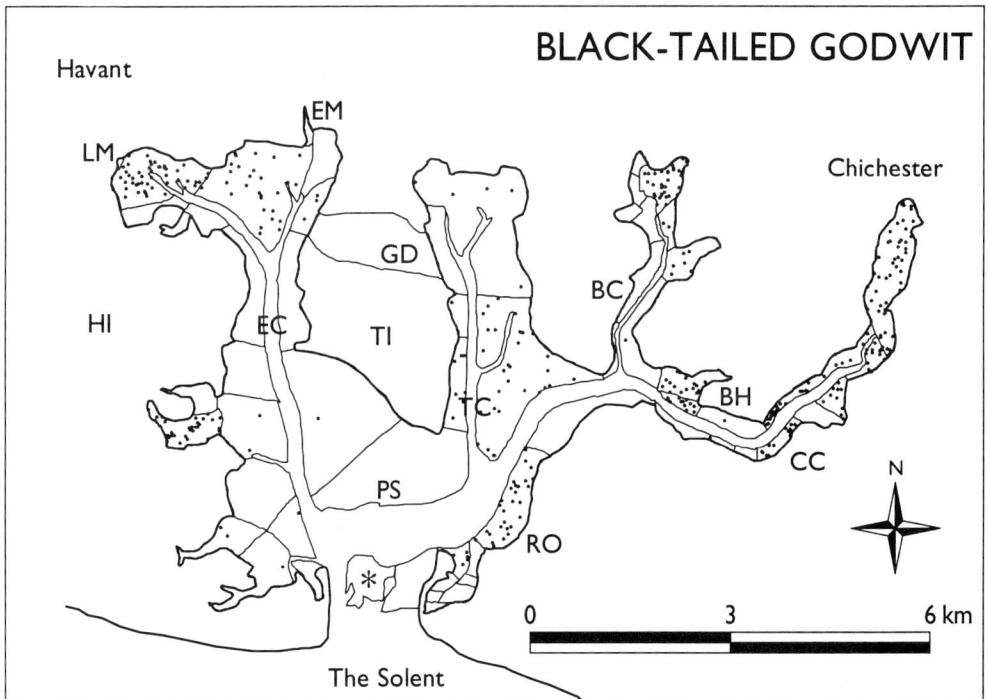

Figure 72. WeBS Low Tide Counts of Bar-tailed Godwit and Black-tailed Godwit at Chichester Harbour, winter 1997-98. (LM=Langstone Mill Pond, EM=Emsworth Mill Pond, GD=Thorney Great Deep, TI=Thorney Island, BC=Bosham Channel, BH=Bosham Hoe, CC=Chichester Channel, RO=Rookwood, PS=Pilsey Sands, HI=Hayling Island, EC=Emsworth Channel, TC=Thorney Channel, *=not counted)

CLEDDAU ESTUARY
Pembrokeshire

Internationally important: None
Nationally important: Little Grebe, Shelduck, Wigeon, Teal, Dunlin, Curlew

Site description

The WeBS site known as the Cleddau Estuary is in reality a series of small estuaries all opening into the sheltered waters of Milford Haven in south-west Wales. In character, the complex is similar to the Fal and Tamar in south-west England, all of these sites being drowned river valleys (or rias) in which mudflats and some areas of saltmarsh have later developed. Most of the main intertidal areas were covered for the WeBS Low Tide Counts, although some of the intervening stretches of rocky coast, as well as the open waters of Milford Haven, were not covered. The *Sea Empress* oil spillage of early 1996 illustrates extremely well one of the principal threats to the area, which is a major centre for oil transport and refining (Prater 1981, Pritchard *et al.* 1992, Davidson 1995a).

Bird distribution

Since this site lacks wide expanses of intertidal habitat, and since there is little other estuarine habitat nearby, it was unsurprising that the overall numbers of birds recorded on Low Tide Counts equated very well to those on the Core Counts, with the sum of the peak counts of all wildfowl and waders at low tide being almost 14,000 birds. The five-year peak mean from Core Counts is about 15,000 birds.

Wigeon were widespread on the Cleddau but highest densities occurred at the western end of the Pembroke River. The overall peak of almost 3,000 birds in November equated very well to the Core Counts. Teal were also recorded in high numbers and were most numerous on the inner parts of the site, especially on the western Cleddau, Sprinkle Pill, Landshipping and on Millin Pill. Mallard were widespread, especially along the western Cleddau and at Millin Pill, and Shelduck were also widespread but favoured Pembroke River and the opposite shore of the Cleddau from Landshipping. Of small numbers of other wildfowl present, Goldeneye were widespread but more concentrated on the pools at the Gann and at the top end of the western Cleddau, with a few Red-breasted Mergansers also present at the latter site. A single Brent Goose wintered in Angle Bay with Canada Geese

found around Landshipping and Llangwm. Little Grebes were widespread, but especially favoured pools at the Gann Estuary. Great Crested Grebes, on the other hand, favoured the open water of Angle Bay and two Great Northern Divers were found on the Pembroke River. Cormorants and Grey Herons favoured the Cleddau Rivers but Little Egrets, of which up to 11 were counted at low tide, were somewhat more widespread.

Dunlin, the most numerous wader species present, was found widely but avoided the upper reaches of the estuary and the highest densities were to be found on Pembroke River. Curlew were distributed more evenly with the highest densities occurring in Sandy Haven Pill. Golden Plovers were virtually absent except in February, when birds were largely to be found at Beggars Reach and on the Carew River; Lapwings numbers were fairly high on all counts and were widespread with the highest densities to be found at Landshipping and along the Cresswell River. Redshank were very widespread with the highest densities at Sprinkle Pill and on the western Cleddau. Oystercatchers, on the other hand, were more numerous towards the lower reaches of the estuarine complex, especially at Angle Bay, Pembroke River, Sandy Haven Pill and the Gann Estuary. Ringed Plovers occurred in small numbers in various localised areas such as the western end of Pembroke River, the lower part of the western Cleddau and the southeast corner of Angle Bay. Turnstone were most widespread on the outer parts of the estuary, such as at the Gann and at Llandstadwell, as well as some parts of Angle Bay and the Pembroke River. Small numbers of Grey Plovers and Bar-tailed Godwits were mostly found on Angle Bay and Pembroke River. In common with estuaries in southwest England, small numbers of Greenshanks and Spotted Redshanks were found around the site. Of the other waders, a few Knot and Sanderling, along with a Green Sandpiper, were noted, but numbers of Snipe were especially notable in the early winter with a peak count of 81 in November; the highest numbers of these were along the western Cleddau. All five common gull species were present with, as usual, Black-headed Gull being the most numerous species but over 300 Lesser Black-backed Gulls in January were also noteworthy.

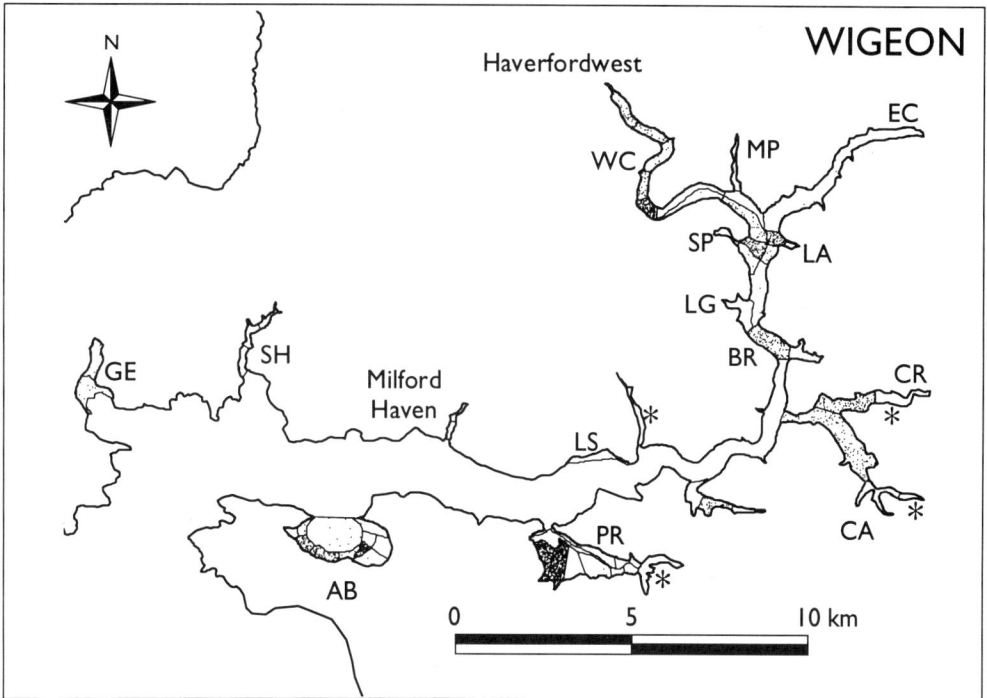

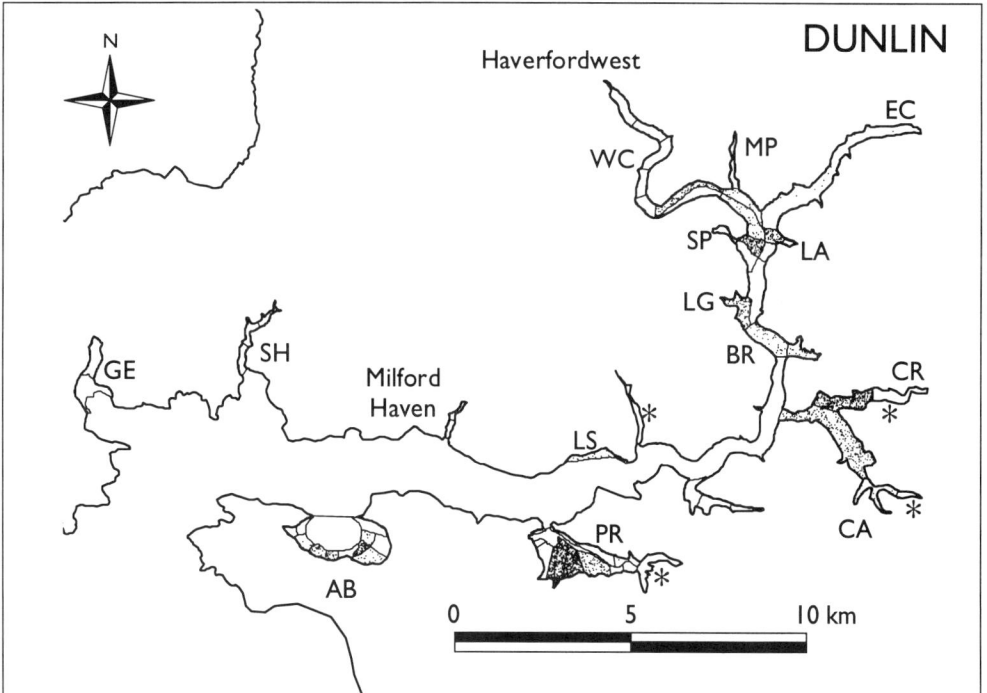

Figure 73. WeBS Low Tide Counts of Wigeon and Dunlin on the Cleddau Estuary, winter 1997-98. (GE=Gann Estuary, SH=Sandy Haven Pill, LS=Llanstadwell, BR=Beggars Reach, LG=Llangwm, SP=Sprinkle Pill, WC=Western Cleddau, MP=Millin Pill, EC=Eastern Cleddau, LA=Landshipping, CR=Cresswell River, CA=Carew River, PR=Pembroke River, AB=Angle Bay, *=not counted)

DEE ESTUARY: NORTH WIRRAL SHORE
Merseyside

Internationally important (whole Dee): Shelduck, Teal, Pintail, Oystercatcher, Grey Plover, Knot, Dunlin, Black-tailed Godwit, Bar-tailed Godwit, Curlew, Redshank, Turnstone

Nationally important (whole Dee): Great Crested Grebe, Cormorant, Wigeon, Sanderling

Site description

The Dee Estuary is one of the most important estuaries in the UK for wintering waterfowl and the whole site was covered as part of the WeBS Low Tide Count scheme during winter 1996-97. One of the most important parts of the Dee, particularly for waders, was confirmed to be the North Wirral Shore, stretching from Red Rocks at the western end to Perch Rock at the mouth of the Mersey. This stretch of coast forms an area of intertidal sand, mudflats and developing saltmarsh approximately 13 km long and about 2 km wide. An important point to note is the proximity of two other WeBS sites: the Mersey Estuary is immediately adjacent to the eastern end of the North Wirral Shore, and this junction of two sites is itself only about 2 km away from the southern flats of the Alt Estuary. There is much interchange between these sites, as discussed below and in the sections describing the Alt and Mersey Estuaries.

Bird distribution

During winter 1997-98, this small stretch of coast, which is at present not part of any SPA or Ramsar site, held low tide peaks of 3,000 Grey Plover, 12,000 Knot, 18,000 Dunlin, 5,000 Bar-tailed Godwit and 2,000 Redshank. All of these counts exceed the level for international importance. In recognition of the critical importance of the area, therefore, it is being proposed as an extension to the existing Dee Estuary Ramsar/SPA site by English Nature.

Many of the most numerous species showed similarities in their distribution along the North Wirral shore. Figure 74 shows the low tide distribution of Dunlin at this site. A two-centred pattern can be seen, with birds favouring Mockbeggar Wharf and East Hoyle Bank, but with very few birds at either the western or eastern ends. There was also a gap between the main two concentrations where very few birds were feeding. The feeding distributions of both Knot and Grey Plover were very similar to that seen for

Dunlin. For all three of these species, the birds occurred more densely on Mockbeggar Wharf than on East Hoyle Bank. Several other common species, such as Oystercatcher, Curlew and Redshank, had a more evenly spread distribution, but all still displayed a tendency towards higher densities on Mockbeggar Wharf.

Bar-tailed Godwits were more unusual, however, in that they showed no interest in East Hoyle Bank at all, being confined exclusively to Mockbeggar Wharf (Figure 74). Lapwings, however, were distributed in a completely different way, roosting almost exclusively in the 'gap' between Mockbeggar and East Hoyle, i.e. away from the feeding concentrations. This may suggest that the feeding distribution of the other species is determined mostly by the food supply and not by excessive disturbance in the central region. Less numerous waders included Ringed Plovers (which were widespread although favouring the western half of the site) and Turnstones. Despite the Dee apparently holding internationally important numbers of Turnstones, with over 900 recorded on Core Counts during winter 1997-98, a peak of only 223 were to be found on the North Wirral shore at low tide during the same winter, mostly towards the eastern end (and last winter's counts showed that the birds were not to be found elsewhere on the Dee apart from about 60 birds on Hilbre). It seems likely that, as with the birds from the Alt, many of the Turnstones roosting on the Dee feed on the adjacent shore at Egremont at the mouth of the Mersey Estuary. Other waders recorded included small numbers of Golden Plover, Sanderling, Black-tailed Godwit and Greenshank.

Despite the major importance of the Dee Estuary as a whole for wintering wildfowl such as Shelduck, Wigeon, Teal and Pintail, very few ducks were to be found on the North Wirral Shore, with just a small number of Shelduck which were largely restricted to the western end. Additionally, small numbers of Mallard, Goldeneye, Red-breasted Merganser and Cormorant were also noted.

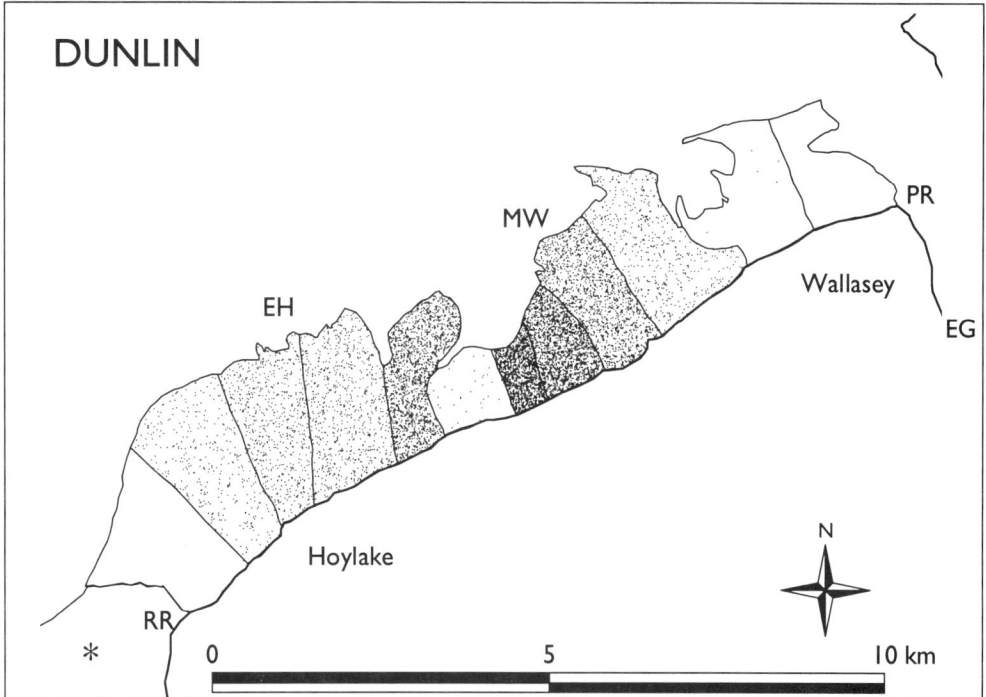

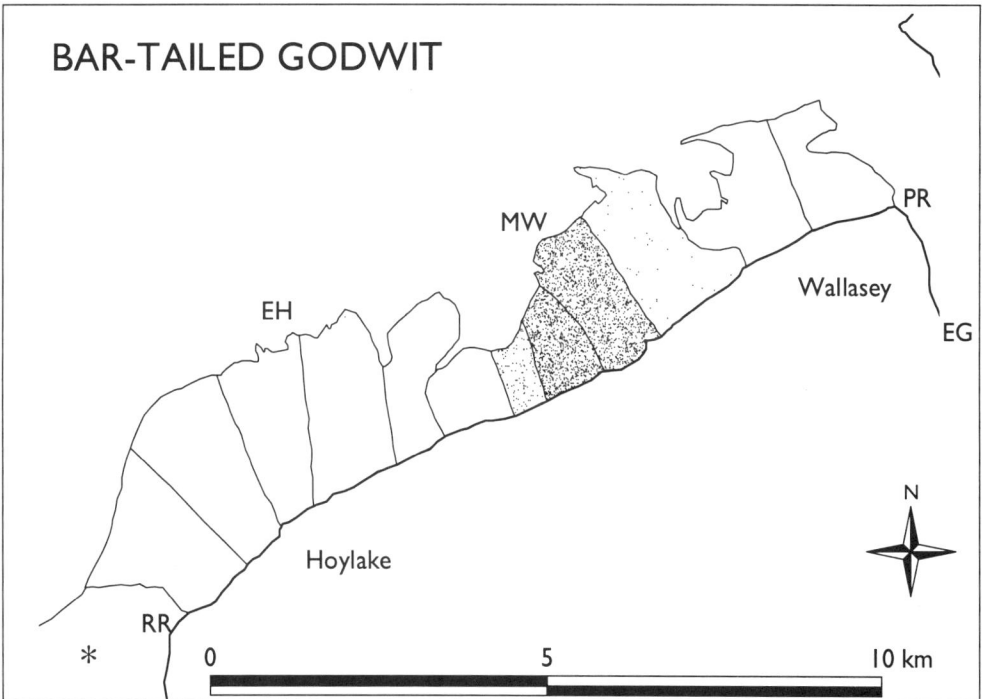

Figure 74. WeBS Low Tide Counts of Dunlin and Bar-tailed Godwit on the North Wirral Shore, winter 1997-98. (RR=Red Rocks, EH=East Hoyle Bank, MW=Mockbeggar Wharf, PR=Perch Rock, EG=Egremont, *=not counted)

HAMFORD WATER

Essex

Internationally important: Dark-bellied Brent Goose, Teal, Grey Plover, Black-tailed Godwit, Redshank

Nationally important: Shelduck, Wigeon, Pintail, Avocet, Ringed Plover, Golden Plover, Dunlin, Bar-tailed Godwit

Site description

Hamford Water is a large, shallow, estuarine basin with an extremely diverse mix of habitat types. The whole site is a mosaic of dissected saltmarshes, islands, channels and mudflats backed by a range of brackish, fresh and reed-fringed marshes. Many of the islands are former saltmarshes embanked and converted to wet grassland, but some have reverted to saltmarsh after sea walls were breached around the end of the 19th century; saltmarsh comprises one third of the whole site. The mouth of the main channel into Pennyhole Bay is flanked on either side by dune-topped shingle spits. The principal causes of disturbance to waterfowl at Hamford Water are light aircraft (more so in the summer months), military helicopter training and explosives testing at a site on the north shore. Much of the surrounding marshland has been converted to arable farmland. As with other sites along this stretch of coast, saltmarsh erosion from rising sea-levels is also a concern (Prater 1981, Pritchard *et al.* 1992, Davidson 1998b).

Bird distribution

Hamford Water is a very important site for wintering waterfowl, supporting an average of about 45,000 birds. WeBS Low Tide Counts were carried out at this site during winter 1992-93 when only very partial coverage was achieved. During 1997-98, coverage was again very patchy (approximately 13% of the intertidal area was covered) primarily because much of the site is difficult to access or view. Consequently, the numbers recorded from the Low Tide Counts were expected to be much lower than from the Core Counts which are made at high tide roosts. However, some of the birds roosting at Hamford do move to the nearby Stour Estuary to feed at low tide (J. Novorol pers. comm.)

The peak winter Core Count of Grey Plovers at Hamford was 3,270 birds whereas the peak Low Tide Count was 1,965. All of the counted areas recorded this species but Dugmore Sands was particularly favoured. This area was also the favoured low tide feeding or roosting area (of those surveyed) for Lapwing, Golden Plover, Bar-

tailed Godwit, Oystercatcher, Ringed Plover, Turnstone and Avocet. Sanderling were found exclusively on Dugmore Sands. Knot favoured the northern parts of Dugmore Sands and Garnham's Island. Dunlin, the most numerous species, was found commonly on all count sections although it displayed a relative avoidance of the Moze Creek area and Dugmore. Curlew were typically evenly distributed. Only two waders, Redshank and Black-tailed Godwit, favoured the inner parts of the site at Landermere, with many Redshank also at Garnham's Island but very few on Dugmore Sands. A count of 28 Ruff in February at Moze Creek was of note. A few Snipe were also found.

A peak of 1,842 Teal was noted in November (although much lower numbers were recorded during the other three months) which is much lower than the Core Count peak of 2,633. Although Teal were found commonly on all covered sections there was a notable concentration at Moze Creek. Wigeon were also widespread although fewer were found on the northern half of Dugmore Sands, an area which was also less favoured by Shelduck; the latter species was found in its highest densities on the southern half of Dugmore Sands. Many Wigeon and Shelduck congregate around shallow pools on this section to preen and roost at low tide (J. Novorol pers. comm.) Very few Pintail were recorded at low tide, with most of those at Garnham's Island, but Pintail numbers were low at Hamford Water on the Core Counts during this winter also. Small numbers of Mallard, Shoveler and Eider were also noted.

Internationally important numbers of Dark-bellied Brent Geese wintering at Hamford Water, although the Core Count winter peak of 4,194 was relatively low for this site. Low Tide Counts found a peak of only 1,829 birds as well as up to two Light-bellied Brents; most of the Brent Geese were presumably elsewhere in the large area of saltmarsh and grazing marsh; they occurred fairly evenly across the recorded parts of the site. Feral Canada and Greylag Geese were also noted (on the southern part of Dugmore Sands) and a flock of 40 Barnacle Geese was recorded in February. No other waterfowl species occurred in noteworthy numbers.

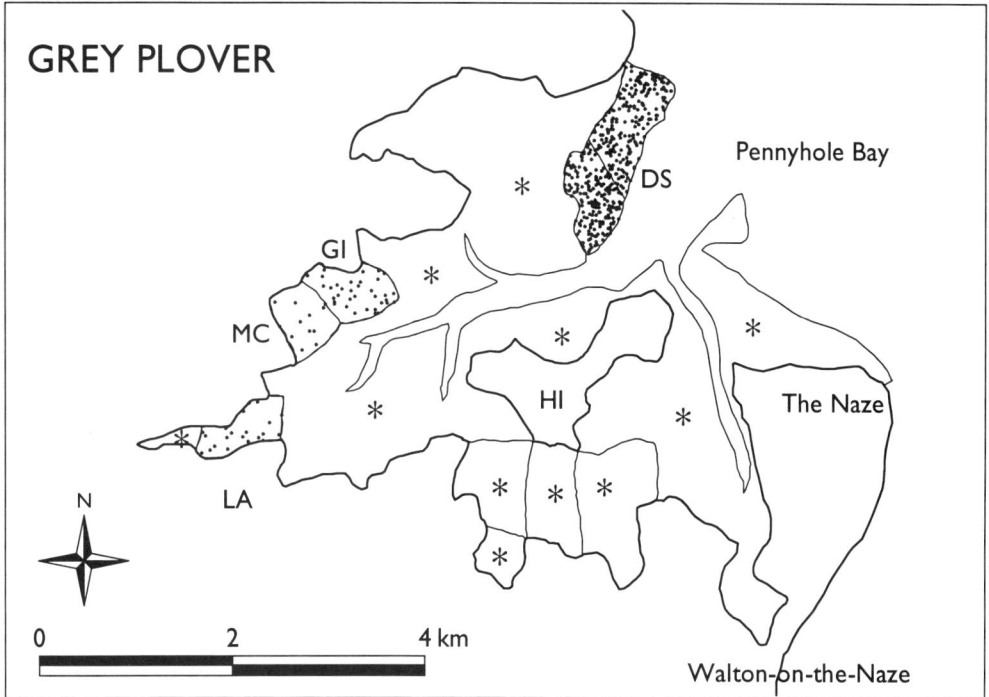

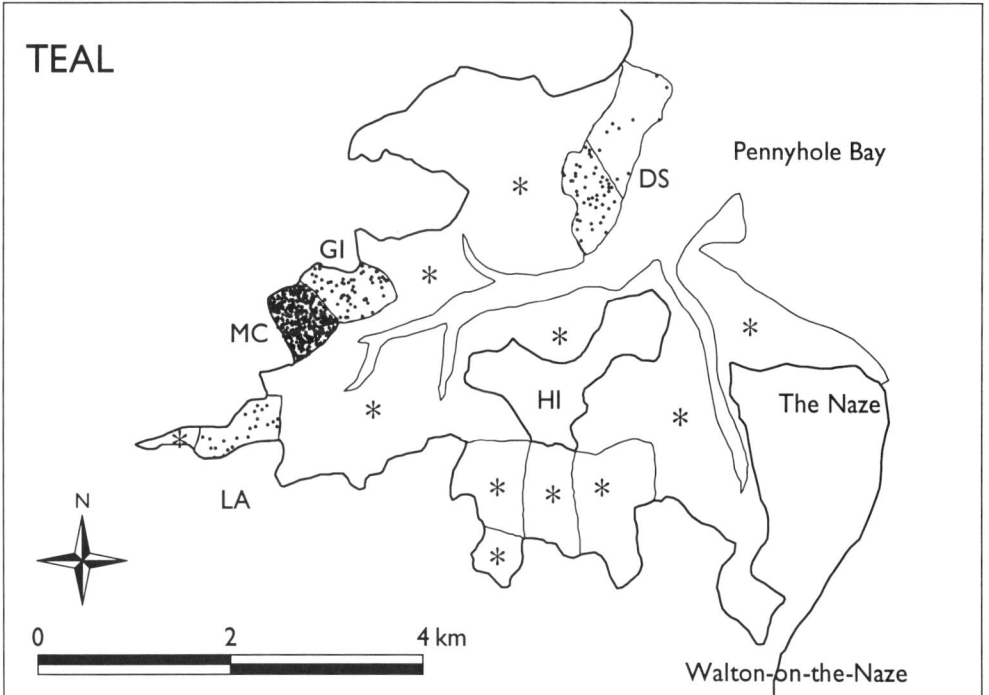

Figure 75. WeBS Low Tide Counts of Grey Plover and Teal at Hamford Water, winter 1997-98. (DS=Dugmore Sands, GI=Garnham's Island, MC=Moze Creek, LA=Landermere, HI=Horsey Island, *=not counted)

MERSEY ESTUARY
Merseyside / Cheshire

Internationally important: Shelduck, Teal, Pintail, Dunlin, Black-tailed Godwit, Redshank
Nationally important: Great Crested Grebe, Wigeon, Golden Plover, Grey Plover, Curlew

Site description

The Mersey is one of the most heavily developed and polluted estuaries in the UK (although pollution levels have lessened somewhat in recent years), with the outer sections of the estuary in particular infringed upon by Liverpool and Birkenhead. The large towns of Widnes, Runcorn and Ellesmere Port are also adjacent to the site. An extensive area of saltmarsh on the southern shore, as well as the important Ince and Stanlow Banks, are protected from disturbance to some degree by the Manchester Ship Canal. The large pools at Frodsham on the south side of the estuary are extremely important as one of the roosts for birds feeding on the estuary. As well as the usual problems which occur on heavily industrialised estuaries, such as pollution and disturbance, a more specific issue which could be detrimental to wintering waterfowl is a proposed second runway for Liverpool Airport to be built on land claimed from the estuary. Additionally, there has been a proposal in recent years for a barrage to generate power from tidal energy, which could resurface in the event of the economics of tidal power being considered more realistic by energy producers (Prater 1981, Pritchard *et al.* 1992, Davidson 1996a).

Bird distribution

The peak Core Count of 52,015 Dunlin during the 1997-98 winter was well matched by a peak Low Tide Count of 48,476. The feeding distribution illustrates clearly the vital importance of the Stanlow Banks to the north of Ellesmere Port for this species. The Dungeon Banks area is also of great importance to Dunlin, but the species was relatively scarce elsewhere in the estuary. Knot similarly favoured Stanlow Banks, but made less use of Dungeon Banks than did Dunlin. There was, however, a concentration of the species at Rock Ferry. Grey Plover also favoured Stanlow Banks, but made greater use of the southern parts of the Ince Banks also. Interestingly, the peak numbers of Grey Plovers noted at low tide were roughly three times higher than those on the Core Counts. Were birds moving into the estuary from outside to feed, or were birds being missed in difficult to view roosts?

Peak numbers of Black-tailed Godwit, Redshank and Curlew were also higher at low tide than at high tide. Black-tailed Godwits, which peaked at 2,655 in February, were quite widespread in the central parts of the estuary, but favoured Dungeon Banks and the flats to the north of Mount Manisty. Curlew were typically widespread around the site although with higher densities at Ellesmere Port. Redshank occurred throughout the site, but the highest densities were to be found on the Egremont and Rock Ferry shores, towards the mouth of the Mersey. It is likely that the Low Tide Counts of Redshank on the Mersey, which peaked at almost 7,000, included some birds which roost on the Alt. Oystercatcher, Turnstone, Sanderling and Ringed Plover were all principally found at Rock Ferry and Egremont. As has been mentioned in the sections on the Alt Estuary and the North Wirral Shore, the numbers of Turnstone feeding at the mouth of the Mersey (peaking at 1,188 in February) were far in excess of the numbers recorded for the Mersey Estuary on WeBS Core Counts (never reaching double figures in the last five years), and it is clear that birds recorded on WeBS Core Counts for the Alt and Dee Estuaries are in fact feeding on the outer Mersey at low tide. Both Lapwing and Golden Plover showed a general preference for the inner parts of the estuary. Only a small number of Bar-tailed Godwits were noted.

Teal were widely distributed in the central parts of the estuary with no single main concentration. For a species which can easily be missed within saltmarshes at low tide, the peak of 8,060 compared quite well with the Core Count peak of 12,065. A peak of over 500 Pintail was noted at low tide, somewhat less than during the Core Counts, with birds quite widespread in the central parts of the estuary but the highest densities to be found at the southern end of the Rock Ferry stretch of mudflats. Wigeon favoured Mount Manisty and Ellesmere Port but Shelduck were somewhat more widespread although still reaching highest densities at Ellesmere Port. Mallard favoured Dungeon Banks, Hill Head and Mount Manisty. Cormorants were widespread but very few Great Crested Grebes were noted at low tide despite the national importance of the site for this species.

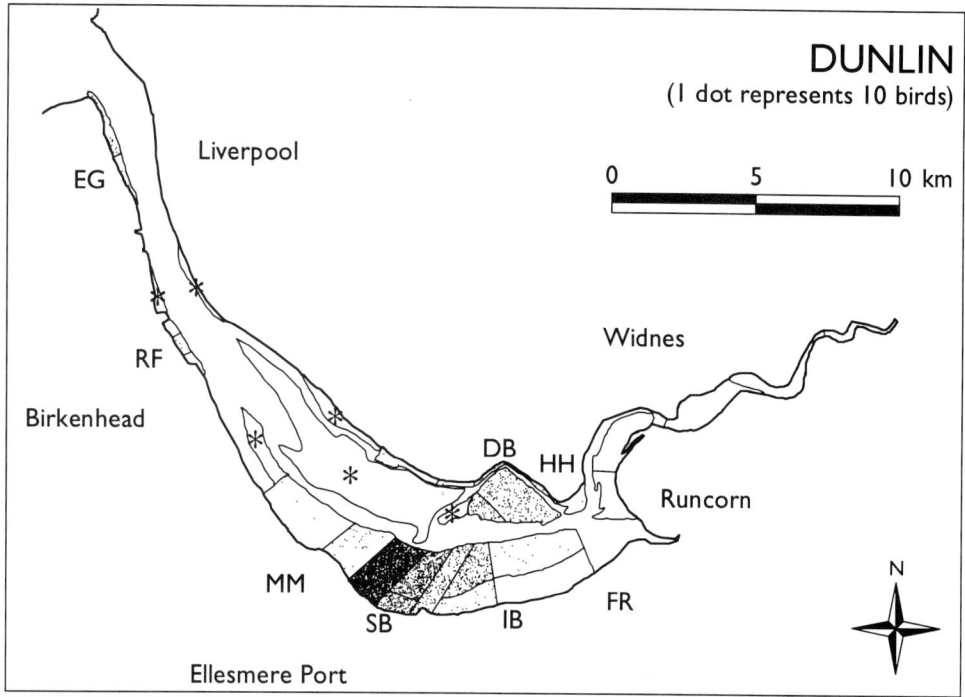

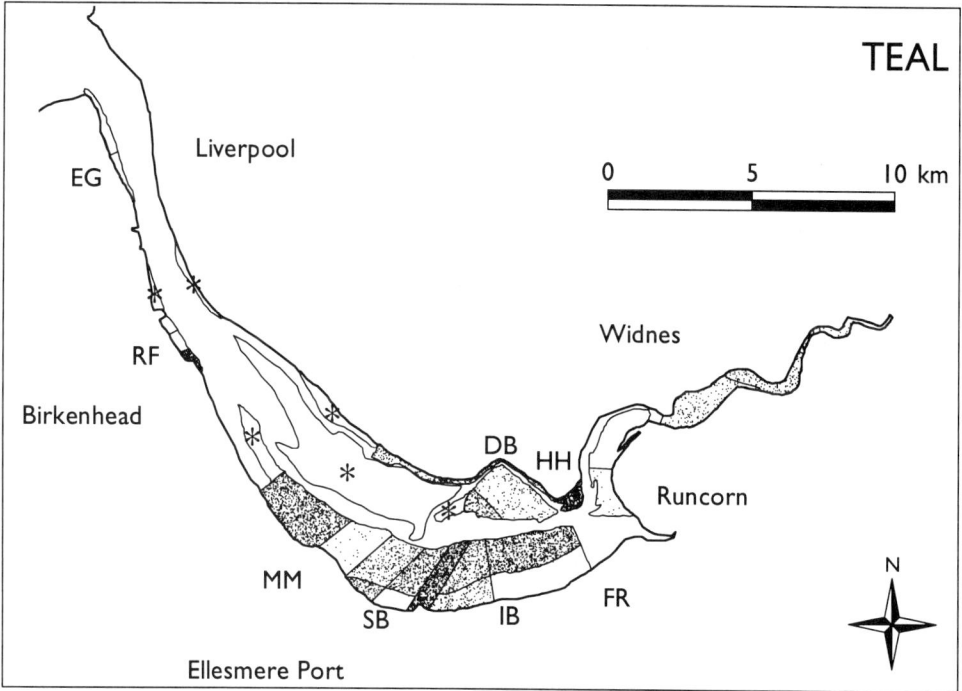

Figure 76. WeBS Low Tide Counts of Dunlin and Teal on the Mersey Estuary, winter 1997-98. (EG=Egremont, RF=Rock Ferry, MM=Mount Manisty, SB=Stanlow Banks, IB=Ince Banks, FR=Frodsham, HH=Hill Head, DB=Dungeon Banks, *=not counted)

MONTROSE BASIN
Angus

Internationally important:	Pink-footed Goose, Knot, Redshank
Nationally important:	Mute Swan, Shelduck, Wigeon, Eider, Red-breasted Merganser, Goosander

Site description

Montrose Basin, the estuary of the South Esk River, is an almost circular basin about 3 km across. The basin is separated from the sea by a broad spit on which the town of Montrose is situated; the river discharges to the sea through a narrow channel at the southern end of this spit. The intertidal flats range from sand to mud and shingle and there are also extensive mussel beds. Eelgrass and algae are also present on the basin, providing a food source for some of the waterfowl. There are areas of saltmarsh on the inner edge of the basin and freshwater grazing fields nearby. Pressure from wildfowling used to be heavy on this site but has been restricted since 1981 when a Local Nature Reserve was created; this led to a dramatic rise in the numbers of waterfowl using the site, particularly Pink-footed Geese. There has also been land-claim for industry and waste disposal during this century (Prater 1981, Pritchard *et al.* 1992, Davidson 1997b).

Bird distribution

Montrose Basin supports an average of about 54,000 wintering waterfowl, of which a large proportion are Pink-footed Geese which roost at the basin in internationally important numbers. Since the geese feed mostly on surrounding farmland, they are not well represented by WeBS Low Tide Counts with over 15,000 noted in November but less than 50 in any other month. Figure 77 shows the low tide distribution of Knot on Montrose Basin with, as can be seen, the birds concentrated into three discrete areas. Peak counts of Knot were about a third lower at low tide than on the WeBS Core Counts on Montrose Basin during the 1997-98 winter. Similarly, Redshank were noted at low tide in about half the numbers recorded at high tide; the species was widespread and fairly evenly distributed although with more of a concentration towards the middle southern sections (Figure 77). Golden Plover were very concentrated in the north-east and south-east corners of the basin, but Lapwing were much more widespread. Dunlin were also widespread, favouring the more open parts of the basin in the north-east and the middle southern sections. Although Oystercatchers were distributed fairly evenly around the basin, there were relatively few at the western end. Curlew were typically distributed very evenly indeed. Both species of godwit peaked at about 80 birds, both favouring southern parts of the estuary, and small numbers of Turnstone were noted around the basin mouth.

Six species of wildfowl occur on Montrose Basin in nationally important numbers. The most numerous, Wigeon, had its main concentrations around Stick's Burn and to the south of the main channel, although smaller numbers were also widespread in the east of the basin. Up to 103 Mute Swans were noted, mostly along the main channel and especially at the western side of the basin. A peak of 1,174 Shelduck at low tide corresponded very well to the core peak of 1,100 with birds found widely but concentrated in the north-east corner. Over 1,000 Eider were found along the lower reaches of the main channel with small numbers of Red-breasted Mergansers. Both this latter species and Goosander (and Eider to a lesser extent) were found in very much lower numbers at low tide than at high tide, probably due to birds moving out of the estuary to feed.

The most numerous other wildfowl species were Mallard (which was mostly found in the south and east of the site), Pintail (in the middle southern parts of the site) and Tufted Duck (with up to 120 in the lower stretches of the main channel). Smaller numbers of Teal and Shoveler were mostly found in the same area as the Pintail and up to 52 Goldeneye and 23 Scaup were found along the main channel. Cormorants were widespread but mostly on the wide flats in the east of the basin. Small numbers of Great Crested Grebes, Grey Herons and Gadwall were also noted, but a Red-throated Diver and two Smew were of particular note. Four species of gull were recorded but there were no Lesser Black-backed Gulls.

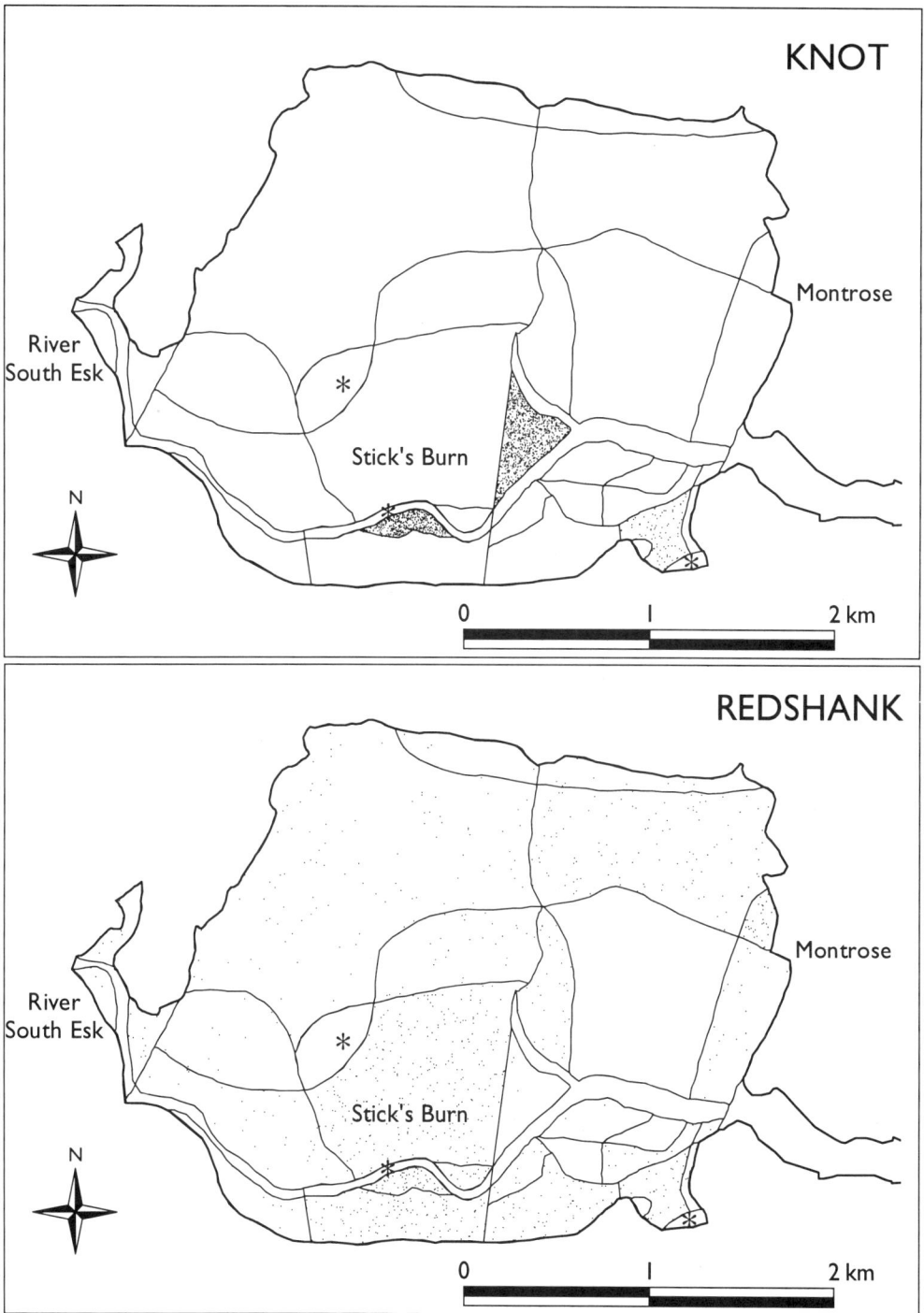

Figure 77. WeBS Low Tide Counts of Knot and Redshank on Montrose Basin, winter 1997-98. (*=not counted)

NORTH NORFOLK COAST
Norfolk

Internationally important: Pink-footed Goose, Dark-bellied Brent Goose, Wigeon, Pintail, Grey Plover, Knot, Bar-tailed Godwit

Nationally important: Red-throated Diver, Little Grebe, Cormorant, European White-fronted Goose, Shelduck, Gadwall, Teal, Shoveler, Scaup, Common Scoter, Goldeneye, Red-breasted Merganser, Avocet, Ringed Plover, Golden Plover, Sanderling, Black-tailed Godwit, Redshank

Site description

The North Norfolk Coast comprises the coastline from the northern edge of Hunstanton in the west to Salthouse Marshes in the east and forms what is arguably the most diverse and complex estuarine system in the UK. There is no single principal river, but several small streams enter the sea here. The coast is the most extensive example of a barrier beach system in the UK, and the large areas of saltmarsh (over 2,000 hectares) are the most diverse in the UK in terms of geomorphology and biology. There is virtually no direct industrial influence on the site, with the main pressures being recreation and exploitation of natural resources. Longer-term threats from sea-level rise may be a more serious problem in the future however (Prater 1981, Pritchard *et al.* 1992, Davidson 1995b, M. Rooney pers. comm.)

Bird distribution

Although Brent Geese are widespread along the coast, they show a clear preference for saltmarsh away from the sea, with highest densities in Blakeney Harbour and north of Burnham Overy Staithe. Few Pink-footed Geese were noted at low tide except for 5,500 in Brancaster Harbour in December; this species uses intertidal habitat principally as a nocturnal roost. No White-fronted Geese were noted, as these birds favour fields at Holkham not covered by the survey. Wigeon showed a similar distribution to Brent Geese, with densest concentrations in Blakeney Harbour. Most Pintail were found in Brancaster and Blakeney Harbours. Teal were numerous in the saltmarshes, particularly near Titchwell, and Mallard were widespread but most concentrated in the harbours. Shelduck numbers exceeded Core Count totals, concentrated in Brancaster and Blakeney Harbours and at Burnham Overy Staithe. Goldeneye and Red-breasted Merganser were both present in similar numbers to Core Counts and favoured the harbours. Up to 70 Little Grebes within the saltmarsh creeks were a significant find, since most birds noted on Core Counts were found on freshwater habitats not covered by this survey. For the size of the site,

the peak of only 16 Grey Herons was low, though up to eight Little Egrets, a Great White Egret and a Bittern were recorded. Cormorants, peaking at 60, were widespread.

Almost all of the Knot occurred on Bob Hall's Sands north of Wells, with smaller numbers in Blakeney Harbour and west of Holme. Bar-tailed Godwits were widespread along the coast, with concentrations at the mouth of Wells channel, on Stiffkey flats and in Blakeney Harbour. Grey Plovers were also widespread, although, as with many other waders, they seemed to avoid Holkham Bay. Whether this is due to the habitat or human disturbance (the area is popular with the public, at least in summer) is uncertain. Relatively few Avocets were noted at low tide, most in Blakeney Harbour. Black-tailed Godwits were similarly few in number, found mostly northeast of Brancaster, in Blakeney Harbour and at Salthouse. Sanderling were spread fairly evenly along all seaward sections and Redshank throughout the whole site, except Holkham Bay and the outer areas at Wells and Stiffkey. Low tide counts of Redshank far exceeded Core Counts and, if maintained over a five year period, suggest not only that the site supports internationally important numbers, but that it may be the sixth most important site in the UK.

Both Turnstones and Ringed Plovers were widespread but favoured Brancaster Harbour and inner Blakeney Harbour. Golden Plover were widespread with particular concentrations on Cockle Bight (north of Brancaster Staithe) and on the inner parts of Blakeney Harbour; the latter was also favoured by Lapwings which were widespread on saltmarsh areas. Oystercatchers and Dunlin were widespread, favouring outer parts of the site, but Curlew favoured areas closer to land, particularly Blakeney Harbour and north of Burnham Deepdale. There was a very high peak of 611 Snipe in January; the size of this saltmarsh population was previously unknown with most birds on Core Counts found on grazing marshes. Less numerous wader species included Spotted Redshank, Greenshank, Purple Sandpiper, Jack Snipe, Woodcock, Whimbrel, Ruff and the long-staying Black-winged Stilt.

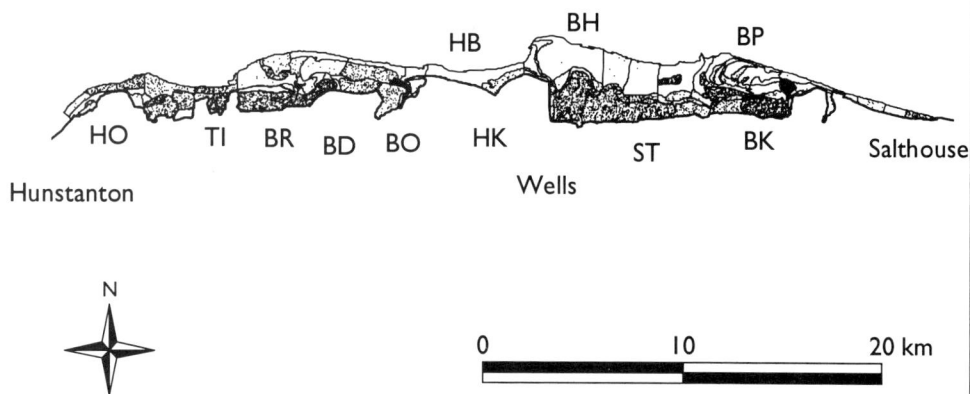

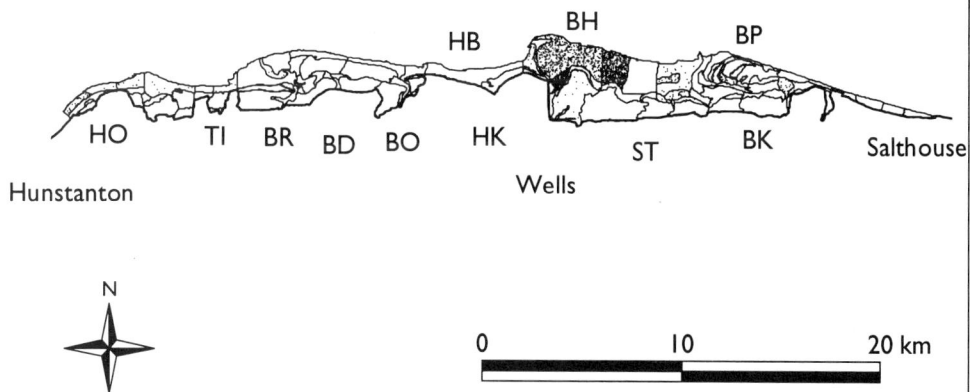

Figure 78. WeBS Low Tide Counts of Brent Goose and Knot on the North Norfolk Coast, winter 1997-98. (HO=Holme, TI=Titchwell, BR=Brancaster, BD=Burnham Deepdale, BO=Burnham Overy Staithe, HK=Holkham, HB=Holkham Bay, BH=Bob Hall's Sands, ST=Stiffkey, BK=Blakeney, BP=Blakeney Point. A total of five small sections were not counted.)

NORTH-WEST SOLENT

Hampshire

Internationally important: None
Nationally important: Dark-bellied Brent Goose, Black-tailed Godwit

Site description

The area known collectively as the North-west Solent includes all of the intertidal parts of the north shore of the Solent from the shingle of Hurst Spit in the west to the shoreline south of Sowley Pond in the east. To the east of here there is only a relatively narrow beach as far as Needs Oar Point and the Beaulieu Estuary (which was described in *Wildfowl and Wader Counts 1996-97*). The site is not far from the Isle of Wight; the birds at Hurst Spit are closer to the Yar Estuary across the Solent than they are to the birds at Pitts Deep. For the purposes of the WeBS Low Tide Counts, no data were received for the brackish lagoons and marshes at Pennington Marshes and Keyhaven Marshes. Although large areas of saltmarsh remain, much of the introduced *Spartina anglica* growth has now died back which, followed by erosion, has increased the area of intertidal flats. There are also areas of *Enteromorpha* which are a favoured food of the Brent Geese. The main threats to the area are from recreational disturbance such as sailing, shooting and walking, and from land-claim for marinas (Prater 1981, Pritchard *et al.* 1992, Davidson 1996c).

Bird distribution

This area of the Solent supports an average of approximately 15,000 wintering waterfowl. Figure 79 depicts the low tide distribution of the nationally important numbers of Black-tailed Godwits in the area. The majority of the godwits were found on the marshes to the south of Keyhaven, although small numbers were also present along the inner parts of the Lymington River and at Oxey Lake. The peak count of 378 Black-tailed Godwits at low tide in December 1997 was far higher than any winter Core Count of this species during the last five years. Bar-tailed Godwits were rather few in number and restricted to the Oxey Lake area. Dunlin, the most numerous wader at low tide with a peak of 5,645 recorded in January, were widespread on muddier parts of the site and also favoured Oxey Lake as well as the area to the south of Keyhaven. The marshes to the south of Keyhaven were the only location where roosting Golden Plover were located and Lapwing were also concentrated here, as well as on the inner part of the Lymington River. Redshank, Curlew, Oystercatcher and Grey Plover were all widespread although Curlew were somewhat more concentrated around the inner Lymington River and there was a concentration of Grey Plovers in Keyhaven Harbour. The few Knot present were to the east of Pitts Deep and most of the Turnstone were recorded at Hurst and south of Keyhaven. Ringed Plover were widespread in small numbers and there were also up to ten Greenshank, two Green Sandpipers and an Avocet recorded.

Numbers of Brent Geese recorded at low tide peaked at 1,852 in February. The species was very widespread but most concentrated to the south of Keyhaven (Figure 79). Shelduck and Wigeon were also widespread but Teal and Pintail were more restricted to the marshes to the south of Keyhaven and the lower Lymington River and marshes to the east of here. Mallard were found mostly along the Lymington River. Most Red-breasted Mergansers and the few Goldeneye recorded were around Hurst Spit. Most of the Cormorants were found to the east of Pitts Deep. Only a single Grey Heron was recorded, in December, compared to widely scattered records of up to five Little Egrets per month. Little Grebes showed a preference for Keyhaven Harbour but there were relatively few Great Crested Grebes noted. Other species of note included single Slavonian Grebe and Long-tailed Duck.

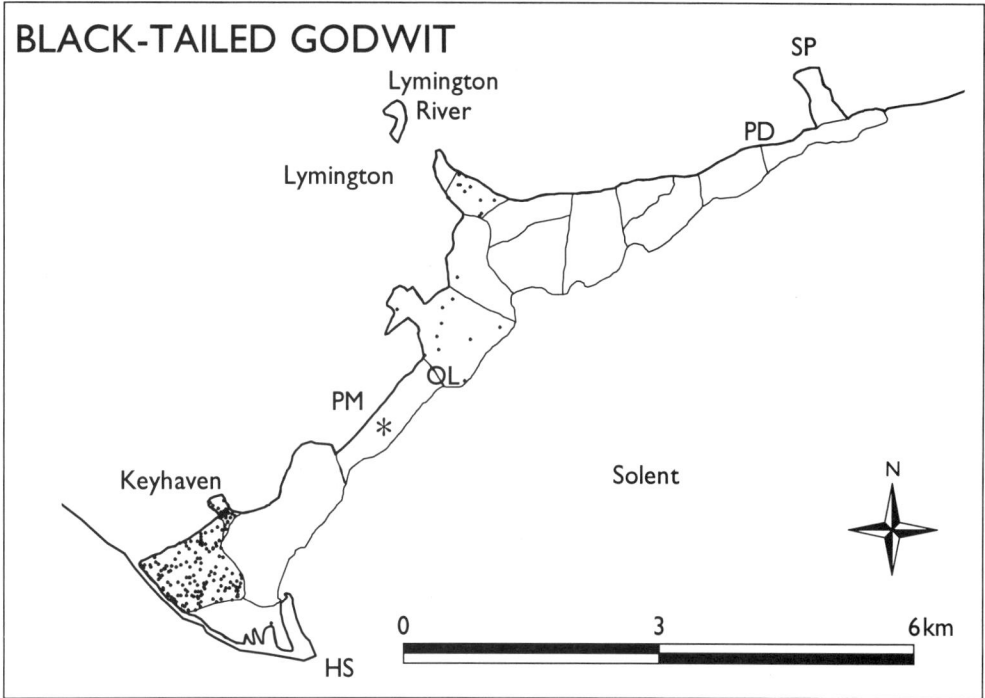

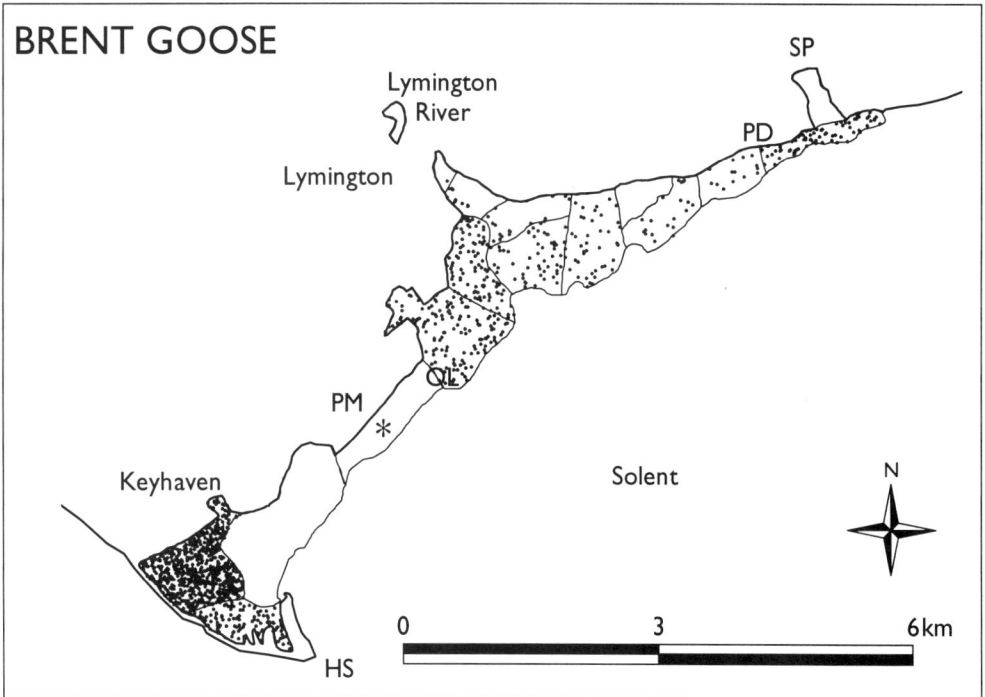

Figure 79. WeBS Low Tide Counts of Black-tailed Godwit and Brent Goose at the North-west Solent, winter 1997-98. (HS=Hurst Spit, PM=Pennington Marshes, PD=Pitts Deep, SP=Sowley Pond, *=not counted)

PAGHAM HARBOUR
West Sussex

Internationally important: Pintail
Nationally important: Cormorant, Dark-bellied Brent Goose, Teal, Grey Plover, Black-tailed Godwit

Site description

Pagham Harbour is a relatively small estuary located just east of Selsey Bill in Sussex. A central area of mudflats and saltmarsh is flanked by brackish marsh and damp pastures. The outlet to the sea is a narrow channel flowing through a shingle beach. There is a brackish lagoon at Pagham and a small pool at Sidlesham Ferry. The area was once claimed as agricultural land but was flooded again early in the 20th century. The harbour is now a designated SPA and Ramsar site. Only a limited amount of sailing takes place in the harbour and fishing is strictly regulated. There are no pressing conservation concerns (Pritchard *et al.* 1992, Buck 1997a, Davidson 1997a).

Bird distribution

Pagham Harbour supports, for its small size, a high number and variety of wintering waterfowl, with roughly 17,000 birds present. The harbour is internationally important for Pintail, with the peak Core Count for winter 1997-98 being 1,087 birds. During the 1997-98 WeBS Low Tide Counts, a peak of 391 was noted (in November) and thus it would appear that a considerable proportion of the wintering Pintail was missed by the Low Tide Counts. This is presumably because, as Figure 80 shows, Pintail favour the inner parts of the harbour where it is relatively easy for them to remain concealed. Another dabbling duck, Teal, which winters on the harbour in nationally important numbers, was similarly counted in fewer numbers at low tide (peak of 525 in December) than at high tide (peak of 969 also in December). Teal were widespread at low tide, especially along the south-west shores. Wigeon were concentrated around the middle parts of the harbour, particularly around the lower parts of White's Creek. Mallard were widespread with highest numbers on Pagham Lagoon and small numbers of Shoveler and Gadwall were also widespread. Brent Geese were recorded in relatively low numbers at low tide, with a peak of only 538 recorded, compared to a peak winter Core Count of 1,071. As was noted in *Wildfowl and Wader Counts 1996-97*, the fields to the north of Pagham Wall were favoured at low tide, with

a few birds scattered about the harbour. Small numbers of Shelduck were fairly widespread in the harbour. Cormorants were mostly found in the outer half of the harbour and both Little and Great Crested Grebes favoured Pagham Lagoon. Seven Slavonian Grebes were recorded in the outer parts of the harbour in November. Up to four Little Egrets and seven Grey Herons were also noted, along with small numbers of Mute Swan, Canada Goose, Red-breasted Merganser and Goldeneye.

Grey Plovers were widespread but obviously favour the southern half of the site. Numbers recorded at low tide were much lower than on the Core Counts, with a peak of only 381 at low tide compared with a Core Count peak of almost 2,500 during the same winter. Some birds were probably moving out of the estuary to feed at low tide but the discrepancy seems large and it may have been that the Core Count peak was due to a relatively short-lived influx into the site which was not recorded by the Low Tide Counts. Black-tailed Godwit, the other nationally important wader wintering at Pagham, was mostly found immediately south of Pagham Wall. Unlike the previous two winters, during the 1997-98 winter the peak Low Tide Count of 124 birds was greater than the peak Core Count of 46. By far the most numerous species of wader in the harbour was Dunlin. This species was widespread but the highest concentrations were along the south-western parts of the site and on the outer shingle bar (where all of the Knot were found); many of the central parts of the harbour were avoided. The outer sections of the harbour were also favoured by Oystercatcher, Ringed Plover, Turnstone and the few Bar-tailed Godwits which were present. Curlew and Redshank were both widespread, although the latter species showed a slight bias towards the inner parts of the harbour. Golden Plover were found in two main concentrations, just to the south of Pagham Wall and to the east of the lower part of White's Creek; Lapwing were a little more widespread. Counts of Avocets at low tide peaked this winter at 14 in January, with the birds favouring the south-west side of the harbour. An high February count of 13 Spotted Redshank was achieved and up to 22 Snipe were also counted, but only a single Sanderling was recorded at low tide.

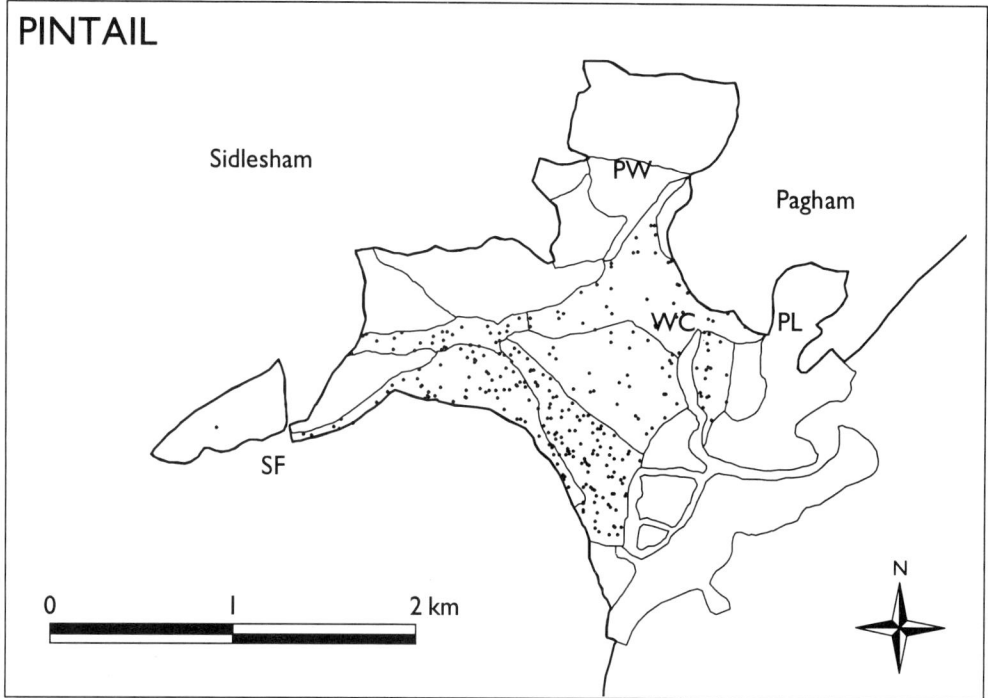

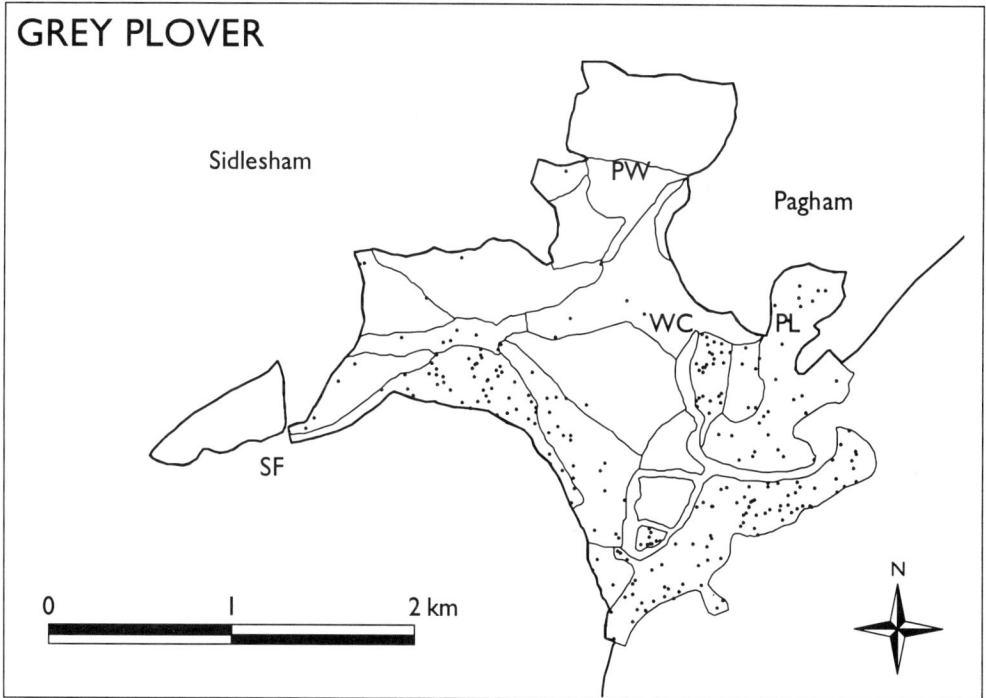

Figure 80. WeBS Low Tide Counts of Pintail and Grey Plover at Pagham Harbour, winter 1997-98. (SF=Sidlesham Ferry, PW=Pagham Wall, WC=White's Creek, PL=Pagham Lake)

PORTSMOUTH HARBOUR
Hampshire

Internationally important: None
Nationally important: Dark-bellied Brent Goose, Black-tailed Godwit

Site description

This large harbour on the Solent lies between Portsmouth to the east and Gosport and Fareham to the west. The main freshwater input is the fairly small Wallington River in the north-west which becomes Fareham Lake; the harbour receives relatively little freshwater. The connection to the sea, via the Solent, is only 200 metres wide at its narrowest point. There is relatively little in the way of saltmarsh but there are extensive areas of eelgrass and algae on the mudflats. The shores of the harbour are highly industrialised with extensive port and housing developments and major naval docks and installations. There have also been problems with land-claim for refuse disposal (Prater 1981, Pritchard *et al.* 1992, Davidson 1996c).

Bird distribution

The numbers of waterfowl wintering on Portsmouth Harbour (recent Core Counts of about 12,000 birds) are relatively low for an estuary of its size, presumably due in part to the built-up nature of the site. A peak of 2,505 Brent Geese was noted at low tide (corresponding very well to the Core Count peak) and the species was widespread around the site with lower densities noted on the outer mudflats and at the top of Fareham Lake (Figure 81). The other species present in nationally important numbers, Black-tailed Godwit, was noted in higher numbers at low tide (peak of 358 birds) than at high tide (peak of 100 birds in 1997-98), implying either movement of birds in and out of the estuary, or birds being hidden when roosting at high tide. Recent colour-ringing studies have shown that many Black-tailed Godwits feeding in the harbour at low tide actually roost at Farlington Marshes, in Langstone Harbour, at high tide (Solent Shorebird Study Group pers. comm.) The main concentration was at Paulsgrove Lake but significant numbers were also present around the upper parts of Fareham Lake (Figure 81).

The most numerous species noted at low tide was the Dunlin, with almost 9,000 counted.

The species was widespread but with the main concentrations on the wider flats, particularly at Whale Island. Both Curlew and Grey Plover were evenly distributed around the harbour. Oystercatcher and Redshank were also both widespread but Oystercatchers occurred in higher densities in the northern part of the harbour and Redshank were more concentrated around the edges of the site, avoiding the centre of the harbour. Lapwings were concentrated in two areas; at Fareham Lake (where small numbers of Golden Plovers were also present) and at Tipner Lake. Turnstone favoured the north-east and north-west corners of the harbour and Ringed Plovers were widespread in small numbers. Knot have decreased greatly on Portsmouth Harbour, with just 11 recorded in February; Prater (1981) reported that Portsmouth Harbour was the third most important site for the species in southern England and peak winter Core Counts of over 1,000 were noted during several winters, but numbers crashed during the 1987-88 winter and have never recovered. The most notable other wader was a single Whimbrel recorded in January and February; this is a very scarce species in the winter in the UK.

With the exception of Brent Geese, wildfowl were not present in large numbers. There were about 100 each of Shelduck, Wigeon and Mallard. Shelduck were widespread, Mallard more confined to the edges of the harbour and the Wigeon were very much confined to Cams Bay, where most of the Teal were also found. Red-breasted Mergansers were relatively common, however, with up to 79 noted at low tide. Both this species and Goldeneye were widespread in the harbour. Mute Swans were also widespread. Of the other species present, Cormorants and Little Grebes were quite common but Great Crested Grebes less so. Little Egrets were roughly as common as Grey Herons and other species noted included Black-throated Diver and Slavonian Grebe. Two Mediterranean Gulls were noted in January, in addition to the five common gull species.

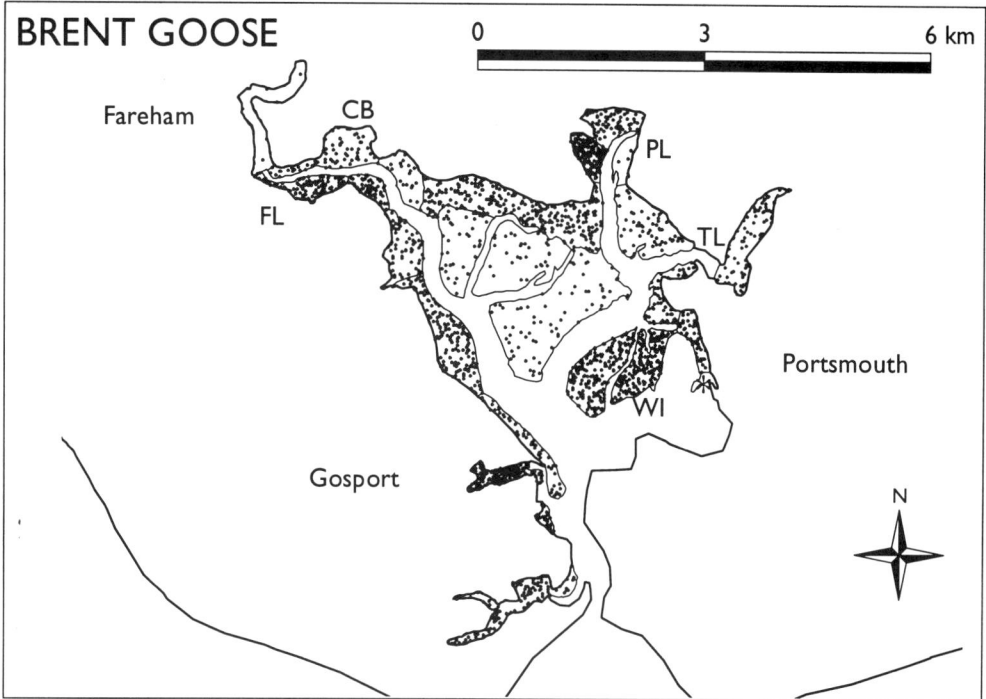

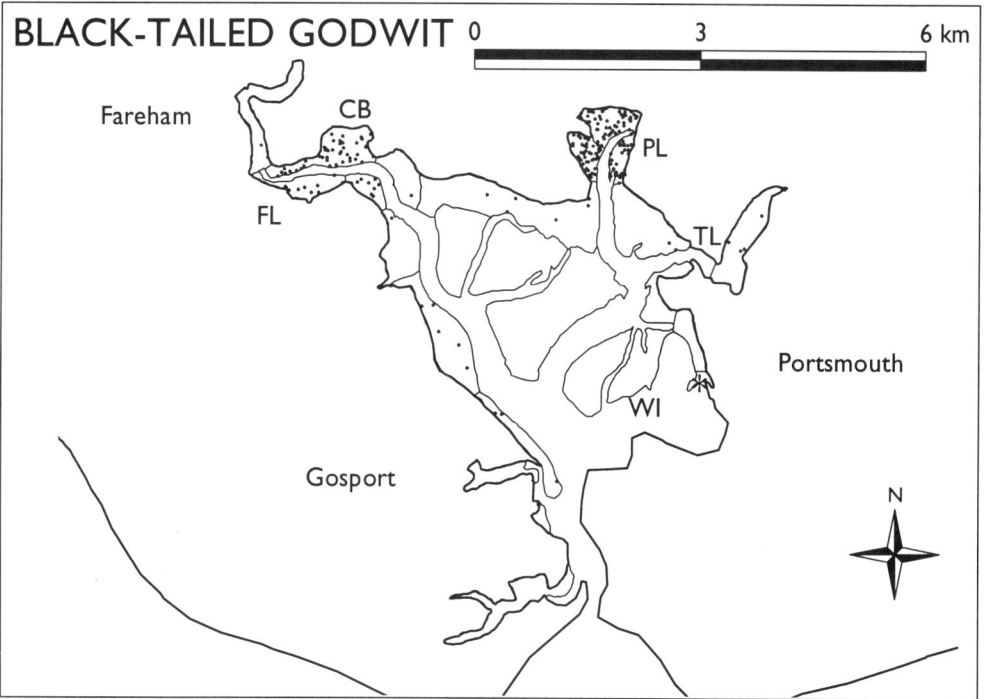

Figure 81. WeBS Low Tide Counts of Brent Goose and Black-tailed Godwit at Portsmouth Harbour, winter 1997-98. (FL=Fareham Lake, CB=Cams Bay, PL=Paulsgrove Lake, TL=Tipner Lake, WI=Whale Island, *=not counted)

RIBBLE ESTUARY
Lancashire / Merseyside

Internationally important: Bewick's Swan, Whooper Swan, Pink-footed Goose, Shelduck, Wigeon, Teal, Pintail, Oystercatcher, Grey Plover, Lapwing, Knot, Sanderling, Dunlin, Black-tailed Godwit, Bar-tailed Godwit, Redshank

Nationally important: Cormorant, Golden Plover, Curlew

Site description

The Ribble Estuary comprises a long, relatively narrow inner estuary, flanked by very large areas of saltmarsh, and a huge area of intertidal flats as the outer regions of the estuary. These flats run south as a wide, sandy shore, past Southport and merge into the area treated above as the Alt Estuary; on the northern side the area extends to the southern outskirts of Blackpool. Current issues concerning the Ribble include the level of sand winning and the use of vehicles on the flats by fisherman. There have also been recent proposals for a landfill site at Hesketh Out Marsh and for a hovercraft service across the estuary, whilst more general disturbance comes from holiday-makers, wildfowlers and the presence of the Warton aerodrome on the north shore. However, disturbance levels are generally low and development pressures are currently light (Prater 1981, Pritchard *et al.* 1992, Davidson 1996a, R.Lambert pers. comm.)

Bird distribution

With the exception of the Wash (which is three times larger), no other UK site holds as many wintering waterfowl as the Ribble. The site also holds internationally important numbers of more species than any other in the UK. The challenge of conducting Low Tide Counts on such a major site was met fairly successfully with the principal exception that the saltmarsh areas were not counted. The intertidal flats were mostly covered using a wide-tyred motorbike. Count sections were much larger than for most other estuaries and consequently the distributional information gathered is of only the broadest nature. As with all low tide counts, but particularly on very large sites, there is a danger that the low tide counts under-represent the importance of some parts of the site which hold very large numbers of feeding waterfowl during the mid-tide period. This was particularly the case for Marshside Sands (R. Lambert pers. comm.) Not surprisingly, the totals of birds counted at low tide were less than those at the high tide roosts, but with a mean Low Tide Count of almost 100,000 birds this was a major advance in our knowledge of this site.

The Ribble is of key significance for Wigeon, with a peak winter Core Count in 1997-98 of 66,197. Up to 46,000 were noted during Low Tide Counts. The highest concentrations were found on Banks Sands but Foulnaze and Southport Sands were also favoured. The main roosting flock, depicted across Banks Sands in Figure 82, was in reality mostly found along the channel separating Banks Sands and Marshside Sands (R. Lambert pers. comm.) Numbers of Teal, peaking at about 1,400, were far fewer than noted on Core Counts, presumably as a result of birds favouring the inner estuary and being missed in the saltmarsh. Most Pintail (peaking at 955 in December) were found on Salter's Bank, with smaller numbers on Banks Sands and Marshside Sands; most are found in the immediate vicinity of the river channel at low tide (R. Lambert pers. comm.) Shelduck occurred at highest densities in the inner estuary and on Banks Sands, but were not found on the sandy shore from Southport southwards. The other three species present in internationally important numbers (the two winter swans and Pink-footed Goose) do no make great use of intertidal areas at low tide and so numbers recorded by the survey were fairly low, but 174 Bewick's Swans on the inner Ribble in February were worthy of note. The nationally important numbers of Cormorants were distributed fairly evenly across the site.

The Ribble Estuary is the most important wintering site in the UK for Bar-tailed Godwits and a peak of over 10,000 were counted at low tide in December. The highest densities occurred on Foulnaze, with large numbers also on Salter's Bank and on Southport and Birkdale Sands. Numbers of Black-tailed Godwits recorded at low tide, however, were much lower than those known to be present from WeBS Core Counts, and occurred mostly on Ainsdale Sands and in the inner estuary. Other important wader species which favoured the inner estuary were Redshank, Curlew, Lapwing and Golden Plover, whereas the distributions of Dunlin, Grey Plover, Knot, Oystercatcher and Sanderling were more concentrated on the outer parts of the site, with Foulnaze particularly important for Dunlin, Knot and Oystercatcher. The Ribble is the only internationally important UK site for wintering Sanderling; these birds favoured Salter's Bank and the Ainsdale/Birkdale Sands stretches. Of the less numerous waterfowl species, a maximum count of 70 Snipe in November was notable.

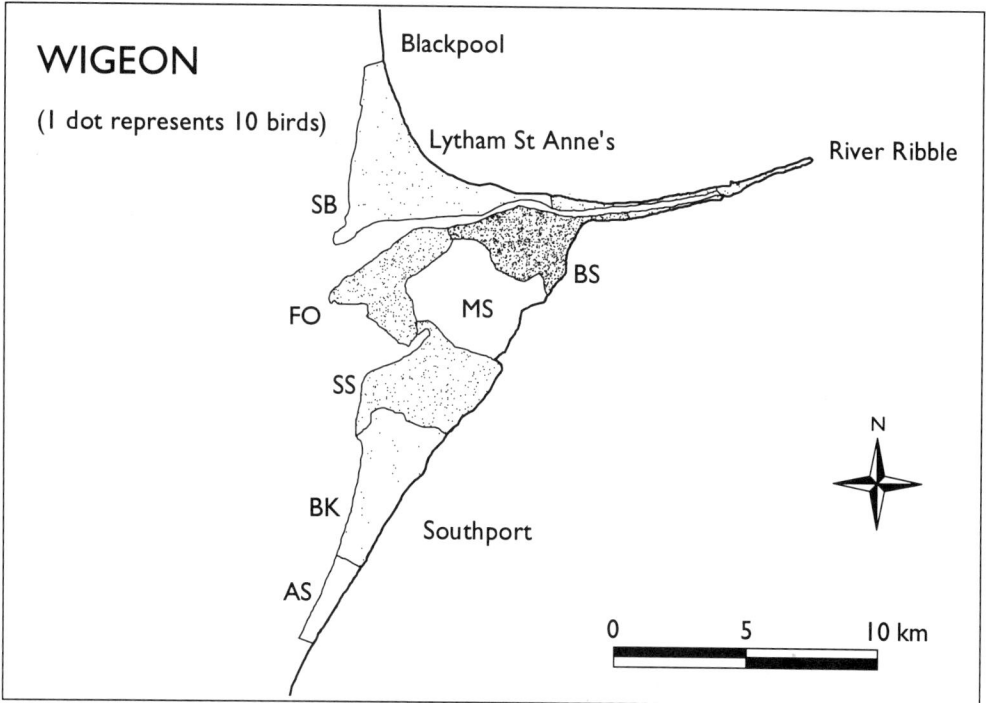

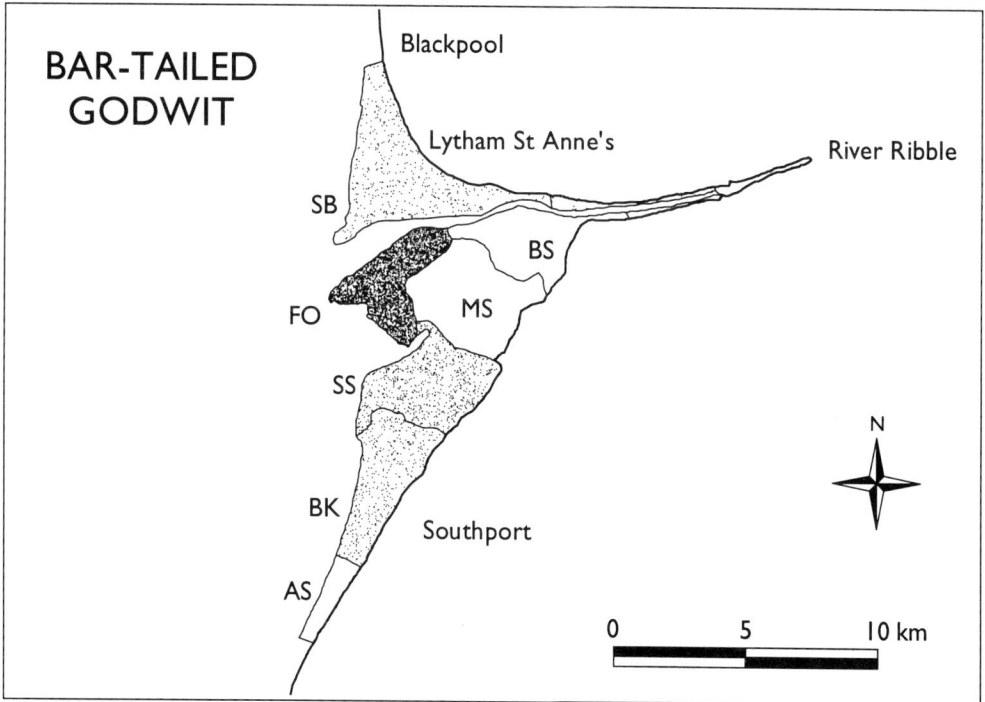

Figure 82. WeBS Low Tide Counts of Wigeon and Bar-tailed Godwit on the Ribble Estuary, winter 1997-98. (SB=Salter's Bank, FO=Foulnaze, SS=Southport Sands, BK=Birkdale Sands, AS=Ainsdale Sands, MS=Marshside Sands, BS=Banks Sands)

SOUTHAMPTON WATER
Hampshire

Internationally important: None
Nationally important: Great Crested Grebe, Dark-bellied Brent Goose, Teal, Black-tailed Godwit

Site description

Southampton Water is part of the Solent complex and lies between the city of Southampton and the New Forest. The three principal rivers entering Southampton Water are the Test, Itchen and Hamble. There are extensive areas of mud on both shores of the estuary, with a large area of *Spartina* saltmarsh along the southern shore. Southampton Water is one of the most heavily developed estuaries in Britain and, as well as being adjacent to a large city, also has important docks, an oil refinery and a power station along its shores. The area is also extremely heavily used by sailing enthusiasts. One of the most significant current development issues is at Dibden Bay, which is actually no longer a bay since dredgings were pumped onto the land here. This area has now dried out and there are plans for further development, which may result in the loss of the remaining intertidal mud (Buck 1997a, Pritchard *et al.* 1992, Davidson 1996c).

Bird distribution

During the 1997-98 winter, Teal were quite widespread but with major concentrations at Cadland Creek and at Bury and Eling Marshes, with smaller numbers at Fawley and on the lower Hamble, an almost exact replica of the distribution noted in 1996-97. Numbers of Wigeon were similar to those of Teal (peaking at about 1,600 each) and occurred in broadly similar areas except for an absence from the Hamble and a separate concentration immediately east of Hythe. All of the Pintail were at Cadland Creek. Mallard were concentrated along the Itchen and on the shore by Titchfield Haven. Small numbers of Gadwall were located along the Fawley to Calshot shore. Brent Geese, peaking at 1,165 in November, were widespread with a general preference for the outer estuary although there were high numbers at Dibden Bay also. The highest densities were found to the north of Calshot and along the Hook to Hill Head stretch. Shelduck occurred principally at Eling Marsh and along the Fawley to Calshot shore. Numbers of other wildfowl species were unremarkable except for a maximum of 104 Mute Swans, concentrated along the Itchen in Southampton.

Dunlin were widespread at low tide but with their highest concentrations at Dibden Bay, on the opposite shore at Weston and at the Hamble Spit. This is the most numerous wader wintering at Southampton Water with a peak of over 7,000 birds at low tide in February. The peak 1997-98 winter Core Count was of only 1,972 and the highest Core Count from the last five winters was of 4,814 in December 1995, whereas similarly high numbers of Dunlin have been recorded at low tide during the previous three winters also. It appears that either the Core Counts are missing some of the Dunlin at the site or that there is immigration into the site at low tide, perhaps from the Beaulieu Estuary. Black-tailed Godwits peaked at 132 during the 1997-98 Low Tide Counts, although comparison with Core Counts is not possible due to a lack of recent core data from the preferred site of Titchfield Haven. The species was most concentrated in the Eling/Bury/Cracknore area with smaller numbers on the lower Hamble and to the east of Fawley. Very few Bar-tailed Godwits were recorded. Curlew were widespread with the highest concentrations at Dibden Bay where there were also held high densities of Grey Plovers and Oystercatchers. Grey Plovers also favoured the east shore at Weston. Lapwing were widespread along the western shore but with concentrations at Eling/Bury Marshes as well as on the lower Hamble. Golden Plovers were virtually all found on the lower Hamble. Redshank and Turnstone were both very widespread and Ringed Plover had concentrations on the Itchen and at Hythe. Only single Knot, Sanderling and Snipe were recorded but there were up to five Greenshank each month and a Common Sandpiper was present all winter.

Among the other species recorded, peak low tide counts of 33 Little Grebes (concentrated along the Itchen), 84 Great Crested Grebes (with the highest concentrations off Dibden and east of Hythe) and 125 Cormorants (quite evenly spread) were notable and up to three Little Egrets were present in the Fawley to Calshot area. All three rarer grebes were recorded as were both Red-throated and Great Northern Divers. Eight species of gull were noted, with maxima of almost 10,000 Black-headed Gulls, six Yellow-legged Gulls and two Mediterranean Gulls.

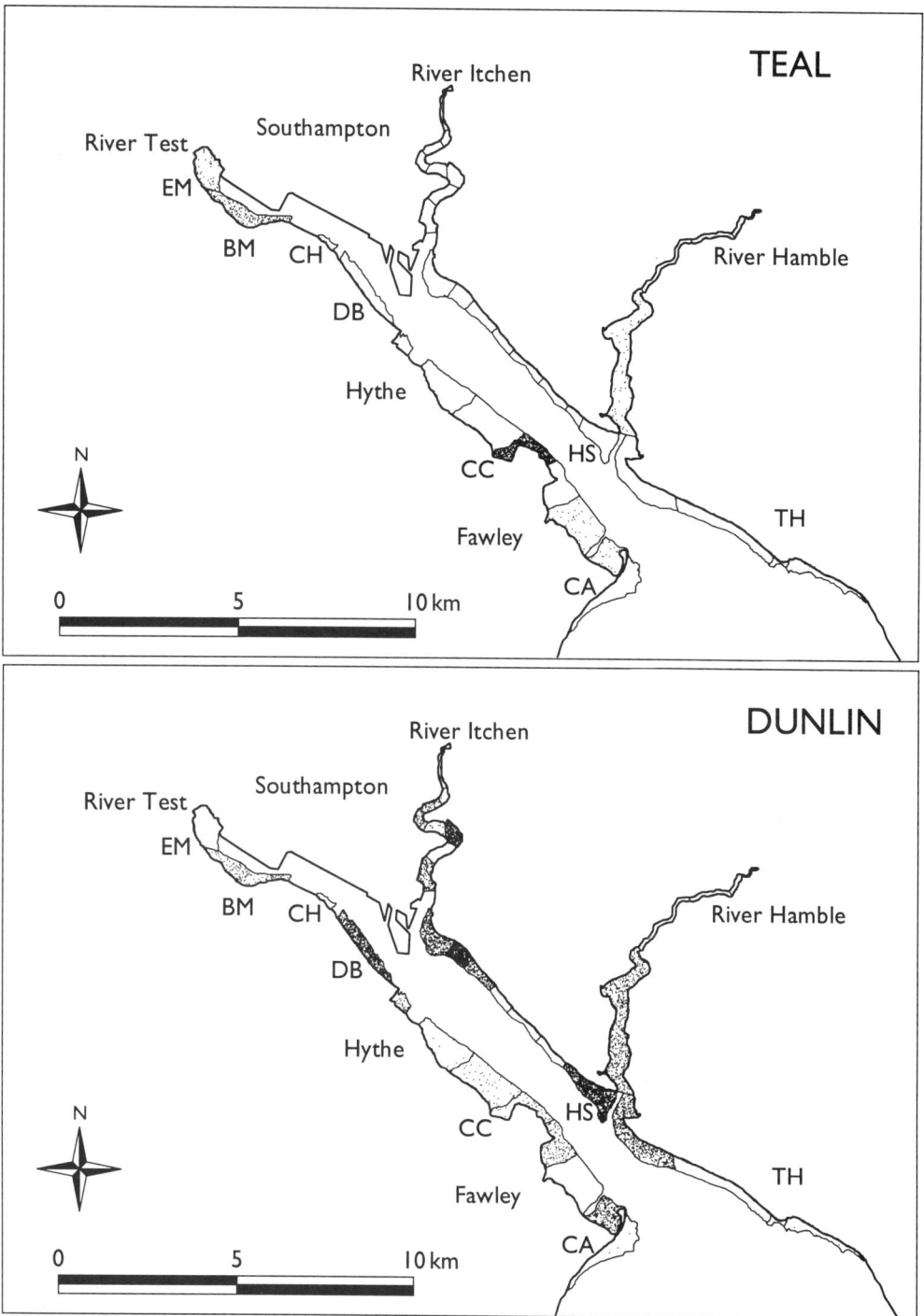

Figure 83. WeBS Low Tide Counts of Teal and Dunlin on Southampton Water, winter 1997-98. (EM=Eling Marsh, BM=Bury Marsh, CH=Cracknore Hard, DB=Dibden Bay, CC=Cadland Creek, CA=Calshot, TH=Titchfield Haven, HS=Hamble Spit)

STRANGFORD LOUGH
Co. Down

Internationally important: Light-bellied Brent Goose, Knot, Bar-tailed Godwit, Redshank
Nationally important: Little Grebe, Great Crested Grebe, Mute Swan, Shelduck, Wigeon, Teal, Mallard, Pintail, Shoveler, Goldeneye, Red-breasted Merganser, Coot, Oystercatcher, Ringed Plover, Golden Plover, Grey Plover, Lapwing, Dunlin, Black-tailed Godwit, Curlew, Turnstone

Site description

Strangford Lough is a large, almost land-locked sea lough situated to the southeast of Belfast. It encompasses extensive tidal flats at the northern end, smaller bays and creeks throughout and numerous small drumlin islands, particularly along the western shore. Principal conservation issues concern large scale recreational use, human population growth around the lough (with resulting increases in eutrophication of the water) and increasing intensification of agriculture. A recent proposal for a tidal barrage at the mouth of the lough was rejected. Much of the lough is managed by the National Trust, which also manages the wildfowling which takes place. The Wildfowl & Wetlands Trust also has a reserve at Castle Espie near Comber (Pritchard *et al.* 1992, Buck & Donaghy 1996, Buck 1997b).

Bird distribution

Bar-tailed Godwits were concentrated in three principal areas in 1997-98, all in the north of the lough. Numbers recorded at low tide were only about two-thirds of the winter Core Count peak. Redshank were much more evenly distributed around the lough although they too were recorded in lower numbers at low tide than at high tide. Knot were concentrated in the northern part of the lough also, with the Ogilby Island area of the northern flats particularly favoured as well as areas around Reagh and Mahee Islands and Greyabbey Bay. Black-tailed Godwits were mostly found on the Newtownards Flats (although higher up the shore than the Bar-tailed Godwits) and in Ardmillan Bay. The peak count of 436 in December was far in excess of the peak winter Core Count. Grey Plover were restricted to the northern mudflats and the Castle Espie area. Golden Plover (which peaked at 12,000 birds in February) also made use of this area as well as the Mahee Island and Greyabbey Bay areas. Lapwing also reached their highest densities in the Greyabbey area but were widespread around the lough. Oystercatcher and Curlew occurred very widely around the site but

were found in highest densities in the north of the lough. The highest densities of Dunlin, which were also widespread, were to the south of Gores Island and in the Ballymoran/Quarterland area. Ringed Plover, Turnstone and smaller numbers of Purple Sandpipers favoured the outer channel in the south of the site but Turnstone were also found east of Castle Espie and Ringed Plovers also occurred south of Greyabbey and around Kircubbin. Of the other wader species, up to 40 Ruff, 24 Greenshank, 11 Snipe and five Common Sandpipers were also recorded.

Light-bellied Brent Geese arrive at the lough in early winter but then disperse to other sites further south. This was reflected in the Low Tide Counts, with 7,813 birds in November decreasing to 2,091 by February. Although widespread, the highest concentrations occurred along the north-east shore from Greyabbey to Newtownards. Shelduck peaked at over 4,000 birds; recorded numbers are typically higher at low tide than high tide at Strangford Lough. Significantly, using Low Tide Counts suggests Strangford Lough qualifies as internationally important for Shelduck with a five-year peak mean of 3,721 birds. The species was widespread but mostly found on the northern mudflats. Wigeon were widespread but quite localised, particularly at the Comber Estuary and to the north of Chapel Island. Teal were also widespread but Mallard were more localised. Most of the Pintail were around the outer parts of the Newtownards mudflats and to the west of Patterson's Hill. Shoveler favoured the Comber Estuary and Mahee Island. Goldeneye showed a preference for the Mahee/Ardmillan area but Red-breasted Mergansers were more widely scattered. Little Grebes were widespread but Great Crested Grebes were more concentrated in the Ringdufferin area. All three divers were recorded as well as a Black-necked Grebe and high count of 11 Slavonian Grebes. Most of the Mute Swans were in Ardmillan Bay and along the north-east shore. Up to 166 Whooper Swans and six Bewick's Swans were present in the Comber Estuary area. Cormorants peaked at 149 in December with the majority in the outer channel from Portaferry/Strangford southwards.

BAR-TAILED GODWIT

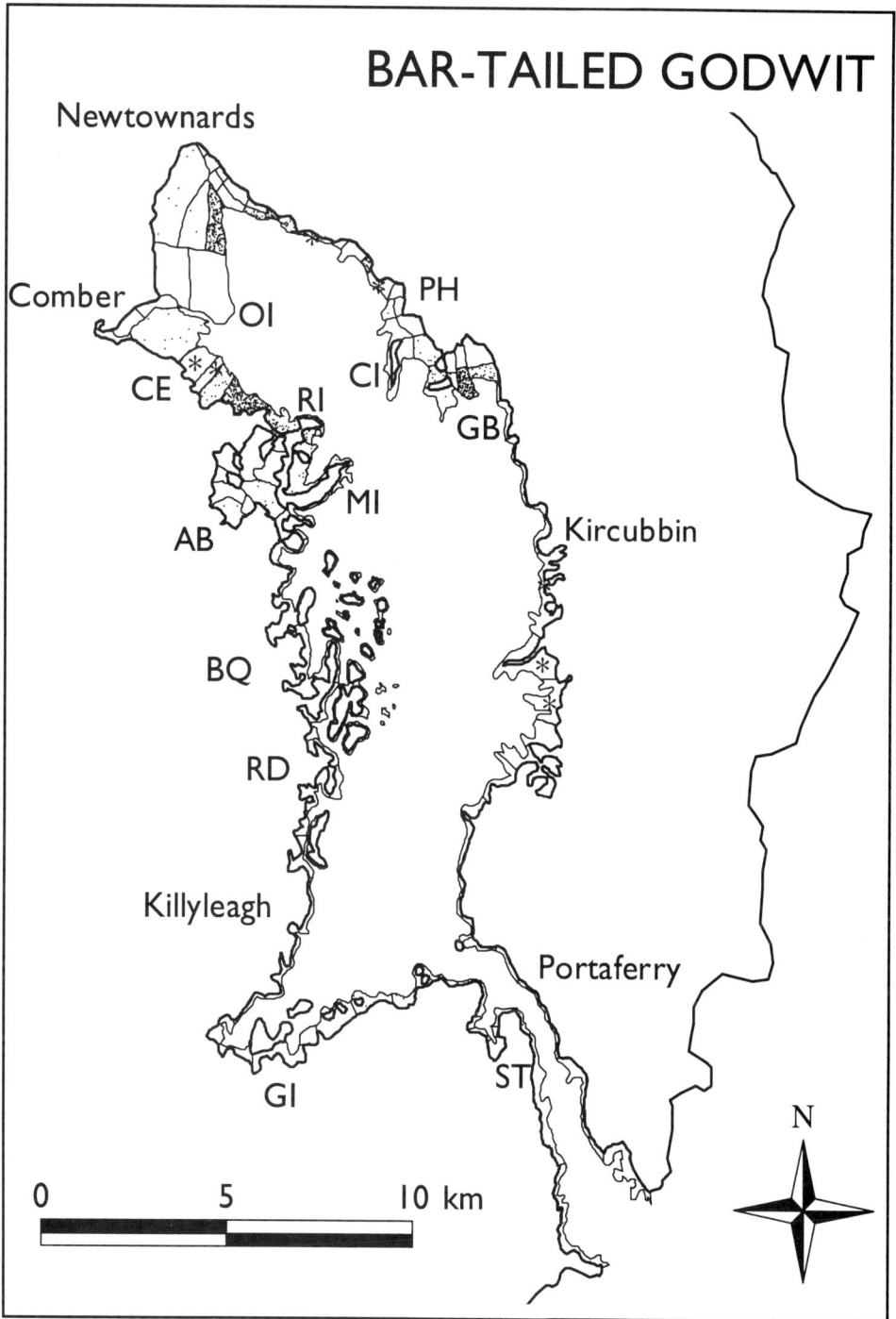

Newtownards

Comber

OI

PH

CE

CI

RI

GB

MI

Kircubbin

AB

BQ

RD

Killyleagh

Portaferry

GI

ST

N

0   5   10 km

Figure 84. WeBS Low Tide Counts of Bar-tailed Godwit on Strangford Lough, winter 1997-98. (ST=Strangford, GI=Gores Island, BQ=Ballymorran/Quarterland, AB=Ardmillan Bay, MI=Mahee Island, RI=Reagh Island, CE=Castle Espie, OI=Ogilby Island, PH=Patterson's Hill, CI=Chapel Island, GB=Greyabbey Bay, *=not counted)

TAMAR COMPLEX
Cornwall / Devon

Internationally important: None
Nationally important: Avocet, Black-tailed Godwit

Site description

The Tamar Complex is the name given to the group of river estuaries (with the exception of the Plym) which reach the sea through Plymouth Sound. The estuaries are the drowned river valleys of the Tamar, Lynher and Tavy, which collectively drain a large area of Devon and Cornwall. The wide area of intertidal flats to the south of Torpoint known as St. John's Lake is also included in the site. The east shore of the lower parts of the complex are adjacent to the city of Plymouth, which has extensive dockyards and naval bases. Apart from the towns of Torpoint and Saltash, most of the west side of the estuary, as well as the upper estuary, is rural in nature. Areas of saltmarsh occur throughout even to the upstream reaches, and stretches of rocky shore also occur unusually far inland. Apart from the Plymouth area, most potential pressures on estuarine waterfowl are concerned with recreational disturbance and there are several proposals for new marinas (Prater 1981, Pritchard *et al.* 1992, Davidson 1996b).

Bird distribution

The Tamar complex was only partially covered for the 1997-98 WeBS Low Tide Counts and this must be taken into account when considering the following details on waterfowl distribution. However, a more detailed study of the site was made by Simon Geary as part of his M.Sc. work (Geary 1997) and his work has been useful in interpreting the results of the current survey.

The complex supports an average of about 13,000 wintering waterfowl based on WeBS Core Counts. In 1997-98, Avocets were restricted to the eastern side of the Tamar between the Tavy and the Tamar Bridges at low tide. However, since the peak count (180) was much lower than the Core Count peak (595) in winter 1997-98, it seems that many birds were feeding in the unsurveyed sections, probably mostly on the Tamar upstream of Saltash (Geary 1997). Black-tailed Godwits showed a similar pattern of occurrence although the Low Tide and Core Count totals for this species corresponded much more closely; Geary found that the Kingsmill Lake and Millbrook areas were also important for

this species in the winter (although the Lynher was used by the species during autumn migration). Oystercatchers were widespread in small numbers as were Curlew although this latter species occurred in higher densities on the Lynher at Sheviock. Dunlin, though widespread, were present in very low numbers on the sections covered, with a peak of less than 100 (compared to a peak of over 3,000 during Core Counts over the same period). The majority of the missing birds will have been on St John's Lake (Geary 1997). Both Golden Plover and Lapwing were concentrated at Sheviock and on the Tavy, with Lapwing also present on Tamerton Lake. Geary also found many Lapwing and Golden Plover on the upper reaches of the Tamar. Redshank were widespread with concentrations on Tamerton Lake and at St Germans. Most other waders were present in small numbers although up to 20 Snipe and three Jack Snipe on the Tavy, as well as up to six Common Sandpipers there, were noteworthy.

The most numerous wildfowl species noted was Canada Goose, present on the Tavy and Lynher but, had the whole site been covered, Shelduck, Wigeon, Teal and Mallard would no doubt have been found to be more abundant. Wigeon were found on the Tamar and Lynher (note that the mid-Lynher, which was not covered, was found to be the most important area for Wigeon by Geary) but Mallard were more widespread with the highest concentration at Polbathic Creek. Mallard were found by Geary to be common also on the upper reaches of the Tamar. Shelduck were distributed quite evenly throughout (note that Geary found many on St John's Lake and mid-Lynher) but most of the Mute Swans were located at Saltash. Despite the fact that several hundred Teal winter on the Tamar, a peak of only two was recorded during the present survey; Geary found most birds on the upper reaches of the Tamar, Kingsmill Lake, the mid-Lynher and Millbrook Lake, none of which were covered during 1997-98. Of the other species present, a peak of 22 Little Egrets was only just exceeded by the peak of 23 Grey Herons. A Spoonbill which had been wintering on the Lynher for several winters was recorded in February. Two Mediterranean Gulls were noted along with the five more common species.

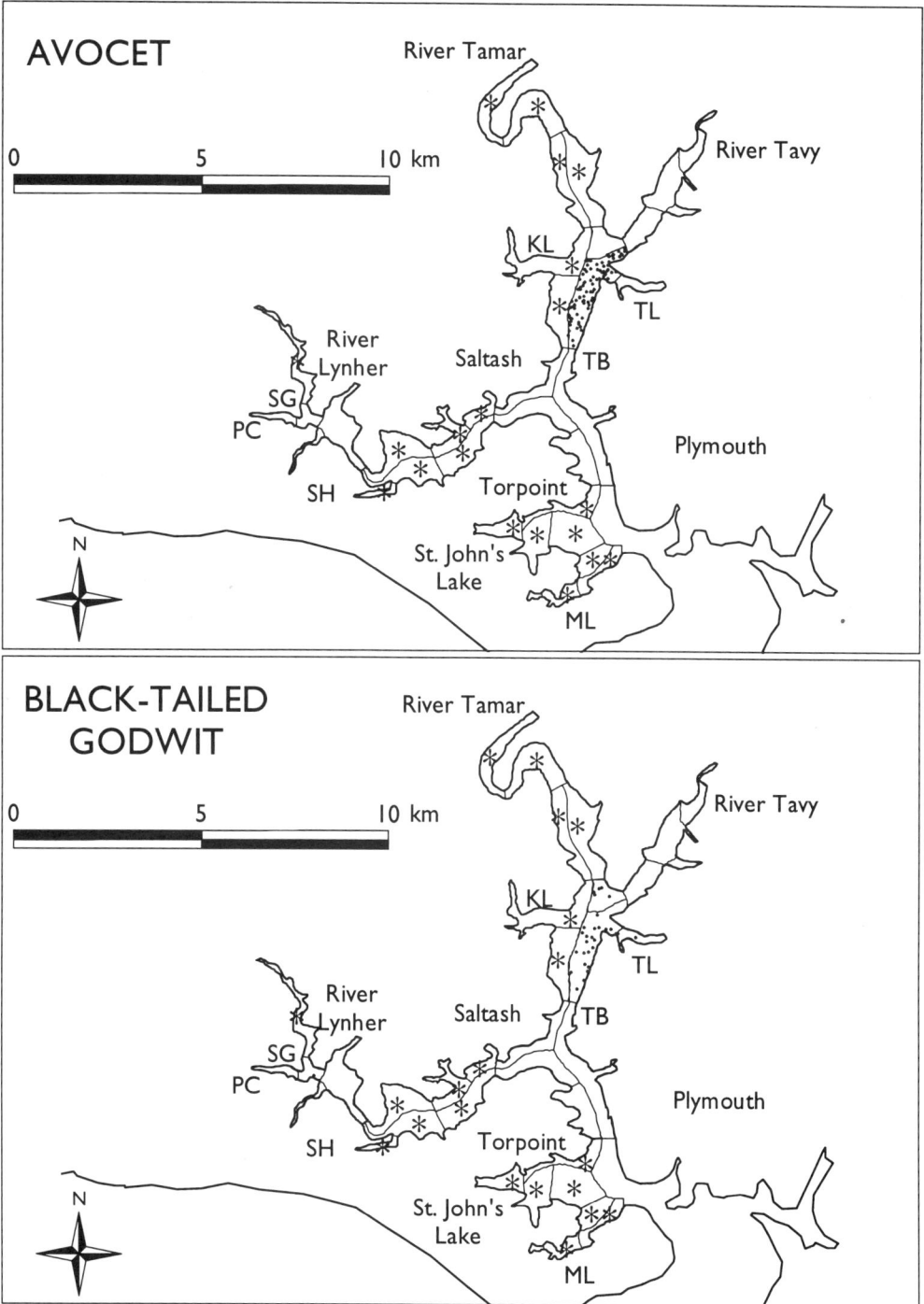

Figure 85. WeBS Low Tide Counts of Avocet and Black-tailed Godwit on the Tamar Complex, winter 1997-98. (KL=Kingsmill Lake, SG=St Germans, PC=Polbathic Creek, SH=Sheviock, ML=Millbrook Lake, TB=Tamar Bridge, TL=Tamerton Lake, *=not counted)

YTHAN ESTUARY
Aberdeenshire

Internationally important: Pink-footed Goose
Nationally important: Eider, Redshank

Site description

The Ythan is a relatively small estuary in north-east Scotland, about ten miles north of Aberdeen. Despite its small size, it is the largest estuary on the coast between Montrose Basin and the Moray Firth and as such is important in a local context. The estuary has a narrow shape and is about 10 km in length. The estuary is shielded from the sea by the nationally important sand dune system known as the Sands of Forvie. The inner estuary is muddy and the outer stretches more sandy, but there is relatively little in the way of saltmarsh. The main human influences on the estuary are through recreation, particularly wildfowling, and through eutrophication of the lower part of the estuary (Prater 1981, Pritchard *et al.* 1992, Davidson 1996d).

Bird distribution

The Ythan Estuary holds a mean winter peak of about 23,000 waterfowl, principally wildfowl. A large proportion of this total is, however, made up of the internationally important numbers of Pink-footed Geese which sometimes roost here, and sometimes at Slains Loch. Since they generally only use the estuary for nocturnal roosting, it was no surprise that no Pink-footed Geese were recorded on the Low Tide Counts.

The estuary is well known for its much-studied Eider flock which is concentrated at low tide around the mouth of the estuary, not reaching much further upstream than the north end of Newburgh (Figure 86). Up to 1,448 Eider were noted during the WeBS Low Tide Counts, approximately half of the corresponding Core Count peak. This is presumably due to birds moving out onto the open coast to feed as the tides goes out. Other wildfowl were noted in smaller numbers with a few hundred Wigeon the next most numerous species; Wigeon occurred widely although relatively few were found in the northern part of the estuary. Small numbers of Tufted Ducks were found with the Eider flock but

Goldeneyes and Red-breasted Mergansers occurred more widely. Shelduck favoured the middle stretches of the estuary around Waterside but Mute Swans were more common at the northern end of the site. Grey Herons were widely scattered throughout but Cormorants favoured the lower stretches. A small number of Mallards and a Little Grebe were also noted. Unfortunately, no King Eiders were noted during the counts at what is probably their most frequented British haunt.

Of the relatively small numbers of waders present, Redshank is the only species for which the site qualifies as nationally important. At low tide, much smaller numbers of this species were noted, with a peak of only 475 birds. Since the estuary is fairly small and does not have extensive areas of saltmarsh for birds to hide in, it seems possible that many birds may move out to the adjacent areas of open coast to feed at low tide. The species was found to be widespread although slightly more numerous at the northern end and around the middle stretches near Waterside (Figure 86).

Of the other waders, most of the Lapwings and all of the Golden Plovers favoured the northern end of the estuary for roosting. Curlews were numerous, especially in February when over 1,500 were present; the greatest concentrations were found around the centre of the estuary. Knot were found exclusively around the mouth of the estuary with the greatest concentration of Oystercatchers, although this species occurred widely. Bar-tailed Godwits and the few Grey Plovers and Turnstones preferred the mouth of the estuary. Dunlin occurred principally in two areas; at the northern end of the estuary and also roughly level with Newburgh. More unusual species included two Greenshanks and a Common Sandpiper, the latter being a rather northerly record for February. Small numbers of Ringed Plover and Snipe were also noted. Four species of gulls were present but no Lesser Black-backed Gulls were recorded.

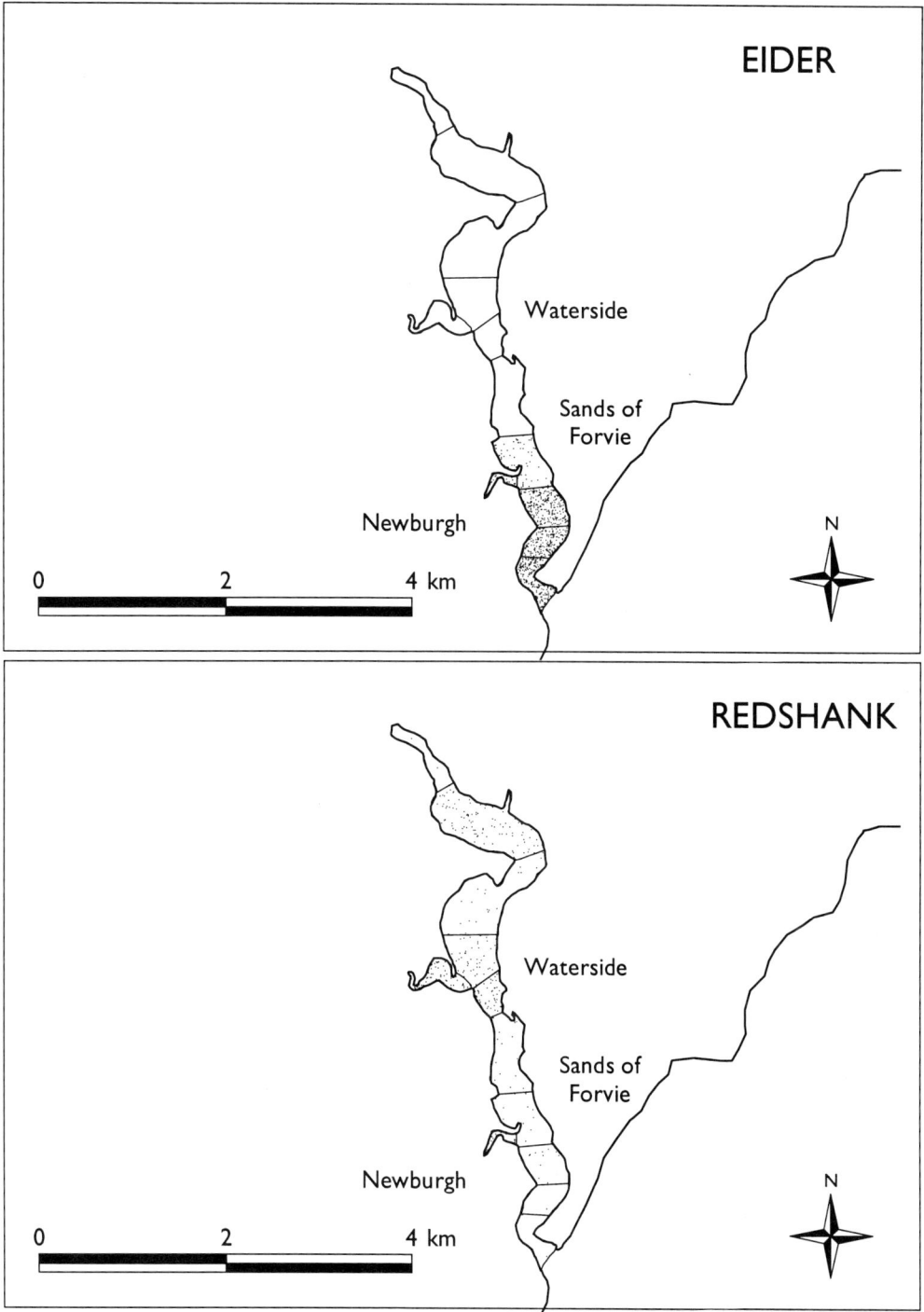

Figure 86. WeBS Low Tide Counts of Eider and Redshank on the Ythan Estuary, winter 1997-98.

REFERENCES

Atkinson, N.K., Davies, M. & Prater, A.J. 1978. The winter distribution of Purple Sandpipers in Britain. *Bird Study* 25: 223-228.

Atkinson-Willes, G.L. (Ed.) 1963. *Wildfowl in Great Britain*. Monographs of the Nature Conservancy Number Three, HMSO.

Bibby, C.J., Burgess, N.D. & Hill, D.A. 1992. *Bird Census Techniques*. Academic Press, London.

BOURC 1999. British Ornithologist's Union Records Committee: 25th Report (October 1998). *Ibis* 141: 175-180.

Bowes, A., Lack, P.C. & Fletcher, M.R. 1984. Wintering gulls in Britain, January 1983. *Bird Study* 31: 161-170.

Bowler, J.M., Butler, L., Liggett, C. & Rees, E.C. 1994. Bewick's and Whooper Swans *Cygnus columbianus bewickii* and *C. cygnus*: the 1993-94 season. *Wildfowl* 45: 269-275.

Buck, A.L. 1997a. *An inventory of UK estuaries. Volume 6. Southern England*. Joint Nature Conservation Committee, Peterborough.

Buck A.L. 1997b. Chapter 4.1 Estuaries. *In: Coasts and seas of the United Kingdom. Region 17 Northern Ireland*, ed. by J.H. Barne, C.F. Robson, S.S. Kaznowska, J.P. Doody, N.C. Davidson & A.L. Buck. Peterborough, Joint Nature Conservation Committee.

Buck, A.L. & Donaghy, A. 1996 *An inventory of UK estuaries. Volume 7. Northern Ireland*. Joint Nature Conservation Committee, Peterborough.

Carp, E. 1972. *Proceedings of the international conference n the conservation of wetlands and waterfowl. Ramsar, Irand, 30 January - 3 February 1971*. IWRB, Slimbridge.

Cayford, J.T. & Waters, R.J. 1996. Population estimates for waders Charadrii wintering in Great Britain, 1987/88-1991/92. *Biol. Conserv.* 77: 7-17.

Clausen, P. & Bustnes, J.O. 1998. Flyways of North Atlantic light-bellied brent geese *Branta bernicla hrota* reassessed by satellite telemetry. In: Mehlum, F., Black, J.M. & Madsen, J. (Eds.) Research on Arctic Geese. Proceedings of the Svalbard Goose Symposium, Oslo, Norway, 23-26 September 1997. *Norsk Polarinst. Skr.* 200: 235-249.

Colhoun, K. 1998. *I-WeBS Report 1996-97: results of the third winter of the Irish Wetland Bird Survey*. BirdWatch Ireland, Dublin.

Colhoun, K. Orr, J. & Delany, S. 1998. *Numbers and distribution of Light-bellied Brent Geese in Ireland: Results of the October 1997 and January 1998 All-Ireland Censuses*. Irish Brent Goose Study Group, Dublin, 9 pp.

Corse, C., Dean, T. & Williams, J. 1998. *Orkney Bird Report 1997*. Orkney Bird Club.

Cranswick, P.A., Kirby, J.S., Salmon, D.G., Atkinson-Willes, G.L., Pollitt, M.S. & Owen, M. 1997 A history of wildfowl counts by The Wildfowl & Wetlands Trust. *Wildfowl* 47: 217-230.

Cranswick, P.A., Stewart, B., Bullock, I., Haycock, R. & Hughes, B. 1998. *Common Scoter Melanitta nigra monitoring in Carmarthen Bay following the Sea Empress oil spill: April 1997 to March 1998*. WWT Wetlands Advisory Service report to CCW, Contract No. FC 73-02-53A, Slimbridge, 25 pp.

Danielsen, F., Skov, H. & Durnick, J. 1993. Estimates of the wintering population of Red-throated Diver *Gavia stellata* and Black-throated Diver *Gavia arctica* in northwest Europe. *Proc. 7th Nordic Congress of Ornithology, 1990.* pp. 18-24.

Davidson N.C. 1995a. Chapter 4.1 Estuaries. *In: Coasts and seas of the United Kingdom. Region 12 Wales: Margam to Little Orme*, ed. by J.H. Barne, C.F. Robson, S.S. Kaznowska & J.P. Doody. Peterborough, Joint Nature Conservation Committee.

Davidson N.C. 1995b. Chapter 4.1 Estuaries. *In: Coasts and seas of the United Kingdom. Region 6 Eastern England: Flamborough Head to Great Yarmouth*, ed. by J.H. Barne, C.F. Robson, S.S. Kaznowska, J.P. Doody & N.C. Davidson. Peterborough, Joint Nature Conservation Committee.

Davidson N.C. 1996a. Chapter 4.1 Estuaries. *In: Coasts and seas of the United Kingdom. Region 13 Northern Irish Sea: Colwyn Bay to Stranraer, including the Isle of Man*, ed. by J.H. Barne, C.F. Robson, S.S. Kaznowska, J.P. Doody & N.C. Davidson. Peterborough, Joint Nature Conservation Committee.

Davidson N.C. 1996b. Chapter 4.1 Estuaries. *In: Coasts and seas of the United Kingdom. Region 10 South-west England: Seaton to the Roseland Peninsula*, ed. by J.H. Barne, C.F. Robson, S.S. Kaznowska, J.P. Doody, N.C. Davidson & A.L. Buck. Peterborough, Joint Nature Conservation Committee.

Davidson N.C. 1996c. Chapter 4.1 Estuaries. *In: Coasts and seas of the United Kingdom. Region 9 Southern England: Hayling Island to Lyme Regis*, ed. by J.H. Barne, C.F. Robson, S.S. Kaznowska, J.P. Doody & N.C. Davidson. Peterborough, Joint Nature Conservation Committee.

Davidson N.C. 1996d. Chapter 4.1 Estuaries. *In: Coasts and seas of the United Kingdom. Region 3 North-east Scotland: Cape Wrath to St Cyrus*, ed. by J.H. Barne, C.F. Robson, S.S. Kaznowska, J.P. Doody & N.C. Davidson. Peterborough, Joint Nature Conservation Committee.

Davidson N.C. 1997a. Chapter 4.1 Estuaries. *In: Coasts and seas of the United Kingdom. Region 8 Sussex: Rye Bay to Chichester Harbour*, ed. by J.H. Barne, C.F. Robson, S.S. Kaznowska, J.P. Doody, N.C.

Davidson & A.L. Buck. Peterborough, Joint Nature Conservation Committee.

Davidson N.C. 1997b. Chapter 4.1 Estuaries. *In: Coasts and seas of the United Kingdom. Region 4 South-east Scotland: Montrose to Eyemouth*, ed. by J.H. Barne, C.F. Robson, S.S. Kaznowska, J.P. Doody, N.C. Davidson & A.L. Buck. Peterborough, Joint Nature Conservation Committee.

Davidson, N.C. 1998a. Compiling estiamtes of East Atlantic Flyway wader populations wintering in coastal Europe in the early 1990s: a summary of the 1996 WSG wader populations workshop. *Wader Study Group Bull.* 86: 18-25.

Davidson N.C. 1998b. Chapter 4.1 Estuaries. *In: Coasts and seas of the United Kingdom. Region 7 South-east England: Lowestoft to Dungeness*, ed. by J.H. Barne, C.F. Robson, S.S. Kaznowska, J.P. Doody, N.C. Davidson & A.L. Buck. Peterborough, Joint Nature Conservation Committee.

Delany, S.N. 1993. Introduced and escaped geese in Britain in summer 1991. *Brit. Birds* 86: 591-599.

Delany, S.N. & Ogilvie, M.A. 1994. *Greenland Barnacle Geese in Scotland, March 1994.* WWT Report to JNCC, Slimbridge, 19 pp.

Forshaw, W.D. 1998. *Report on wild geese and swans in Lancashire, 1997/98.* Unpubl. report, 9 pp.

Forster, S. 1998. *COUNTer proDUCKtive News. Issue 1.* Newsletter for WeBS counters in the Lough Neagh area, EHS Northern Ireland.

Fox, A.D. & Francis, I. 1998. *Report of the 1997/98 national census of Greenland White-fronted Geese in Britain.* GWGS Report to WWT, Kalø, Denmark, 16 pp.

Fox, A.D., Glahder, C., Mitchell, C.R., Stroud, D.A., Boyd, H. & Frikke, J. 1996. North American Canada Geese (*Branta canadensis*) in west Greenland. *Auk* 113: 231-233.

Garner, M. 1999. Identification of Common Merganser. *Birding World* 12: 31-33.

Geary S. 1997. The abundance and low tide distribution of waterfowl on the Tamar Complex 1989-1996. Unpublished M.Sc. thesis, University of Plymouth.

Gilbert, G., Gibbons, D.W. & Evans, J. 1998. *Bird Monitoring Methods.* RSPB, Sandy.

Goss-Custard, J.D., Ross, J., McGrorty, S., Durell, S.E.A. LeV. dit, Caldow, R.W.G. & West, A.D. 1998. Locally stable wintering numbers in the Oystercatcher *Haematopus ostralegus* where carrying capacity has not been reached. *Ibis* 140: 104-112.

Green, M. & Elliott, D. 1993. *Surveys of wintering birds and cetaceans in northern Cardigan Bay, 1990-93.* Report by Friends of Cardigan Bay, 32 pp.

Hearn, R.D. 1998. *The 1997 national census of Pink-footed Geese and Icelandic Greylag Geese in Britain.* WWT Report to JNCC, Slimbridge, 14 pp.

Heubeck, M. 1993. Moult flock surveys indicate a continued decline in the Shetland Eider population, 1984-92. *Scott. Birds* 17: 77-84.

Heubeck, M. 1998. *SOTEAG ornithological monitoring programme: 1997 summary report.* SOTEAG, Aberdeen, 27 pp.

Hjort, C. 1995. Brent Geese in northeasternmost Greenland. Dansk Ornitologisk Forenings Tidsskrift 89: 89-91.

Holloway, L.G. 1998. *The Sussex Bird Report No. 50 1997.* The Sussex Ornithological Society.

Holmes, J.S. & Stroud, D.A. 1995. Naturalised birds: feral, exotic, introduced or alien? *Brit. Birds* 88: 602-603.

Holmes, J.S., Marchant, J., Bucknell, N., Stroud, D.A. & Parkin, D.T. 1998. The British List: new categories and their relevance to conservation. *Brit. Birds* 92: 2-11.

Hughes, B., Criado, J., Delany, S.N., Gallo-Orsi, U., Green, A.J., Grussu, M., Perennou, C. & Torres, J.A. 1999. *The status of the North American Ruddy Duck* Oxyura jamaicensis *in the Western Palearctic: towards an action plan for eradication.* Report by The Wildfowl & Wetlands Trust to the Council of Europe.

Kershaw, M. 1998. *Long term population trends in wintering Pintail (*Anas acuta*) in Great Britain 1966-95.* WWT report to JNCC, Slimbridge.

Kirby, J.S. 1995. Winter population estimates for selected waterfowl species in Britain. *Biol. Cons.* 73: 189-198.

Kirby, J.S., Salmon, D.G. & Atkinson-Willes, G.L. & Cranswick, P.A. 1995. Index numbers for waterbird populations, III. Long-term trends in the abundance of wintering wildfowl in Great Britain, 1966/67 to 1991/2. *J. Appl. Ecol.* 32: 536-551.

Lloyd, C., Tasker, M.L., & Partridge, K. 1991. *The Status of Seabirds in Britain and Ireland.* T & AD Poyser, London.

Lock, L. & Cook, K. 1998. The Little Egret in Britain: a successful colonist. *Brit. Birds* 91: 273-280.

Madsen, J. 1991. Status and trends of goose populations in the Western Palearctic in the 1980s. In: Fox, A.D., Madsen, J. & van Rhijn, J. (Eds.) Western Palearctic Geese. Proc. IWRB Symp. Kleve, Germany, February 1989. *Ardea* 79: 113-122.

Madsen, J., Cracknell, G. & Fox, A.D. (Eds.). 1999. *Goose populations of the Western Palearctic. A review of status and distribution.* Wetlands International Publ. No. 48, Wetlands International, Wageningen, The Netherlands. National

Environmental Research Institute, Rönde, Denmark, 34 pp.

Marquiss, M. 1993. Grey Heron. In: Gibbons, D.W., Reid, J.B. & Chapman, R. (Eds.) 1993. *The New Atlas of Breeding Birds in Britain and Ireland: 1988-1991.* T. & A.D. Poyser, London: 48-49.

Mathers, R.G., Portig, A.A., & Montgomery, W.I. 1998. Distribution and abundance of Pale-bellied Brent Geese and Wigeon on Strangford Lough, Northern Ireland. *Bird Study* 45: 18-34.

Mayhew, P. & Houston D. 1999. Effects of winter and early spring grazing by Wigeon *Anas penelope* on their food supply. *Ibis* 141: 80-84.

Meek, E.R. 1993. Population fluctuations and mortality of Mute Swans on an Orkney loch system in relation to a Canadian Pondweed growth cycle. *Scott. Birds* 17: 85-92.

Meltofte, H., Blew, J., Frikke, J., Rösner, H.-U., & Smit, C.J. 1994. Numbers and distribution of waterbirds in the Wadden Sea. Results and evaluation of 36 simultaneous counts in the Dutch-German-Danish Wadden Sea 1980-91. *IWRB Publ. 34/Wader Study Group Bull 74, Special issue.*

Milne, P., & O'Sullivan, O. 1996. Forty-fourth Irish Bird Report, 1996. *Irish Birds* 6: 61-90.

Mitchell, C.R., MacDonald, R. & Boyer, P.R. 1995. *Greylag Geese on the Uists.* WWT report to JNCC, Slimbridge, 6 pp.

Mitchell, C.R., Patterson, D.J., Price, D.J. & Kerr, S. 1997. *Aerial counts of Barnacle Geese on proposed SPA sites in north and west Scotland.* WWT Report to SNH, Contract No. SNH/RASD/043/017/N2K, 19 pp.

Moss, S. 1998. Predictions of the effects of global climate change on Britain's birds. *Brit. Birds* 91: 307-325.

Murray, R.D. (Ed) 1998. *Scottish Bird Report 1996.* Scottish Ornithologists' Club, Edinburgh.

Nightingale, B. & Allsopp, K. 1998. The ornithological year 1997. *Brit. Birds.* 91: 527-539.

Norris, K., Bannister, R.C.A. & Walker, P.W. 1998. Changes in the number of oystercatchers *Haematopus ostralegus* wintering in the Burry Inlet in relation to the biomass of cockles *Cerastoderma edule* and its commercial exploitation. *J. Appl. Ecol.* 35: 75-85.

Owen, M., Atkinson-Willes, G.L. & Salmon, D.G. 1986. *Wildfowl in Great Britain.* 2nd Edition. University Press, Cambridge.

Percival, S.M., Sutherland, W.J. & Evans, P.R. 1998. Intertidal habitat loss and wildfowl numbers: applications of a spatial depletion model. *J. App. Ecol.* 35: 57-63.

Pirot, J.-Y., Laursen, K., Madsen, J. & Monval, J.-Y.

1989. Population estimates of swans, geese, ducks and Eurasian Coot *Fulica atra* in the Western Palearctic and Sahelian Africa. In: Boyd, H. & Pirot, J.-Y. (Eds). Flyways and Reserve Networks for Water Birds. *IWRB Spec. Publ. 9,* IWRB, Slimbridge: 12-23.

Prater, A.J. 1975. The wintering population of the Black-tailed Godwit. *Bird Study* 22:169-176.

Prater, A.J. 1981. *Estuary birds of Britain and Ireland.* T. & A.D. Poyser, Carlton.

Pritchard, D.E., Housden, S.D., Mudge, G.P., Galbraith, C.A. & Pienkowski, M.W. (Eds.) 1992. *Important Bird Areas in the UK including the Channel Islands and the Isle of Man.* RSPB, Sandy.

Prŷs-Jones, R. P., Underhill, L.G. & Waters, R.J. 1994. Index numbers for waterbird populations. II Coastal wintering waders in the United Kingdom, 1970/71 - 1990/91. *J. Appl. Ecol.* 31: 481-492.

Rafe, R. 1998. *The Harrier* No. 114. Suffolk Ornithologists' Group.

Ramsar Convention Bureau 1998. *Convention on Wetlands of International Importance especially as Waterfowl Habitat.* Proceedings of the third meeting of the Conference of the Contracting Parties, Regina, Canada, 1987. Ramsar, Switzerland.

Rees, E.C. & Bacon, P.J. 1996. Migratory tradition in Bewick's Swans (*Cygnus columbianus bewickii*). In: Proceedings of the Anatidae 2000 Conference, Strasbourg, France, 5-9 December 1994, Birkan, M., van Vessem, J., Havet, P., Madsen, J., Trolliet, B. & Moser, M. (Eds.) *Gibier Faune Sauvage, Game Wild.* 13: 407-420.

Rees, E.C. & Bowler, J.M. 1997. Fifty years of swan research and conservation by The Wildfowl & Wetlands Trust. *Wildfowl* 47: 248-263.

Rees, E.C & White, G.T. 1998. *Whooper Swans wintering in the Black Cart Floodplain: 1994-95 to 1997-98 perspective.* WWT Wetland Advisory Service Report to Renfrewshire Enterprise, Slimbridge.

Ridgill, S.C. & Fox, A.D. 1990. Cold weather movements of waterfowl in Western Europe. *IWRB Spec. Publ. 13.* IWRB, Slimbridge.

Rose, P.M. 1995. *Western Palearctic and South-West Asia Waterfowl Census 1994.* IWRB Publ. 29. IWRB, Slimbridge, UK. 102 pp.

Rose, P.M. & Scott, D.A. 1997. *Waterfowl Population Estimates - Second Edition.* Wetlands International Publ. 44, Wageningen, The Netherlands.

Rose, P.M. & Stroud, D.A. 1994. Estimating international waterfowl populations: current activity and future directions. *Wader Study Group Bull.* 73: 19-26.

Scott, D.A. & Rose, P.M. 1996. *Atlas of Anatidae Populations in Africa and Western Eurasia.* Wetlands International Publ. No. 41, Wetlands International,

Wageningen, The Netherlands.

Simpson, J. & Maciver, A. 1997. *Population and Distribution of Bean Geese in the Slamannan Area, year 1996/97*. The Bean Goose Working Group.

Smit, C.J. & Piersma, T. 1989. Numbers, midwinter distribution and migration of wader populations using the East Atlantic flyway. In: Boyd, H. & Pirot, J.-Y. (Eds.) Flyways and reserve networks for waterbirds. *IWRB Spec. Publ. 9*, Slimbridge: 24-64.

Smith, T., Bainbridge, I. & O'Brien, M. 1994. *Distribution and habitat use by Bean Geese in the Slamannan area*. RSPB Report to SNH, 71 pp.

Stenning, J. 1994. *Moray Firth Monitoring: winter 1993-94*. RSPB report, 36 pp.

Stenning, J. 1998. *Moray Firth monitoring: winter 1997-98*. RSPB report to Talisman Energy, 4 pp.

Stewart, B., Hughes, B., Bullock, I. & Haycock, R. 1997. *Common scoter* Melanitta nigra *monitoring in Carmarthen Bay following the Sea Empress oil spill*. WWT Wetlands Advisory Service report to the *Sea Empress* Environmental Evaluation Committee, Contract No. FC 73-02-53, Slimbridge.

Stroud, D.A. 1998. Breeding introduced Trumpeter Swans in Northamptonshire: first wild breeding in Britain. *Northamptonshire Bird Report 1998*: 55.

Stroud, D.A., Mudge, G.P. & Pienkowski, M.W. 1990. *Protecting internationally important bird sites: a review of the EEC Special Protection Area network in Great Britain*. NCC, Peterborough, 230 pp.

Syroechkovski, E.E., Zockler, C. & Lappo, E. 1998. Status of Brent Goose in northwest Yakutia, East Siberia. *Brit. Birds* 91: 565-572.

Tomkovich, P.S. & Zharikov, Y.V. 1998. Waer breeding conditions in the Russian tundras in 1997. *Wader Study Group Bull.* 87: 30-42.

Tucker, G.M. & Heath, M.F. 1994. *Birds in Europe: their conservation status*. BirdLife International, Cambridge.

Underhill, L.G. 1989. Indices for waterbird populations. *BTO Research Report 52*.

Underhill, L.G. & Prŷs-Jones, R. 1994 Index numbers for waterbird populations. I. Review and methodology. *J. Appl. Ecol.* 31: 463-480.

Vinicombe, K., Marchant, J. & Know, A. 1993. Review of status and categorization of feral birds on the British List. *Brit. Birds* 75: 1-11.

Watson, A., Marquiss, M. & Cosgrove, P.J. 1998. Northeast Scottish counts of Goldeneye, Goosander, Red-breasted Merganser and Cormorant in 1944-50 compared with 1988-97. *Scott. Birds* 19: 249-258.

Way, L.S., Grice, P., MacKay, A., Galbraith, C.A., Stroud, D.A. & Pienkowski, M.W. 1993. *Ireland's internationally important bird sites: a review of sites for the EC Special Protection Area network*. JNCC, Peterborough, 231 pp.

Wilson, J.R., Czajkowski, M.A. & Pienkowski, M.W. 1980. The migration through Europe and wintering in West Africa of Curlew Sandpipers. *Wildfowl* 31: 107-122.

GLOSSARY

The terms listed below are generally restricted to those that have been adopted specifically for use within WeBS or more widely for monitoring.

Autumn For waders, autumn comprises July to October inclusive. Due to differences in seasonality between species (see *Monthly Fluctuations*), a strict definition of autumn is not used for wildfowl.

British Trust for Ornithology (BTO) The BTO is a well respected organisation, combining the skills of professional scientists and volunteer birdwatchers to carry out research on birds in all habitats and throughout the year. Data collected by the various surveys form the basis of extensive and unique databases which enable the BTO to objectively advise conservation bodies, government agencies, planners and scientists on a diverse range of issues involving birds.

Complex site A *WeBS site* that consists of two or more *sectors*.

Core Counts The basic WeBS counts that monitor all wetlands throughout the UK once per month on priority dates. Used to determine population estimates and trends and identify important sites.

Local Organiser Person responsible for co-ordinating counters and counts at a local level, normally a county or large estuary, and the usual point of contact with WeBS partner HQs.

Incomplete counts When presenting counts of an individual species, a large proportion of the number of birds was suspected to have been missed, e.g. due to part coverage of the site or poor counting conditions, or when presenting the total number of birds of all species on the site, a significant proportion of the total number was missed.

I-WeBS An independent but complementary scheme operating in the Republic of Ireland to monitor non-breeding waterfowl, organised by the IWC Birdwatch Ireland, the National Parks and Wildlife Service (Ireland) and The Wildfowl & Wetlands Trust.

Joint Nature Conservation Committee (JNCC) JNCC is the statutory body constituted by the Environmental Protection Act 1990 to be responsible for research and advice on nature conservation at both UK and international levels. The committee is established by English Nature, Scottish Natural Heritage and the Countryside Council for Wales, together with independent members and representatives from the Countryside Commission and Northern Ireland, and is supported by specialist staff.

Low Tide Counts (LTC) WeBS counts made at low tide to assess the relative importance of different parts of individual estuaries as feeding areas for intertidal waterfowl.

Royal Society for the Protection of Birds (RSPB) The RSPB is the charity that takes action for wild birds and the environment in the UK. The RSPB is the national BirdLife partner in the UK.

Spring For waders, spring comprises April to June inclusive. Due to differences in seasonality between species (see *Monthly Fluctuations*), a strict definition of spring is not used for wildfowl.

Waterfowl WeBS follows the definition adopted by Wetlands International. This includes a large number of families, those occurring regularly in the UK being divers, grebes, cormorants, herons, storks, ibises and spoonbills, wildfowl, cranes, rails, waders and gulls and terns. Note that, due to differences in coverage, not all families may be included in the 'waterfowl totals' given in this report, although the species excluded and the reasons for this will be given in each case.

WeBS count sector The unit of division of large *sites* into areas which can be counted by one person in a reasonable time period. They are often demarcated by geographic features to facilitate recognition of the boundary by counters. The finest level at which data are recorded.

WeBS count site A biologically meaningful area that represents a discrete area used by waterfowl such that birds regularly move within but only occasionally between sites. The highest level at which count data are stored.

WeBS count sub-site A grouping of *sectors* within a *site* to facilitate co-ordination. In most cases, sub-sites also relate to biologically meaningful units for describing waterfowl distribution.

WeBS count unit The area/boundary within which a count is made. The generic term for *sites*, *sub-sites* and *sectors*.

Wetland Advisory Service (WAS) The environmental consultancy wing of The Wildfowl & Wetlands Trust.

The Wildfowl & Wetlands Trust (WWT) Founded by Sir Peter Scott in 1946, WWT is the only wildlife conservation charity specialising in wetlands and the wildlife they support. It has pioneered the bringing together of people and wildlife for the benefit of both and seeks to raise awareness of the value of wetlands, the threats they face and the actions needed to save them. To this end, WWT has eight centres throughout the UK and is dedicated to saving wetlands for wildlife and people.

Winter For waders, winter comprises November to March inclusive. Due to differences in seasonality between species (see *Monthly Fluctuations*), a strict definition of winter is not used for wildfowl.

Winter (five-year) peak mean Calculated by averaging the peak count in each season for a particular species at an individual site (i.e. the right hand column of figures in the table in each species account). Normally calculated using the most recent five years' data, this figure is compared with the respective *1% thresholds* to determine if the site qualifies as nationally or internationally important.

1% criterion The Ramsar Convention has established site selection criteria. One such criterion (currently numbered Criterion 3c) indicates that a site is identified as being of international importance if it holds 1% or more of a population of waterfowl A change in the 1% criterion would be if the selection threshold changes to, say, 2% of a population (the 2% criterion) or 0.5% of a population (0.5% criterion). The term thus relates to the proportion (1%) that is used as a criterion for internationally important site selection.

1% threshold This logically derives from the *1% criterion* and relates to the number of birds that are used as the nominal 1% of the population for the purposes of site selection. Thus, an international population of 75,215 Shelduck has a derived 1% threshold (adopting rounding conventions) of 750.

APPENDIX I. INTERNATIONAL DESIGNATIONS

The Ramsar Convention on Wetlands of International Importance especially as Waterfowl Habitat requires each Contracting Party to designate suitable wetlands, selected on account of their international significance in terms of ecology, botany, zoology, limnology or hydrology, for inclusion in a List of Wetlands of International Importance (known as Ramsar sites) (Carp 1972). The Directive on the Conservation of Wild Birds (EC/79/409) lays emphasis on the need to conserve bird habitats as a means of maintaining populations and that this, in part, should be achieved by the establishment of a network of protected areas termed Special Protection Areas (SPAs) (Stroud *et al.* 1990). Ramsar Sites and SPAs may be identified using a number of criteria, including a number of numeric selection critieria (see Appendix 2) which draw heavily upon waterfowl counts, especially WeBS and the other data presented in this report.

Between 1 January 1998 and 31 March 1999, a total of 38 SPAs and 23 Ramsar sites were designated by the UK, with 17 of these sites receiving dual designation. These new designations are indicated in Table A1 by 'N'.

This represents excellent progress during the period. By 31 March 1999, 137 Ramsar sites and 199 SPAs had been designated in the UK, with a further three UK Ramsar sites in Dependent Territories. The total area designated as Ramsar sites has reached 661,217.27 ha, and that as SPAs is 943,584.23 ha. Since many sites are designated as both a Ramsar site and SPA, this represents a total area designated of just under one million square kilometres.

Table A1. Ramsar Sites and SPAs designated in the UK as of 31 March 1999. (R) = Ramsar site only; (S) = SPA only; the remainder have dual designation. Sites designated during the period 1 January 1998 to 31 March 1999 are denoted by 'N'.

	Abberton Reservoir	Castle Loch, Lochmaben	N	Eoligarry (S)
	Abernethy Forest (S)	Castlemartin Coast (S)		Esthwaite Water (R)
	Achanalt Marshes (S)	Chesil Beach and The Fleet		Exe Estuary
	Ailsa Craig (S)	Chew Valley Lake (S)		Fair Isle (S)
	Alde-Ore Estuary	Chichester and Langstone		Fala Flow
	Alt Estuary	Harbours		Farne Islands (S)
	Ashdown Forest (S)	Chippenham Fen (R)		Fetlar (S)
N	Auskerry (S)	Claish Moss (R)		Feur Lochain (part of Rinns of
N	Avon Valley	Coll		Islay)
N	Ballochbuie (S)	Colne Estuary (Mid-Essex Coast		Firth of Forth Islands (S)
N	Ballynahone Bog (R)	Phase 2)		Flamborough Head and Bempton
N	Beinn Dearg (S)	Copinsay (S)		Cliffs (S)
N	Belfast Lough	Coquet Island (S)		Flannan Isles (S)
N	Ben Wyvis (S)	Cors Caron (R)		Foula (S)
	Benacre to Easton Bavents (S)	Cors Fochno and Dyfi (R)		Foulness (Mid-Essex Coast Phase
	Benfleet and Southend Marshes	N Corsydd Môn a Llyn /Anglesey		5)
N	Berwyn (S)	and Llyn Fens (R)		Fowlsheugh (S)
	Blackwater Estuary (Mid-Essex	N Creag Meagaidh (S)	N	Garron Plateau (R)
	Coast Phase 4)	N Cromarty Firth (Moray Basin		Gibraltar Point
	Bowland Fells (S)	Firths and Bays)		Glac na Criche (part of Rinns of
	Breydon Water	N Crouch and Roach Estuaries		Islay)
	Bridgend Flats, Islay	(Mid-Essex Coast Phase 3)		Gladhouse Reservoir
	Bridgwater Bay (part of Severn	Crymlyn Bog (R)		Glannau Aberdaron and Ynys
	Estuary) (R)	N Cuilcagh Mountain (R)		Enlli/Aberdaron Coast and
N	Broadland	Deben Estuary		Bardsey Island (S)
N	Buchan Ness to Collieston Coast	Dengie (Mid-Essex Coast Phase		Glannau Ynys Gybi /Holy Island
	(S)	1)		Coast (S)
	Bure Marshes (part of	Dersingham Bog (R)	N	Glas Eileanan (S)
	Broadland) (R)	Derwent Ings (part of Lower		Glen Tanar (S)
	Burry Inlet	Derwent Valley)		Grassholm (S)
	Caenlochan (S)	Din Moss - Hoselaw Loch		Great Yarmouth North
	Cairngorm Lochs (R)	Dornoch Firth and Loch Fleet		Denes (S)
	Cairngorms (S)	N Dorset Heathlands		Greenlaw Moor
N	Caithness and Sutherland	Drumochter Hills (S)		Gruinart Flats, Islay
	Peatlands	N Duddon Estuary		Hamford Water
N	Caithness Lochs	East Caithness Cliffs (S)		Handa (S)
N	Calf of Eday (S)	N East Devon Heaths (S)		Hermaness and Saxa Vord (S)
	Cameron Reservoir	East Sanday Coast		Hickling Broad and Horsey Mere
N	Canna and Sanday (S)	Eilean na Muice Duibhe (Duich		(part of Broadland) (R)
	Cape Wrath (S)	Moss) Islay		Holburn Lake and Moss
N	Carlingford Lough	Elenydd-Mallaen (S)		Hornsea Mere (S)

Humber Flats, Marshes and
   Coast (Phase 1)
N Inner Moray Firth (Moray Basin
   Firths and Bays)
   Irthinghead Mires (R)
N Kilpheder to Smerclate, South
   Uist (S)
N Kintyre Goose Roosts
   Laggan, Islay (S)
   Larne Lough
   Leighton Moss
   Lindisfarne
   Llyn Idwal (R)
   Llyn Tegid (R)
   Loch an Duin (R)
   Loch Ashie (S)
   Loch Druidibeg, Loch a' Machair
   and Loch Stilligarry (R)
   Loch Eye
   Loch Flemington (S)
   Loch Ken and River Dee
   Marshes
   Loch Knockie and Nearby
   Lochs (S)
   Loch Leven (R)
   Loch Lomond
   Loch Maree
N Loch of Inch and Torrs Warren
   Loch of Kinnordy
   Loch of Lintrathen
   Loch of Skene
   Loch of Strathbeg
   Loch Ruthven
   Loch Spynie
   Loch Vaa (S)
N Lochnagar (S)
   Lochs Druidibeg, a' Machair (S)
   Lochs of Spiggie and Brow (S)
N Lough Foyle
   Lough Neagh and Lough Beg
   Lower Derwent Valley
   Malham Tarn (R)
   Martin Mere
   Marwick Head (S)
   Medway Estuary and Marshes
   Mersey Estuary
   Midland Meres and Mosses
   Phase 0
   Midland Meres and Mosses
   Phase 2 (R)
   Mingulay and Berneray (S)
   Minsmere - Walberswick
N Mointeach Scadabhaigh (S)
   Monach Isles (S)

   Montrose Basin
   Moor House (S)
   Moray and Nairn Coast
   Morecambe Bay
   Mousa (S)
   Nene Washes
N Ness & Barvas, Lewis (S)
   New Forest
   North Caithness Cliffs (S)
   North Colonsay and Western
   Cliffs (S)
   North Harris Mountains (S)
   North Inverness Lochs (S)
   North Norfolk Coast
N North Sutherland Coastal
   Islands (S)
N North Uist Machair and Islands
   Noss (S)
   Old Hall Marshes (part of
   Blackwater Estuary)
   Orfordness-Havergate (part of
   Alde-Ore Estuary) (S)
   Ouse Washes
   Pagham Harbour
   Papa Westray (North Hill and
   Holm) (S)
   Pentland Firth Islands (S)
   Pettigoe Plateau
N Pevensey Levels (R)
N Poole Harbour
   Porton Down (S)
   Portsmouth Harbour
   Priest Island (Summer Isles) (S)
   Ramna Stacks and Gruney (S)
   Ramsey and St David`s Peninsula
   Coast (S)
   Rannoch Moor (R)
N Rathlin Island (S)
   Redgrave and South Lopham
   Fens (R)
   Ribble and Alt Estuaries
   (Phase 2)
   Ribble Estuary (S)
   Rinns of Islay
   River Spey - Insh Marshes
   Rockcliffe Marsh (part of Upper
   Solway Flats and Marshes)
   Ronas Hill - North Roe and
   Tingon
   Rostherne Mere (R)
   Roydon Common (R)
   Rum (S)
   Rutland Water

   Salisbury Plain (S)
   Severn Estuary
   Sheep Island (S)
   Shiant Isles (S)
   Silver Flowe (R)
   Skomer (S)
N Solent and Southampton Water
   Somerset Levels and Moors
   South Pennine Moors Phase 1 (S)
   South Pennine Moors Phase 2 (S)
   South Tayside Goose Roosts
   South Uist Machair and Lochs
   St Abb's Head to Fast Castle (S)
   St Kilda (S)
   Stodmarsh
   Stour and Orwell Estuaries
N Strangford Lough
   Sule Skerry and Sule Stack (S)
   Sumburgh Head (S)
   Swan Island (S)
   Tamar Estuaries Complex (S)
   Teesmouth and Cleveland Coast
   Thanet Coast and Sandwich Bay
   The Dee Estuary
   The Swale
   The Wash
   Thursley and Ockley Bog (R)
   Thursley, Hankley and Frensham
   Commons (Wealden Heaths
   Phase 1) (S)
   Traeth Lafan/Lavan Sands,
   Conway Bay (S)
   Treshnish Isles (S)
   Troup, Pennan and Lion's
   Heads (S)
   Upper Lough Erne
   Upper Severn Estuary (part of
   Severn Estuary)
   Upper Solway Flats and Marshes
   Walmore Common
N Wealden Heaths Phase 2 (S)
   West Westray (S)
   Westwater
   Wicken Fen (R)
   Woodwalton Fen (R)
   Ynys Feurig, Cemlyn Bay and
   The Skerries (S)
N Ythan Estuary and Meikle
   Loch (R)
N Ythan Estuary, Sands of Forvie
   and Meikle Loch (S)

APPENDIX 2. INTERNATIONAL AND NATIONAL IMPORTANCE

Any site recognised as being or international ornithological importance is considered for classification as a Special Protection Area (SPA) under the EC Directive on the Conservation of Wild Birds (EC/79/409), whilst a site recognised as an internationally important wetland qualifies for designation as a Ramsar site under the Convention on Wetlands of International Importance especially as Waterfowl Habitat. Criteria for assessing the international importance of wetlands have been agreed by the Contracting Parties to the Ramsar Convention on Wetlands of International Importance (Ramsar Convention Bureau 1988). Under one criterion, a wetland is considered internationally important if it regularly holds at least 1% of the individuals in a population of one species or subspecies of waterfowl, while any site regularly holding a total of 20,000 or more waterfowl also qualifies. Britain and Ireland's wildfowl belong, in most cases, to the northwest European population (Pirot et al. 1989), and the waders to the east Atlantic flyway population (Smit & Piersma 1989).

A wetland in Britain is considered nationally important if it regularly holds 1% or more of the estimated British population of one species or subspecies of waterfowl, and in Northern Ireland important in an all-Ireland context if it holds 1% or more of the estimated all-Ireland population.

The 1% thresholds for British, all-Ireland and international waterfowl populations, where known, are listed in Table A2. Thus, any site regularly supporting at least this number of birds potentially qualifies for designation under national legislation, or the EC Bird's Directive or Ramsar Convention. The international population for each species and sub-species is also specified in the table. However, it should be noted that, where 1% of the national population is less than 50 birds, 50 is normally used as a minimum qualifying threshold for the designation of sites of national or international importance.

1% thresholds have not been derived for introduced species since, for these species, protected sites (e.g. SSSIs) would not be identified on the basis of numbers for these birds.

Sources of qualifying levels represent the most up-to-date figures following recent reviews: for British wildfowl see Kirby (1995); for British waders see Cayford & Waters (1996); for all-Ireland importance for divers see Danielsen et al. (1993) and for other waterfowl see Whilde (in prep.) cited in Way et al. (1993). International criteria follow Smit & Piersma (1989) or Scott & Rose (1996).

It was agreed at the meeting of the Ramsar Convention in Brisbane that population estimates will be reviewed by Wetlands International every three years and 1% thresholds revised every nine years (Rose & Stroud 1994; Ramsar Resolution VI.4).

The third edition of *Waterfowl Population Estimates*, presented to the Seventh Meeting of the Contracting Parties to the Ramsar Convention in Costa Rica in May 1999, includes revisions for a number of goose populations (following Madsen et al. 1999) and of a number of east Atlantic flyway wader populations (see Davidson 1998a). The next revision of British population sizes will be undertaken in the year 2000.

Table A2. 1% thresholds for national and international importance.

	Great Britain	all-Ireland	International	Population
Red-throated Diver	50	10 *	750	Europe/Greenland
Black-throated Diver	7 *	1 *	1,200	Europe/W Siberia
Great Northern Diver	30 *	?	50	Europe
Little Grebe	30 *	?	?	W Palaearctic
Great Crested Grebe	100	30 *	?	NW Europe
Red-necked Grebe	1 *	?	330	NW Europe
Slavonian Grebe	4 *	?	50	NW Europe
Black-necked Grebe	1 *	?	1,000	W Palaearctic
Cormorant	130	?	1,200	NW Europe
Little Egret	?	?	800	W Mediterranean
Grey Heron	?	?	4,500	Europe/N Africa
Mute Swan	260	55	2,400	NW Europe
Bewick's Swan	70	25 *	170	W Siberia/NW Europe
Whooper Swan	55	100	160	Iceland/UK/Ireland
Bean Goose	4 *	+ *	800	NE & NW Europe
Pink-footed Goose: Iceland/Greenland	1,900	+ *	2,250	E Greenland/Iceland/UK
European White-fronted Goose	60	+ *	6,000	NW Siberia/NE & NW Europe
Greenland White-fronted Goose	140	140	300	Greenland/Ireland/UK
Greylag Goose: Iceland	1,000	40 *	1,000	Iceland/UK/Ireland
Hebrides/N Scotland	50	n/a	50	NW Scotland

	Great Britain	all-Ireland	International	Population
Barnacle Goose: Greenland	270	75	320	E Greenland/ Ireland/Scotland
Svalbard	120	+ *	120	Svalbard/SW Scotland
Dark-bellied Brent Goose	1,000	+ *	3,000	bernicla
Light-bellied Brent Goose: Canada	+ *	200	200	Canada/Ireland
Svalbard	25 *	+ *	50	Svalbard/Denmark/UK
Shelduck	750	70	3,000	NW Europe
Wigeon	2,800	1,250	12,500	NW Europe
Gadwall	80	+ *	300	NW Europe
Teal	1,400	650	4,000	NW Europe
Mallard	5,000	500	20,000 **	NW Europe
Pintail	280	60	600	NW Europe
Garganey	+ *	+ *	20,000 **	Europe/W Africa
Shoveler	100	65	400	NW Europe/Central Europe
Red-crested Pochard	+ *	+ *	250	C & SW Europe/W Mediterranean
Pochard	440	400	3,500	NW Europe
Tufted Duck	600	400	10,000	NW Europe
Scaup	110	30 *	3,100	NW Europe
Eider	750	20 *	20,000 **	Europe
Long-tailed Duck	230	+ *	20,000 **	Iceland/Greenland/ NW Europe
Common Scoter	350	40 *	16,000	W Siberia/W Europe/ NW Africa
Velvet Scoter	30 *	+ *	10,000	W Siberia/NW Europe
Goldeneye	170	110	3,000	NW & Central Europe
Smew	2 *	+ *	250	NW & Central Europe
Red-breasted Merganser	100	20 *	1,250	NW & Central Europe
Goosander	90	+ *	2,000	NW & Central Europe
Coot	1,100	250	15,000	NW Europe
Oystercatcher	3,600	500	9,000	Europe/W Africa (win)
Avocet	10 *	+ *	700	Europe/NW Africa (bre)
Little Ringed Plover	?	?	?	Europe/W Africa
Ringed Plover	290	125	500	Europe/NW Africa (win)
passage	300			
Golden Plover	2,500	2,000	18,000	NW Europe (bre)
Grey Plover	430	40 *	1,500	E Atlantic
Lapwing	20,000 **	2,500	20,000 **	Europe/W Africa
Knot C. c. islandica	2,900	375	3,500	W Europe/Canada
C. c. canutus			5,000	W Africa/W Siberia
Sanderling	230	35 *	1,000	E Atlantic
passage	300			
Little Stint	?	?	2,100	W Africa/Europe
Curlew Sandpiper	?	?	4,500	W Africa/SW Europe(win)
Purple Sandpiper	210	10 *	500	E Atlantic
Dunlin C. a. arctica			150	Greenland (bre)
C. a. schinzii (Icelandic)			8,000	Iceland/Greenland (bre)
C. a. schinzii (temperate)			200	UK/Ireland/Baltic
C. a. alpina	5,300	1,250	14,000	Europe (bre)
passage	2,000			
Ruff	7 *	+ *	10,000	W Africa (win)
Jack Snipe	?	250	?	Europe/W Africa (win)
Snipe	?	?	10,000	Europe/W Africa (bre)
Woodcock	?	?	20,000 **	Africa/Europe
Black-tailed Godwit	70	90	700	Iceland (bre)
Bar-tailed Godwit	530	175	1,000	W Europe (win)
Whimbrel	+ *	+ *	6,500	Europe/W Africa (win)
passage	50			
Curlew	1,200	875	3,500	Europe/NW Africa
Spotted Redshank	+ *	+ *	1,500	Europe/W Africa
Redshank T. t. totanus	1,100	245	1,500	Europe/W Africa (win)
T. t. robusta	1,100		1,500	NW Europe (win)
passage	1,200			
Greenshank	+ *	9 *	3,000	Europe/W Africa
Green Sandpiper	?	?	?	Europe (bre)

	Great Britain	all-Ireland	International	Population
Common Sandpiper	?	?	?	Europe (bre)
Turnstone	640	225	700	Europe (win)
Little Gull	?	?	750	Cent/E Europe (bre)
Black-headed Gull	?	?	20,000 **	NW Europe
Common Gull	?	?	16,000	NW Europe
Lesser Black-backed Gull	?	?	4,500	W Europe
Herring Gull	?	?	13,000	W Europe/Iceland
Great Black-backed Gull	?	?	4,800	W Atlantic
Kittiwake	?	?	20,000 **	E Atlantic
Sandwich Tern	?	?	1,500	W Europe/W Africa
Common Tern	?	?	6,000	N/E Europe
Little Tern	?	?	340	E Atlantic
Black Tern	?	?	2,000	Europe/Asia

?   *Population size not accurately known*
+   *Population too small for meaningful figure to be obtained*
*   *Where 1% of the British or all-Ireland wintering population is less than 50 birds, 50 is normally used as a minimum qualifying level for national or all-Ireland importance respectively*
** *A site regularly holding more than 20,000 waterfowl qualifies as internationally important by virtue of absolute numbers*

APPENDIX 3. ANALYSES

This appendix provides additional detail about the analyses used in this report to that presented in *Analyses* and lists the index values used to produce the graphs of annual and monthly indices in the species accounts.

Data availability

The count scheme first begun in 1947 has developed considerably over time (see Cranswick *et al.* 1997). In particular, coverage of species and area has expanded during this time. The first year for which data for certain species or areas are available for use in analyses are given below:

Table A3. First year of availability of WeBS Core Count data for different species and areas

Wildfowl in GB	1960 on computer (collected since 1947)
Waders in UK	1969-70
Great Crested Grebe	1982-83
Coot	1982-83
Little Grebe	1985-86
Comorant	1986-87
Wildfowl in Northern Ireland	1986-87
Rare grebes, divers, rarities	1993-94

National totals for goose populations

Figures presented in Tables 1 & 2 and in Appendices 4-9 for total counts of the various goose populations are derived initially from WeBS Core Counts, but are replaced by results of dedicated censuses (see *Survey Methods*, *Analyses* and *Coverage* for appropriate references, methods and dates) where these provide better counts. Several goose populations are identified according to location (and totals derived by summing counts from particular WeBS regions) where they cannot be separated in the field by appearance.

Table A4. Use of WeBS Core Count and goose census data to compile national totals for goose populations in 1997-98.

Bean Goose	WeBS Core Counts in all months
Pink-footed Goose	October and November counts replaced by summed counts from the co-ordinated national censuses
European White-fronted Goose	WeBS Core Counts in all months
Greenland White-fronted Goose	November and March counts replaced by summed counts from the co-ordinated late autumn and late spring international censuses, respectively
Greylag Goose:	
Iceland	WeBS Core Counts from all WeBS regions in Scotland except those on the west coast (see NW Scotland population) plus Northumberland and North Cumbria. October and November counts replaced by summed counts from the co-ordinated national censuses.
NW Scotland	WeBS Core Counts from WeBS regions Islay/Jura/Colonsay, Mull/Lismore/Coll/Tiree Skye, Highland Southwest and North and South Outer Hebrides. August and February counts replaced by summed counts from co-ordinated censuses of Outer Hebrides. August 1997 count replaced by total from full national survey
naturalised	WeBS Core Counts for all sites in Wales and England, except for Northumberland and North Cumbria

Note that Icelandic and NW Scotland populations overlap in WeBS regions Orkney and North Highland. NW Scotland birds counted by WeBS in these regions will be included Icelandic population totals. Note also that up to 2,340 naturalised birds occur in Scotland (Delany 1992) and others in Northumberland and North Cumbria which are therefore incorrectly included in totals of Icelandic birds in Appendices 4-9

Canada Goose	WeBS Core Counts in all months
Barnacle Goose:	
Greenland	WeBS Core Counts from all WeBS regions on Scottish west coast, plus Shetland and Orkney. November and March counts replaced by summed counts from the co-ordinated late autumn and late spring censuses in Argyll plus the monthly maximum count from Hoy, Orkney.
Svalbard	WeBS Core Counts from WeBS regions Dumfries & Galloway, North Cumbria, Northumberland, Borders, Lothians, Central, Fife, Perth & Kinross, Angus, Grampian, Moray and SE Highland. Dumfries & Galloway and North Cumbria WeBS Core Counts replaced by Solway-wide counts and censuses between October and March.
naturalised	WeBS Core Counts for all WeBS regions in Wales and England, except for Northumberland and North Cumbria
Dark-bellied Brent Goose	WeBS Core Counts, plus additional counts of inland areas in January and February
Light-bellied Brent Goose:	

| Canada | | | WeBS Core Counts for sites in Northern Ireland, Wales and WeBS regions Shetland, Orkney, Highland North, Western Isles, Skye, Higland Southwest, Islay/Jura/Colonsay, Argyll West Mainland, Mull/Lismore/Coll/Tiree, Dumbarton/SE Argyll, Renfrew, Lanarkshire/Strathkelvin, Ayrshire & Arran, Dumfries & Galloway West, Cornwall, Devon and the Channel Islands. October count in Northern Ireland replaced by counts in Northern Ireland made during all-Ireland census. |
| Svalbard | | | WeBS Core Counts from regions not used to compile Canada population totals (see above) |

Annual indices

Underhill index values are derived from sites where at least 50% of the maximum possible number of counts, bearing in mind that different months are used for different species, were complete. Index values provided extend back to 1966-67 for wildfowl and 1971-72 for waders, representing the first years in which coverage was deemed sufficient for data to be included in the calculation of the index. A number of species were only first included in WeBS in the 1980s, whilst counts of wildfowl in Northern Ireland only began in earnest in 1985-86.

Underhill (1989) recommends that, where possible, the index is based on counts from more than one month. The months chosen for each species are given below. The most appropriate grouping of months on which to base the annual index for waders is December, January and February, the period when the wintering population in Britain and Northern Ireland is most stable (Prŷs-Jones et al. 1994). However, the peak abundance of different wildfowl occur in different months according to species, and thus different months and different numbers of months were selected for each (Kirby et al. 1995).

The selection of months for calculating indices for wildfowl and their allies was made by first calculating monthly index values for all months September to March, and selecting that with the highest index value and any adjacent months with overlapping consistency intervals. Data from all years from 1966-67 onwards were used for calculating the index for each of these species, as recommended in Kirby et al. (1995), or from the years in which data were first available for species added to the scheme subsequently (see above). Caution is urged in particular regarding the first few years' index values for these species only recently included in the scheme; missing counts may have been incorrectly recorded as nil counts, giving rise to anomalous index values. The parameters used for indexing each species follow Kirby et al. (1995).

Due to more stable populations of waders during the winter, the months December to February are chosen for calculation of index values for all waders for which there are suitable data. Due to the small number of sites in Northern Ireland, data are combined for analysis at the UK level.

Table A5. Months used in calculating indices for wildfowl species in Great Britain and Northern Ireland (indicated using the first letter of the months September to March)

Species	GB	NI	Species	GB	NI
Little Grebe	SO	SON	Shelduck	JF	DJFM
Great Crested Grebe	SON	SONDJFM	Wigeon	J	SONDJFM
Cormorant	SONDJFM	SOND	Gadwall	SONDJFM	SONDJ
Mute Swan	SONDJFM	SONDJ	Teal	D	DJ
Bewick's Swan	JF	NDJF	Mallard	D	SO
Whooper Swan	ND	ONDJFM	Pintail	ONDJ	ONDJFM
Pink-footed Goose	O or N	-	Shoveler	SO	SONDJFM
European Whitefront	JF	-	Pochard	NDJ	NDJF
Greenland Whitefront	N or M	N or M	Tufted Duck	NDJF	ONDJFM
Greylag Goose: Icelandic	O or N	-	Goldeneye	F	DJFM
naturalised	S	-	Red-breasted Merganser	ONDJFM	SONDJFM
Canada Goose	S	-	Goosander	DJF	-
Barnacle Goose: Svalbard	any month	-	Ruddy Duck	SONDJFM	-
Dark-bellied Brent	DJF	-	Coot	SONDJ	SONDJFM
Light-bellied Brent	-	SONDJFM			

Table A6. Great Britain annual index values for wildfowl

Year	LG	GG	CA	MS	BS	WS	PG	EW	NW	I-GJ	N-GJ	CG	DB	SU	WN	GA	T.	MA	PT	SV	PO	TU	GN	RM	GD	RY	CO
1966-67				64	40	85	32	117		76	3	18	22	67	67	5	25	163	32	45	127	71	83	27	41	0	
1967-68				64	43	46	28	223		67	3	16	27	78	77	4	18	142	45	49	134	75	71	24	45	0	
1968-69				58	33	54	28	303		76	2	22	22	73	67	6	31	146	55	68	117	80	86	76	52	1	
1969-70				59	32	81	31	328		78	2	25	25	102	70	5	35	118	50	55	106	80	59	52	68	—	
1970-71				55	70	55	31	263		81	3	31	30	87	73	12	32	172	73	48	105	90	86	56	52	—	
1971-72				56	50	85	28	110		80	5	27	30	92	57	14	36	147	69	73	127	101	103	42	61	—	
1972-73				58	30	82	29	69		92	6	38	34	80	64	13	38	144	90	70	142	110	108	38	46	2	
1973-74				56	49	63	35	202		96	5	33	45	81	65	11	47	133	154	83	136	109	133	43	103	3	
1974-75				56	38	96	38	55		87	6	33	38	77	57	14	43	143	134	52	152	105	129	55	88	3	
1975-76				53	53	75	31	112		79	5	25	53	93	53	12	53	128	126	52	154	111	78	49	64	6	
1976-77				51	78	59	30	143		70	8	37	49	87	72	14	50	145	118	67	120	89	80	56	57	12	
1977-78				51	89	89	29	98		84	12	39	52	76	54	21	54	112	142	78	136	96	85	58	50	11	
1978-79				54	96	67	33	186		96	12	47	62	118	81	21	65	139	130	81	112	95	84	61	91	13	
1979-80				55	94	97	34	97		102	14	52	78	99	66	25	62	138	130	83	110	103	89	59	75	20	
1980-81				54	123	86	41	116		113	16	56	63	112	64	28	74	141	159	74	96	106	75	68	100	34	
1981-82				59	119	60	38	140		120	20	61	69	124	74	33	94	137	138	74	109	94	96	76	114	30	
1982-83		62		60	105	99	38	117	35	101	25	70	92	94	48	36	79	148	151	81	116	95	79	74	110	33	90
1983-84		71		54	125	74	43	107	40	103	30	67	94	102	46	40	82	157	150	78	114	91	80	67	90	44	98
1984-85		84		54	188	75	36	145	46	78	40	81	92	96	116	39	65	134	184	85	101	93	94	76	81	49	103
1985-86	19	72		57	162	107	54	138	55	134	40	86	98	114	95	43	67	165	143	90	112	96	83	108	116	50	99
1986-87	29	71	29	61	210	122	58	125	56	128	41	92	98	119	105	47	64	144	100	82	103	101	108	118	128	53	98
1987-88	26	73	63	69	96	112	73	157	59	132	49	104	95	100	80	53	64	166	152	78	119	93	92	94	130	57	93
1988-89	70	94	87	74	128	160	75	140	62	137	52	93	108	100	80	60	78	147	138	80	111	109	91	96	124	62	106
1989-90	76	93	88	78	174	152	78	122	70	105	55	106	85	104	77	57	98	139	136	92	110	87	86	66	79	64	107
1990-91	81	90	101	88	193	170	81	98	74	144	61	117	115	112	76	63	86	135	118	103	105	97	115	82	101	66	93
1991-92	74	101	92	83	205	118	99	136	79	111	58	107	141	116	99	57	82	129	141	100	100	91	108	77	89	68	92
1992-93	70	91	91	81	140	114	84	33	74	123	65	111	96	100	95	58	71	127	100	77	105	93	107	82	86	64	96
1993-94	88	99	106	87	124	102	95	115	83	125	67	99	122	110	97	70	88	116	99	82	106	103	109	94	82	68	109
1994-95	103	100	114	92	98	121	111	83	94	108	80	115	98	95	108	79	85	115	106	86	124	99	111	114	81	65	110
1995-96	124	99	116	93	193	107	85	88	105	104	88	127	92	111	111	87	91	115	115	107	124	95	101	88	131	76	113
1996-97	103	96	105	96	185	151	99	102	102	100	104		99	114	130	87	81	110	119	102	104	104	145	89	162	93	112
1997-98	100	100	100	100	100	100	100	100	100	100	100	100	100	100	100	100	100	100	100	100	100	100	100	100	100	100	100

Wildfowl species codes

LG	Little Grebe	EW	European White-fronted Goose
GG	Great Crested Grebe	NW	Greenland White-fronted Goose
CA	Cormorant	I-GJ	Icelandic Greylag Goose
MS	Mute Swan	N-GJ	Naturalised Greylag Goose
BS	Bewick's Swan	CG	Canada Goose
WS	Whooper Swan	DB	Dark-bellied Brent Goose
PG	Pink-footed Goose	PB	Light-bellied Brent Goose

SU	Shelduck	PO	Pochard
WN	Wigeon	TU	Tufted Duck
GA	Gadwall	GN	Goldeneye
T.	Teal	RM	Red-breasted Merganser
MA	Mallard	GD	Goosander
PT	Pintail	RY	Ruddy Duck
SV	Shoveler	CO	Coot

Table A7. Northern Ireland annual index values for wildfowl

Year	LG	GG	CA	MS	BS	WS	PB	SU	WN	GA	T.	MA	PT	SV	PO	TU	GN	RM	CO
1986-87	48	44	3	64	351	98	38	68	108	80	68	118	153	197	110	54	186	57	75
1987-88	20	55	32	71	519	130	89	60	114	113	112	86	80	133	117	104	212	96	90
1988-89	112	68	72	78	498	134	85	103	145	117	110	104	117	152	167	110	177	110	85
1989-90	114	53	84	86	889	103	87	70	103	169	120	123	63	129	151	116	164	104	114
1990-91	87	47	60	82	976	118	105	82	130	133	138	116	101	110	154	109	209	101	105
1991-92	97	54	50	85	485	107	108	67	120	155	103	115	128	141	164	123	226	81	104
1992-93	110	68	71	90	235	111	77	61	92	183	71	97	97	104	119	114	199	105	115
1993-94	118	43	64	80	398	97	70	72	74	146	73	106	95	170	105	116	137	78	62
1994-95	139	95	68	96	130	108	86	78	84	192	76	118	64	137	100	116	147	98	99
1995-96	148	85	76	100	189	93	85	93	76	141	82	145	70	141	142	138	151	110	138
1996-97	106	68	67	85	329	116	88	86	84	143	89	102	97	127	109	111	119	82	97
1997-98	100	100	100	100	100	100	100	100	100	100	100	100	100	100	100	100	100	100	100

Table A8. UK annual index values for waders

Year	OC	AV	RP	GV	KN	SS	DN	BW	BA	CU	RK	TT
1970-71	68	4	79	16	127	80	82	17	113	56	60	86
1971-72	74	4	77	17	155	115	98	22	104	67	89	109
1972-73	72	3	94	20	121	104	110	34	96	62	91	99
1973-74	76	4	107	29	120	124	131	26	113	69	94	103
1974-75	77	3	109	31	92	147	125	36	90	84	102	98
1975-76	99	3	119	30	92	131	129	31	99	78	101	115
1976-77	96	1	101	34	92	95	132	31	121	69	92	120
1977-78	87	4	101	23	62	107	99	33	98	59	79	117
1978-79	86	3	96	31	75	132	102	34	111	59	85	115
1979-80	95	4	99	37	85	145	101	34	130	75	90	113
1980-81	102	3	91	51	96	108	98	33	105	68	82	101
1981-82	101	8	78	41	83	90	86	28	137	66	73	100
1982-83	92	9	89	44	96	102	85	38	116	68	71	98
1983-84	96	7	93	48	83	99	89	41	104	65	79	122
1984-85	99	13	98	56	89	92	81	36	128	68	80	124
1985-86	111	15	100	61	116	109	85	50	130	73	86	148
1986-87	107	13	101	63	98	102	74	37	133	79	85	137
1987-88	117	15	106	86	106	118	77	45	109	83	108	159
1988-89	113	23	135	91	111	114	96	60	106	78	104	135
1989-90	112	25	123	83	104	95	98	66	104	82	105	141
1990-91	119	32	113	92	105	102	116	51	126	76	90	130
1991-92	109	34	98	96	110	122	109	60	101	91	97	143
1992-93	105	48	107	90	118	100	95	63	101	93	92	126
1993-94	96	60	102	104	94	77	96	83	91	83	92	123
1994-95	101	78	106	126	102	112	108	90	90	102	108	125
1995-96	93	65	108	109	95	86	98	83	117	74	85	107
1996-97	120	61	96	128	111	157	127	105	163	80	94	106
1997-98	100	100	100	100	100	100	100	100	100	100	100	100

Wader species codes

OC	Oystercathcer
AV	Avocet
RP	Ringed Plover
GV	Grey Plover
KN	Knot
SS	Sanderling
DN	Dunlin
BW	Black-tailed Godwit
BA	Bar-tailed Godwit
CU	Curlew
RK	Redshank
TT	Turnstone

Table A10. Great Britain monthly index values for wildfowl.

	Sep	Oct	Nov	Dec	Jan	Feb	Mar
BS	0.0	5.9	23.9	97.1	96.7	100.0	1.7
CA	100.0	94.3	96.9	81.8	84.5	84.5	76.9
CO	88.7	90.2	100.0	93.9	71.1	58.7	39.2
DB	0.3	65.6	88.9	100.0	99.2	82.9	50.1
EW	0.1	1.3	10.1	49.3	89.7	100.0	14.0
GA	58.1	65.2	100.0	93.1	75.5	70.5	43.7
GD	24.7	21.9	52.6	91.5	95.4	100.0	55.7
GG	100.0	94.1	96.5	90.6	76.8	79.2	81.1
GN	2.3	19.0	54.0	92.9	83.2	100.0	90.1
LG	100.0	88.8	74.8	76.2	61.9	59.3	60.4
MA	85.9	88.1	100.0	91.5	79.6	56.3	35.9
PO	32.7	45.5	73.0	90.1	100.0	75.3	29.8
PT	36.7	71.2	60.4	100.0	54.1	47.2	17.0
SU	55.7	92.9	91.0	100.0	78.5	80.8	66.5
SV	83.7	100.0	81.7	72.5	69.1	69.7	66.5
T.	46.4	68.1	100.0	97.0	70.7	48.0	29.3
TU	80.3	77.2	93.8	100.0	85.6	78.7	69.8
WN	15.7	66.1	95.0	95.0	100.0	86.6	58.6
WS	2.3	42.6	100.0	92.9	78.8	90.9	76.6

Table A11. Northern Ireland monthly index values for wildfowl.

	Sep	Oct	Nov	Dec	Jan	Feb	Mar
BS	0.0	13.2	34.9	53.5	86.6	90.5	36.7
CA	86.7	100.0	70.6	67.3	47.3	58.5	43.1
CO	84.2	89.5	100.0	79.0	44.4	41.0	42.7
GA	84.9	50.0	66.5	81.6	60.5	78.3	100.0
GG	100.0	42.9	33.8	27.1	33.2	29.2	74.5
GN	1.2	9.3	100.0	72.4	74.4	75.5	92.5
LG	99.1	66.5	100.0	93.1	51.3	38.4	27.5
MA	100.0	79.6	75.0	70.4	43.8	39.9	23.6
PB	85.9	100.0	62.4	37.9	20.9	23.0	15.0
PO	1.0	8.8	47.9	100.0	97.7	41.4	10.5
PT	2.8	8.1	7.3	88.8	33.2	100.0	29.6
SP	38.7	0.1	36.1	13.3	85.4	100.0	60.1
SU	2.2	19.4	61.5	80.1	100.0	83.3	56.3
SV	10.8	48.7	37.8	100.0	49.7	35.7	31.9
T.	16.0	45.2	54.4	100.0	75.9	66.8	52.1
TU	30.4	38.1	100.0	87.8	85.6	68.9	44.9
WN	18.9	100.0	88.1	87.2	29.6	33.4	31.2
WS	2.0	50.5	58.3	32.3	71.6	59.7	100.0

APPENDIX 4. TOTAL NUMBERS OF WATERFOWL RECORDED BY WeBS IN ENGLAND, 1997-98

	Apr	May	Jun	Jul	Aug	Sep	Oct	Nov	Dec	Jan	Feb	Mar
Sectors	*1,081*	*904*	*817*	*856*	*950*	*1,330*	*1,624*	*1,704*	*1,717*	*1,787*	*1,774*	*1,713*
Sites	*595*	*563*	*522*	*544*	*559*	*768*	*934*	*981*	*983*	*1018*	*999*	*957*
Red-t. Diver	89	15	0	26	4	47	82	168	182	193	161	226
Black-t. Diver	2	0	0	0	0	0	6	8	12	5	4	5
Great N. Diver	2	0	0	0	0	0	1	8	29	30	17	21
Pied-billed Grebe	0	0	1	0	0	0	0	0	0	0	0	0
Little Grebe	807	464	500	707	1,388	2,471	2,507	2,425	2,385	1,873	1,859	1,933
Great C. Grebe	3,571	2,710	2,785	3,116	4,533	6,171	7,050	7,415	6,797	6,187	6,576	6,527
Red-n. Grebe	2	1	2	3	6	6	20	13	31	28	21	10
Slavonian Grebe	5	1	0	0	0	11	13	27	90	104	50	72
Black-n. Grebe	5	10	7	4	15	23	23	31	47	48	13	28
Cormorant	5,003	3,547	2,416	3,349	5,755	9,105	10,343	10,072	9,412	9,271	9,046	8,080
Bittern	1	0	0	1	0	0	8	7	14	12	10	6
Little Bittern	0	0	1	0	0	0	0	0	0	0	0	0
Little Egret	146	21	47	189	405	373	408	259	255	259	250	299
Great White Egret	1	0	0	0	0	0	0	0	0	2	0	1
Grey Heron	1,346	1,033	1,245	1,648	2,177	2,765	2,742	2,517	2,392	2,174	2,222	2,186
White Stork	1	1	1	1	1	1	3	3	3	3	2	2
Spoonbill	0	2	2	4	2	1	5	5	11	7	12	6
Chilean Flamingo	1	0	1	0	0	0	0	0	0	0	0	0
Lesser Flamingo	0	0	0	0	1	1	1	1	1	1	0	1
Greater Flamingo	1	1	1	0	1	1	1	2	1	1	0	0
Fulv. Whist. Duck	0	0	0	0	0	2	0	0	0	0	0	1
Mute Swan	6,825	5,209	6,499	7,869	9,288	10,673	12,241	14,187	13,207	12,782	11,856	11,308
Black Swan	11	6	11	9	12	18	15	21	23	12	11	21
Trumpeter Swan	0	0	0	0	0	3	3	3	2	3	3	2
Bewick's Swan	4	0	0	3	0	0	127	370	5,409	4,099	4,883	100
Whooper Swan	181	5	4	5	3	6	454	691	1,599	863	2,389	1,578
Swan Goose	0	0	11	13	14	16	33	34	26	38	30	25
Bean Goose	2	1	1	1	0	0	1	5	49	41	34	3
Pink-f. Goose	1,247	90	10	9	11	3,862	25,263	39,683	87,543	50,066	29,431	18,360
White-f. Goose	0	0	1	0	0	0	0	0	0	0	0	0
Euro. Whitefront	3	0	0	1	1	2	52	474	2,150	4,091	5,341	659
Greenl. Whitefront	0	0	0	0	0	0	0	1	1	2	6	0
Lesser WF Goose	0	0	1	1	1	0	1	1	0	2	2	0
Greylag (Iceland)	141	107	597	616	244	634	1210	2,083	1,842	769	978	1,328
Greylag (natur.)	4,426	3,092	5,661	6,386	11,810	15,175	16,697	15,206	15,356	12,815	12,869	9,911
Bar-headed Goose	5	4	5	7	8	15	7	14	14	6	9	14
Snow Goose	27	10	24	25	4	14	67	63	32	44	62	64
Ross's Goose	1	0	0	0	5	0	0	1	1	1	0	0
Emperor Goose	0	0	1	1	2	2	3	4	6	5	3	3
Canada Goose	10,576	8,751	17,942	18,112	26,336	38,125	38,839	41,330	36,630	34,708	28,832	21,940
Barnacle (natur.)	83	43	31	98	144	332	609	548	268	388	362	433
Brent Goose	0	0	0	0	0	0	0	0	2	10	0	0
Dark-b. Brent	16,667	2,012	41	26	41	260	52,178	83,603	98,792	92,112	77,862	51,644
Black Brant	0	0	0	0	0	0	0	1	0	1	1	0
Light-b. Brent (Sva)	5	0	0	0	0	1,659	2,383	2,576	2,495	729	429	29
Light-b. Brent (Canada)	0	0	0	0	0	0	0	1	1	1	0	2
Red-b. Goose	0	0	0	0	0	0	0	0	0	0	0	1
Egyptian Goose	27	17	29	93	373	219	232	167	163	152	163	101
Feral/hybrid Goose	32	42	45	56	39	47	169	159	250	361	242	194
Unidentified Goose	14	0	0	0	0	0	0	0	0	0	0	0
Ruddy Shelduck	3	0	2	6	2	9	7	6	8	8	5	7
Cape Shelduck	0	2	2	4	0	0	1	0	0	0	0	0
Paradise Shelduck	0	0	0	0	0	0	1	0	0	1	0	0
Shelduck	25,179	11,906	11,162	14,817	25,063	29,480	50,499	53,656	65,247	56,887	56,135	44,044
Muscovy Duck	10	13	11	15	12	20	101	116	127	99	93	35
Wood Duck	3	0	0	1	1	0	4	8	8	2	5	9
Mandarin	76	52	120	72	47	155	174	314	258	227	170	179
Wigeon	9,356	178	112	1,040	552	28,282	136,663	213,676	247,170	268,598	224,884	160,192
Am. Wigeon	0	0	0	0	0	0	0	1	1	0	1	3
Chiloe Wigeon	0	0	0	1	0	0	2	4	1	3	4	4
Falcated Duck	0	0	0	0	0	0	0	0	0	0	0	1
Gadwall	2,176	1,143	1,203	1,030	3,728	6,483	7,937	12,749	12,438	10,720	10,640	6,680
Teal	13,522	274	353	1,606	11,502	43,547	64,932	104,632	112,235	94,341	64,046	41,266
Speckled Teal	0	1	0	1	1	1	0	0	0	0	1	2

	Apr	May	Jun	Jul	Aug	Sep	Oct	Nov	Dec	Jan	Feb	Mar
Mallard	21,806	16,818	25,454	32,137	58,735	74,648	86,930	108,284	99,486	90,778	65,029	43,003
Chestnut Teal	0	0	0	0	0	0	0	1	0	0	0	0
Pintail	602	10	13	63	105	6,029	11,717	14,467	18,780	12,460	12,969	7,304
Bahama Pintail	0	0	0	0	0	1	0	0	0	0	0	0
Cape Teal	0	0	0	0	0	0	1	1	1	0	0	0
Garganey	12	25	6	19	37	17	4	4	0	0	1	5
Shoveler	2,607	308	270	598	3,109	6,997	8,363	7,840	7,064	6,400	7,051	6,689
Ringed Teal	0	0	0	0	1	0	0	0	0	0	0	0
Red-c. Pochard	0	1	2	0	6	15	46	94	70	82	39	44
Pochard	984	431	745	1,796	6,252	6,935	10,570	17,687	23,672	35,379	25,573	9,586
Ring-necked Duck	0	0	0	0	1	1	0	1	1	1	2	3
Ferruginous Duck	1	1	0	0	1	1	2	1	2	1	2	1
Tufted Duck	16,094	5,627	6,454	13,936	24,975	27,543	29,769	38,945	43,013	40,588	37,301	31,688
Scaup	832	20	2	3	11	42	50	100	101	189	800	529
Lesser Scaup	1	0	0	0	0	0	0	0	0	0	0	0
Eider	6,570	3,006	3,627	4,303	5,499	6,783	9,993	3,989	6,074	3,183	3,492	5,097
Long-tailed Duck	0	0	0	1	0	0	14	14	86	63	48	19
Common Scoter	1,058	80	54	47	70	544	493	1,615	2,502	1,637	1,062	646
Velvet Scoter	0	0	0	0	0	0	22	2	11	4	0	0
Unid. scoter sp.	0	0	0	0	0	0	0	1	0	0	0	0
Goldeneye	1,816	23	5	8	12	71	761	3,166	5,829	6,549	6,907	5,363
Hooded Merganser	0	0	0	0	1	1	0	0	0	0	0	0
Smew	5	0	0	0	0	0	2	11	138	278	214	63
Red-b. Merganser	779	143	250	109	142	590	979	1,670	2,225	2,465	2,316	2,664
Goosander	270	82	69	160	184	196	207	819	2,156	2,467	2,655	1,383
Ruddy Duck	1,194	520	442	493	1,164	1,766	2,183	2,513	3,259	2,668	2,902	2,517
Feral/hybrid duck	59	116	110	90	90	136	131	137	129	162	129	102
Hybrid Aythya	1	0	0	0	0	1	0	1	1	0	1	1
Water Rail	82	45	34	41	54	103	167	223	276	212	231	212
Spotted Crake	0	0	2	0	1	2	0	0	0	0	0	0
Moorhen	4,226	2,749	2,311	3,760	5,333	7,574	9,243	10,345	9,597	9,449	10,530	9,697
Coot	18,109	10,524	15,990	22,510	41,185	64,552	76,027	91,144	85,403	70,387	58,140	40,410
Crane	0	0	0	0	0	0	0	6	4	6	6	0
Oystercatcher	55,394	28,250	23,880	40,421	121,197	171,640	161,287	187,903	157,448	133,140	151,180	97,137
Black-winged Stilt	0	0	0	0	1	0	0	1	1	1	1	1
Avocet	1,035	227	237	453	819	1,550	2,554	2,409	3,859	3,464	3,232	2,469
Black-w. Pratincole	0	0	0	0	1	0	0	0	0	0	0	0
Little R. Plover	192	278	249	139	52	17	5	0	0	0	0	9
Ringed Plover	3,154	6,864	793	1,262	19,874	9,984	6,357	7,378	6,130	5,051	5,556	4,063
Kentish Plover	1	0	0	0	0	0	0	0	0	0	0	0
Dotterel	1	4	0	0	0	0	0	0	1	0	0	0
Golden Plover	7,650	387	140	1,781	25,367	31,462	52,824	149,119	162,598	126,829	131,645	40,375
Grey Plover	32,886	16,269	947	4,776	26,256	34,431	31,367	43,249	32,330	42,938	44,680	40,997
Unid. wader	0	0	0	0	0	0	0	0	0	0	1	0
Lapwing	8,623	3,325	12,098	47,746	57,422	81,510	145,399	360,853	416,469	404,283	309,958	31,186
Knot	70,113	8,217	3,079	12,738	70,088	134,362	182,062	286,487	241,799	154,199	211,009	128,648
Sanderling	7,368	11,089	131	9,113	7,146	7,670	7,552	5,512	5,749	5,228	4,634	6,639
Little Stint	1	12	5	8	28	41	15	5	2	1	1	2
Pectoral Sandpiper	0	0	0	0	0	3	0	0	0	0	0	0
Curlew Sandpiper	0	7	2	16	39	93	36	1	0	0	1	0
Purple Sandpiper	512	12	0	12	38	34	161	257	379	553	349	562
Dunlin	94,790	83,189	1,212	43,027	75,522	79,633	218,289	343,674	401,699	379,789	351,978	162,524
Ruff	276	93	4	136	418	498	211	182	261	380	420	395
Jack Snipe	9	1	0	0	0	3	56	42	52	47	90	64
Snipe	753	113	33	90	817	1,245	3,513	4,487	5,948	3,639	4,339	3,057
Great Snipe	0	0	0	0	1	0	0	0	0	0	0	0
Long-b. Dowitcher	0	0	0	1	0	0	1	1	0	0	0	0
Woodcock	2	2	0	0	0	1	10	17	38	18	15	5
Black-t Godwit	9,983	1,473	1,264	3,240	16,578	16,019	12,796	11,997	12,368	11,108	10,498	14,180
Bar-t Godwit	3,612	1,055	1,078	7,807	16,594	25,922	18,216	37,666	30,779	39,954	36,733	14,805
Whimbrel	111	1,343	128	620	550	281	41	11	2	1	4	9
Curlew	26,671	3,823	9,406	41,546	59,179	67,032	63,359	45,138	54,003	54,549	59,548	49,466
Spot. Redshank	67	22	14	160	205	178	125	174	90	48	129	52
Redshank	31,400	2,656	2,961	14,306	40,523	54,582	58,523	54,354	52,108	51,658	61,649	57,183
Greenshank	89	93	14	628	1,617	1,435	575	204	127	465	94	121
Lesser Yellowlegs	0	0	0	0	0	0	0	0	0	0	2	0
Green Sandpiper	55	5	37	212	447	241	150	156	123	72	97	84
Wood Sandpiper	3	1	1	12	43	16	1	0	0	0	0	0
Common Sand.	42	323	235	707	1,217	272	40	27	24	18	43	39

	Apr	May	Jun	Jul	Aug	Sep	Oct	Nov	Dec	Jan	Feb	Mar
Turnstone	6,589	2,008	237	1,057	4,219	7,745	10,694	9,918	9,023	7,917	7,638	8,913
Grey Phalarope	0	0	0	1	1	1	3	0	0	2	0	0
Mediterranean Gull	17	8	6	28	30	15	48	29	15	69	59	46
Little Gull	47	26	5	21	28	9	17	15	0	4	0	1
Sabine's Gull	0	0	0	0	1	3	0	0	0	0	0	0
Black-h. Gull	48,713	25,917	30,942	65,680	97,271	110,095	109,268	216,106	217,789	204,935	163,136	133,797
Ring-b. Gull	0	0	0	0	0	1	2	1	3	2	5	3
Common Gull	4,075	2,184	1,750	2,727	9,246	11,155	10,193	46,049	48,359	42,522	50,999	23,617
Lesser B-b Gull	33,731	37,573	46,545	56,750	19,505	10,800	9,025	19,001	7,245	5,029	17,108	28,996
Herring Gull	27,283	26,021	28,260	29,148	18,748	29,040	31,378	47,583	35,233	46,276	45,655	39,168
Iceland Gull	1	2	0	0	0	0	0	0	0	2	2	3
Glaucous Gull	3	0	0	0	0	1	0	0	1	5	7	2
Great B-b Gull	1,164	1,250	1,335	1,600	3,669	4,850	4,783	11,062	7,173	8,135	3,552	1,794
Kittiwake	884	787	308	580	1,257	164	297	42	31	108	15	325
Unidentified gull	0	0	0	0	3,500	1,260	1,017	499	4,165	990	280	480
Sandwich Tern	347	1,326	684	2,154	2,739	652	51	3	0	1	0	6
Roseate Tern	0	0	0	0	2	0	0	0	0	0	0	0
Common Tern	118	1,575	2,235	2,888	3,038	597	16	1	0	0	0	0
Arctic Tern	6	112	131	170	50	23	8	2	0	0	0	0
Little Tern	3	267	216	84	65	23	0	0	0	0	0	0
Black Tern	0	23	2	0	21	4	1	0	0	0	0	0
White-w Black Tern	0	0	0	0	2	0	0	0	0	0	0	0
Unidentified tern	0	16	16	31	63	20	0	0	0	0	0	0
Kingfisher	58	61	78	117	145	225	243	180	188	119	162	166

TOTALS

	Apr	May	Jun	Jul	Aug	Sep	Oct	Nov	Dec	Jan	Feb	Mar
WATERFOWL	540,558	252,434	163,914	373,061	796,759	1,132,466	1,658,374	2,463,652	2,628,308	2,375,942	2,185,269	1,219,557
GULLS	115,918	93,768	109,151	156,534	153,255	167,393	166,028	340,387	320,014	308,077	280,818	228,232
TERNS	474	3,319	3,284	5,327	5,980	1,319	76	6	0	1	0	6

APPENDIX 5. TOTAL NUMBERS OF WATERFOWL RECORDED BY WeBS IN SCOTLAND, 1997-98

	Apr	May	Jun	Jul	Aug	Sep	Oct	Nov	Dec	Jan	Feb	Mar
Sectors	277	233	201	209	212	482	657	624	641	651	661	638
Sites	230	190	163	175	170	417	541	540	532	539	550	554
Red-t. Diver	232	48	4	114	23	192	129	210	360	235	339	135
Black-t. Diver	14	14	0	0	3	1	22	29	18	8	14	17
Great N. Diver	32	20	2	2	1	16	19	15	43	11	13	22
Little Grebe	110	83	94	123	254	656	551	350	309	345	290	281
Great C. Grebe	249	147	152	300	586	910	812	435	592	544	555	740
Red-n. Grebe	10	2	0	18	64	29	27	18	10	8	17	17
Slavonian Grebe	46	3	1	1	3	25	116	118	219	136	139	81
Black-n. Grebe	1	3	3	4	3	1	1	0	0	0	1	0
Cormorant	800	513	544	688	978	1,521	2,578	2,128	1,925	2,364	1,678	1,539
Grey Heron	178	108	185	259	329	524	706	458	677	654	537	479
Mute Swan	1,282	1,053	1,375	1,500	1,636	2,698	3,941	3,515	3,580	3,534	3,169	2,634
Black Swan	0	0	2	1	1	1	5	4	7	3	4	4
Bewick's Swan	0	0	2	0	0	0	0	3	0	10	0	0
Whooper Swan	224	7	6	4	2	41	1,144	1,614	1,515	1,524	1,340	1,441
Bean Goose	2	0	0	0	0	0	35	159	0	4	21	0
Pink-f. Goose	18,071	554	5	10	249	60,127	209,936	142,449	32,831	17,909	42,915	35,811
White-f. Goose	0	0	0	0	0	1	0	0	0	0	0	0
Euro. Whitefront	230	0	0	0	0	0	0	0	5	4	4	5
Greenl. Whitefront	79	0	0	0	0	0	497	20,572	525	587	100	19,647
Greylag (Iceland)	5,330	209	341	462	488	2,505	38,189	77,394	26,393	15,556	12,869	11,684
Greylag (NW Scot)	52	4	7	9	9,793	242	228	183	129	229	2,535	293
Snow Goose	0	0	0	0	0	1	3	5	3	5	2	5
Canada Goose	139	188	379	472	624	1,292	1,192	624	611	435	472	423
Barnacle (Greenland)	0	0	0	0	3	37	85	35,123	34	48	28	33,841
Barnacle (Svalbard)	3599	1	1	1	1	631	19,882	22,244	15,719	18,138	21,777	23,863
Dark-b. Brent	0	0	0	0	0	2	7	0	0	1	1	0
Light-b. Brent (Svalbard)	1	0	0	0	0	1	2	7	12	24	2	1
Light-b. Brent (Canada)	0	0	0	0	0	0	10	3	11	3	10	7
Feral/hybrid Goose	0	0	0	0	0	112	107	102	88	78	68	65
Unid. goose	65	0	0	0	0	0	0	9	0	0	0	0
Shelduck	2,688	1,852	3,350	744	5,824	3,349	5,212	6,442	5,871	5,875	5,508	3,734
Muscovy Duck	7	6	5	0	0	0	0	0	0	0	0	1
Mandarin	0	1	1	0	2	1	5	0	7	7	2	2
Wigeon	1,802	249	98	63	318	6,266	62,361	30,453	54,226	48,262	35,576	18,873
Am. Wigeon	0	0	0	0	0	0	0	0	0	0	2	1
Falcated Duck	0	0	0	0	0	0	0	0	0	0	0	1
Gadwall	24	11	4	3	15	445	503	167	112	61	110	145
Teal	1,109	89	111	113	1,685	6,110	18,103	11,412	17,581	15,554	10,509	5,565
Mallard	2,738	2,354	3,316	4,508	7,722	20,444	29,755	26,397	29,919	27,317	19,493	9,678
Pintail	105	6	0	4	57	1,002	2,024	1,419	3,771	1,608	1,660	466
Shoveler	78	19	15	16	86	647	560	249	250	173	165	198
Pochard	89	22	46	161	236	927	4,392	5,091	4,520	5,081	3,072	1,353
Ring-n. Duck	0	0	0	0	0	1	0	0	0	0	0	0
Tufted Duck	2,538	977	1,061	2,610	3,225	8,970	8,038	7,630	7,476	8,339	7,219	7,402
Scaup	905	5	3	6	26	167	1,004	1,303	7,411	4,182	3,364	778
Eider	13,467	11,198	11,814	11,891	16,048	16,124	14,542	15,087	13,223	12,135	10,991	10,809
King Eider	0	0	0	0	0	1	1	0	0	0	0	0
Long-tailed Duck	1,186	22	0	0	0	1	362	656	1,705	1,451	683	807
Common Scoter	2,988	1,186	482	454	805	2,889	3,227	3,529	4,637	2,914	2,185	1,365
Surf Scoter	3	0	0	0	0	0	1	3	6	3	2	3
Velvet Scoter	576	172	101	18	84	61	289	742	781	386	454	528
Goldeneye	2,163	139	77	124	88	169	2,624	4,875	10,156	7,548	7,674	7,381
Smew	1	2	0	0	0	0	0	4	6	13	13	7
Red-b. Merganser	1,322	484	493	523	846	1,193	1,520	1,180	1,759	1,434	1,395	1,368
Goosander	221	372	356	822	845	610	741	1,220	1,017	975	903	724
Ruddy Duck	47	54	61	62	61	151	156	97	69	60	35	42
Feral/hybrid duck	2	2	2	2	1	3	4	11	6	2	3	6
Unid. duck	0	0	0	0	0	2	3	0	0	0	0	0
Water Rail	14	0	10	6	18	11	12	21	29	21	21	8
Moorhen	295	180	149	205	354	897	984	890	925	924	852	791
Coot	1,290	793	957	1,794	2,368	6,254	7,579	7,907	6,899	6,407	6,260	3,802
Oystercatcher	17,044	7,260	6,797	13,759	29,005	29,952	52,242	43,474	57,625	58,874	55,429	37,539
Little R. Plover	0	6	0	0	0	0	0	1	0	0	0	0
Ringed Plover	883	2,358	340	270	2,220	1,767	2,525	1,661	1,991	1,810	1,521	527

	Apr	May	Jun	Jul	Aug	Sep	Oct	Nov	Dec	Jan	Feb	Mar
Golden Plover	843	7	26	579	2,435	5,149	16,322	11,786	11,728	9,895	10,886	2,480
Grey Plover	1,058	527	87	80	1,029	1,058	1,637	1,350	1,663	2,292	1,806	1,537
Lapwing	882	424	871	6,113	10,620	17,345	24,729	30,509	27,016	21,916	29,346	5,627
Knot	1,211	72	27	15	442	1,529	2,261	7,001	13,428	23,353	13,323	2,475
Sanderling	238	202	0	74	153	521	762	185	656	489	444	475
Western Sandpiper	0	0	0	0	1	0	0	0	0	0	0	0
Little Stint	0	0	0	0	0	2	0	0	0	0	0	0
Curlew Sandpiper	0	0	0	0	2	9	3	0	0	0	0	0
Purple Sandpiper	190	10	0	0	2	18	67	288	621	506	303	55
Dunlin	2,620	8,495	476	1,269	2,950	5,209	11,633	18,023	40,347	48,437	35,916	5,452
Ruff	6	1	0	1	59	88	4	1	1	4	4	4
Jack Snipe	3	0	0	0	0	0	7	10	23	11	10	4
Snipe	86	23	12	50	115	237	1,100	972	662	368	428	323
Woodcock	0	0	1	0	1	0	0	1	1	2	3	4
Black-t. Godwit	292	102	89	120	180	360	497	413	202	185	265	147
Bar-t. Godwit	638	239	233	1,177	627	2,957	2,719	2,922	5,290	7,984	5,743	1,538
Whimbrel	5	102	11	5	14	3	2	0	0	0	0	0
Curlew	3,797	832	1,091	6,035	8,866	11,253	14,527	10,040	18,827	21,992	23,227	13,638
Spot. Redshank	0	0	0	1	3	3	2	2	0	1	1	1
Redshank	6,919	851	690	2,898	7,977	11,624	20,171	14,842	19,439	17,153	14,906	12,062
Greenshank	9	4	11	42	101	181	60	32	34	34	39	29
Green Sandpiper	0	1	0	1	4	1	0	2	3	2	1	1
Wood Sandpiper	0	2	0	0	2	0	0	0	0	0	0	0
Common Sand.	1	108	78	133	38	9	0	2	0	0	0	0
Turnstone	1,305	415	114	208	1,065	1,555	2,915	2,575	3,266	2,395	2,186	1,905
Mediterranean Gull	0	0	0	0	0	0	1	0	0	3	0	0
Little Gull	0	4	0	18	0	0	1	0	0	4	0	0
Black-h. Gull	3,242	4,589	2,416	5,523	10,072	13,879	18,803	24,562	17,805	17,993	16,574	16,112
Ring-b. Gull	0	0	0	0	0	0	0	0	0	0	0	1
Common Gull	4,068	1,088	1,333	2,079	2,827	10,869	22,532	23,721	20,197	25,563	33,787	14,529
Lesser B-b Gull	1,439	686	1,851	1,735	1,502	2,029	1,668	751	134	107	220	1,007
Herring Gull	5,685	4,225	4,876	6,984	11,381	13,038	9,640	15,067	11,292	11,892	12,186	6,655
Iceland Gull	0	0	0	0	0	0	0	2	0	1	3	0
Glaucous Gull	1	0	0	0	0	1	0	0	0	3	1	2
Great B-b Gull	431	319	234	853	1,337	1,200	1,346	2,264	1,581	1,202	794	856
Kittiwake	413	290	77	513	2,864	728	591	195	167	73	1	12
Unid. gull	19	310	53	305	242	356	370	431	2,330	2,252	5,480	120
Sandwich Tern	25	807	171	1,100	2,553	1,803	36	0	0	0	0	0
Common Tern	6	495	220	1,078	643	297	20	0	0	0	0	0
Arctic Tern	1	137	254	837	1,287	54	0	0	0	0	0	0
Little Tern	0	90	55	32	70	0	1	0	0	0	0	0
Black Tern	0	0	0	0	1	2	0	0	0	0	0	0
Unid. tern	0	2	0	414	1	0	0	0	0	0	0	0
Kingfisher	3	5	3	4	3	18	17	14	11	9	9	15

TOTALS

	Apr	May	Jun	Jul	Aug	Sep	Oct	Nov	Dec	Jan	Feb	Mar
WATERFOWL	104,481	45,193	36,569	60,927	123,667	239,091	598,412	580,698	460,901	430,832	402,838	294,696
GULLS	15,298	11,511	10,840	18,010	30,225	42,100	54,952	66,994	53,506	59,093	69,046	39,294
TERNS	32	1,531	700	3,461	4,555	2,156	57	0	0	0	0	0

APPENDIX 6. TOTAL NUMBERS OF WATERFOWL RECORDED BY WeBS IN WALES, 1997-98

	Apr	May	Jun	Jul	Aug	Sep	Oct	Nov	Dec	Jan	Feb	Mar
Sectors	111	108	105	108	99	173	202	191	195	194	187	194
Sites	66	69	64	64	65	102	115	113	115	115	108	110
Red-t. Diver	2	0	0	0	0	1	27	17	29	3	1	1
Black-t. Diver	0	0	0	0	0	1	0	2	5	2	3	2
Great N. Diver	0	0	0	0	0	0	0	1	1	2	5	1
Little Grebe	33	22	22	44	69	194	215	248	309	229	204	155
Great C. Grebe	117	82	81	120	118	173	129	128	312	168	169	152
Red-n. Grebe	0	0	0	0	0	0	0	0	0	0	0	2
Slavonian Grebe	4	0	0	0	0	0	0	1	6	2	3	3
Black-n. Grebe	2	0	0	0	0	1	0	1	0	0	0	0
Cormorant	173	188	468	296	563	730	730	626	594	541	414	392
Bittern	0	0	0	0	0	0	0	0	0	0	0	1
Little Egret	3	2	3	5	9	16	22	18	34	19	22	25
Grey Heron	82	47	90	131	175	270	232	148	219	127	168	156
Spoonbill	0	0	0	0	0	1	0	0	0	1	0	0
Mute Swan	209	181	231	293	294	471	501	468	577	447	388	422
Black Swan	0	0	0	0	1	1	1	1	1	1	0	1
Bewick's Swan	16	0	0	0	0	0	0	0	32	1	17	4
Whooper Swan	0	1	0	0	0	0	21	50	57	63	52	46
Bean Goose	0	0	0	0	0	1	0	0	0	0	0	0
Pink-f. Goose	0	0	0	0	0	0	0	0	1	4	55	0
Euro. Whitefront	0	0	0	0	1	1	0	0	0	0	0	0
Greenl. Whitefront	110	0	0	0	0	8	48	81	74	3	97	109
Greylag (naturalised)	236	250	734	563	596	802	600	216	377	135	193	157
Bar-headed Goose	0	0	0	0	2	0	2	0	0	0	2	2
Snow Goose	0	0	0	0	0	1	0	0	0	0	0	0
Ross's Goose	1	1	1	0	1	0	0	0	1	0	1	1
Canada Goose	270	189	1,062	934	985	978	1,755	1,271	2,401	1,878	1,160	617
Barnacle (naturalised)	0	0	0	0	3	4	10	36	36	0	32	2
Brent Goose	0	0	0	0	0	2	2	1	1	1	0	1
Dark-bellied Brent	6	0	0	0	0	0	8	800	253	456	1,165	438
Light-b. Brent (Canada)	4	0	0	0	0	0	3	11	11	18	25	17
Feral/hybrid Goose	1	0	9	7	5	8	3	2	1	0	0	1
Ruddy Shelduck	0	0	0	1	1	1	1	1	1	1	0	0
Shelduck	2,191	1,316	1,230	313	125	936	2,249	3,041	3,226	3,762	4,150	3,906
Muscovy Duck	8	10	12	12	1	12	0	0	0	0	0	0
Mandarin	1	1	1	0	0	0	1	1	1	1	1	0
Wigeon	128	7	1	2	148	2,648	9,010	13,844	13,316	10,173	8,141	4,035
Chiloe Wigeon	0	0	0	0	0	0	1	0	0	0	0	0
Gadwall	53	23	38	23	91	85	84	157	189	175	106	71
Teal	346	11	2	21	278	1,927	3,627	6,913	7,938	7,054	5,610	2,240
Mallard	1,024	831	1,651	1,175	3,468	5,928	6,374	5,524	5,228	4,842	3,633	1,996
Pintail	26	0	0	0	0	190	756	1,586	1,966	766	1,133	226
Garganey	0	1	0	0	0	0	0	0	0	0	0	0
Blue-winged Teal	0	0	0	0	0	0	0	0	0	0	0	0
Shoveler	109	17	6	0	47	126	345	541	544	837	524	370
Red-c. Pochard	0	1	1	1	2	1	0	0	1	3	2	2
Pochard	138	61	94	78	148	240	553	1,298	1,540	1,631	1,233	376
Tufted Duck	606	200	240	611	945	1,244	982	1,379	1,515	1,610	1,038	1,025
Scaup	0	0	0	1	0	6	2	6	17	26	9	7
Eider	20	17	43	33	30	58	43	61	52	63	86	72
Long-tailed Duck	0	0	0	0	0	0	0	0	2	0	1	0
Common Scoter	10	2	0	0	14	0	151	96	1,426	937	170	260
Goldeneye	137	4	0	0	1	1	70	194	370	421	478	395
Smew	0	0	0	0	0	0	0	0	5	9	9	2
Red-b. Merganser	239	77	128	69	44	181	264	184	184	117	186	238
Goosander	28	1	6	4	2	11	7	40	72	71	70	59
Ruddy Duck	85	54	49	90	107	270	164	177	257	264	168	97
Feral/hybrid duck	30	17	26	24	53	29	35	17	23	17	0	23
Hybrid Aythya	0	0	0	0	0	0	0	0	0	0	0	1
Water Rail	1	2	0	1	0	10	21	26	40	23	15	19
Moorhen	232	190	170	175	237	399	483	566	592	446	461	400
Coot	789	503	880	1,428	2,523	3,303	3,586	3,456	3,090	2,098	1,605	1,176
Oystercatcher	8,113	5,689	4,042	8,600	21,700	24,450	22,778	19,999	24,434	21,194	20,462	14,189
Little R. Plover	2	4	0	1	0	1	0	0	0	0	0	0
Ringed Plover	299	127	58	70	1,069	829	484	567	428	447	257	191

	Apr	May	Jun	Jul	Aug	Sep	Oct	Nov	Dec	Jan	Feb	Mar
Golden Plover	0	1	0	0	0	2	1,504	3,772	1,119	1,437	4,397	1,380
Grey Plover	11	12	0	0	90	30	208	485	546	462	290	125
Lapwing	183	181	143	770	727	1,360	2,659	7,374	20,981	9,289	7,972	478
Knot	0	3	0	0	25	210	224	537	2,170	1,665	678	8
Sanderling	143	38	6	31	310	240	709	785	308	199	195	174
Little Stint	0	0	0	0	0	4	0	0	0	0	0	0
Curlew Sandpiper	0	0	0	0	9	14	3	0	0	0	0	0
Purple Sandpiper	5	0	0	0	0	3	0	0	3	2	6	5
Dunlin	166	567	26	503	1,638	2,995	2,541	14,881	20,536	20,569	17,196	1,221
Ruff	0	0	0	2	3	1	0	0	1	0	0	0
Jack Snipe	0	0	0	0	0	0	2	3	3	2	3	6
Snipe	25	2	3	0	16	47	214	567	793	614	271	161
Long-b. Dowitcher	0	0	0	0	0	0	1	0	0	0	0	0
Woodcock	0	0	0	0	0	0	0	1	1	9	7	0
Black-t. Godwit	241	133	249	119	186	333	1,755	1,538	629	954	1,759	1,134
Bar-t. Godwit	21	12	1	0	6	119	111	93	337	375	285	82
Whimbrel	1	290	84	87	36	12	1	1	1	2	1	0
Curlew	2,047	469	1,126	4,564	5,376	8,013	7,733	6,759	8,702	9,141	8,845	5,068
Spot. Redshank	4	1	0	5	13	24	20	3	8	6	12	7
Redshank	2,045	78	430	1,626	3,256	5,237	5,961	4,408	4,185	3,979	4,064	3,173
Greenshank	7	2	1	42	112	169	78	25	31	28	25	30
Green Sandpiper	0	0	0	3	1	3	7	3	6	4	2	3
Wood Sandpiper	0	0	0	0	1	1	0	0	0	0	0	0
Common Sand.	4	30	11	72	81	21	3	0	2	2	1	2
Turnstone	150	20	2	2	199	412	477	430	431	432	439	532
Mediterranean Gull	0	3	7	2	0	4	1	6	0	1	6	2
Little Gull	0	0	0	0	0	0	0	1	0	1	0	0
Black-h. Gull	1,519	1,746	1,775	11,043	13,429	16,859	9,884	10,797	18,164	9,430	10,239	4,408
Ring-b. Gull	1	0	1	0	0	0	0	0	0	0	0	0
Common Gull	556	223	124	308	1,507	1,719	385	1,907	4,393	2,004	1,740	2,061
Lesser B-b Gull	689	562	715	600	2,404	2,072	1,777	2,176	1,122	1,095	3,240	6,963
Herring Gull	2,197	2,041	3,265	3,801	2,641	6,937	4,648	4,571	4,590	4,999	3,747	3,718
Iceland Gull	1	0	0	0	0	0	0	0	0	0	0	0
Great B-b Gull	116	86	237	351	229	406	201	498	260	162	101	145
Kittiwake	0	0	18	13	30	1	6	0	0	2	0	0
Sandwich Tern	14	88	21	603	426	220	0	0	0	0	0	0
Common Tern	100	22	39	57	15	10	4	0	0	0	0	1
Arctic Tern	0	0	0	0	0	1	0	0	0	0	0	0
Little Tern	0	100	161	120	3	0	0	0	0	0	0	0
Black Tern	0	0	0	0	2	0	0	0	0	0	0	0
Kingfisher	2	0	0	2	5	17	20	16	11	11	7	7

TOTALS

	Apr	May	Jun	Jul	Aug	Sep	Oct	Nov	Dec	Jan	Feb	Mar
WATERFOWL	20,937	11,968	13,461	22,952	45,941	65,800	80,591	105,467	132,582	110,260	100,172	47,645
GULLS	5,079	4,661	6,142	16,118	20,240	27,998	16,902	19,956	28,529	17,694	19,073	17,297
TERNS	114	210	221	780	446	231	4	0	0	0	0	1

	Apr	May	Jun	Jul	Aug	Sep	Oct	Nov	Dec	Jan	Feb	Mar
Sectors	*0*	*0*	*0*	*0*	*0*	*1*	*1*	*1*	*1*	*1*	*1*	*1*
Sites	*0*	*0*	*0*	*0*	*0*	*1*	*1*	*1*	*1*	*1*	*1*	*1*
Cormorant	0	0	0	0	0	3	7	4	1	1	2	0
Grey Heron	0	0	0	0	0	0	4	0	6	1	2	6
Mute Swan	0	0	0	0	0	0	0	0	0	0	0	4
Shelduck	0	0	0	0	0	1	0	0	8	10	26	32
Wigeon	0	0	0	0	0	0	65	57	109	66	65	0
Mallard	0	0	0	0	0	16	11	8	42	18	13	6
Eider	0	0	0	0	0	0	1	0	0	0	0	3
Goldeneye	0	0	0	0	0	0	0	0	0	0	3	2
Oystercatcher	0	0	0	0	0	57	88	34	73	70	125	39
Ringed Plover	0	0	0	0	0	0	1	52	0	63	6	0
Grey Plover	0	0	0	0	0	0	0	0	4	4	0	0
Lapwing	0	0	0	0	0	0	7	0	0	0	2	0
Dunlin	0	0	0	0	0	0	1	10	0	25	0	0
Snipe	0	0	0	0	0	0	0	0	1	0	0	0
Curlew	0	0	0	0	0	60	2	23	35	10	17	19
Redshank	0	0	0	0	0	7	4	7	19	10	4	4
Greenshank	0	0	0	0	0	0	0	1	0	0	0	0
Turnstone	0	0	0	0	0	0	13	3	20	10	0	0
Black-h. Gull	0	0	0	0	0	89	74	109	163	109	84	5
Common Gull	0	0	0	0	0	1	1	0	1	1	2	0
Lesser B-b Gull	0	0	0	0	0	0	0	0	0	0	1	0
Herring Gull	0	0	0	0	0	149	500	209	279	78	202	293
Great B-b Gull	0	0	0	0	0	43	12	26	34	7	9	10
Kittiwake	0	0	0	0	0	1	0	0	0	0	0	0
TOTALS												
WATERFOWL	0	0	0	0	0	144	204	19	318	288	265	115
GULLS	0	0	0	0	0	283	587	344	477	195	198	308

APPENDIX 8. TOTAL NUMBERS OF WATERFOWL RECORDED BY WeBS IN THE CHANNEL ISLANDS, 1997-98

	Apr	May	Jun	Jul	Aug	Sep	Oct	Nov	Dec	Jan	Feb	Mar
Sectors	*0*	*0*	*0*	*0*	*0*	*2*	*9*	*9*	*24*	*24*	*24*	*7*
Sites	*0*	*0*	*0*	*0*	*0*	*2*	*9*	*9*	*11*	*11*	*11*	*7*
Little Grebe	0	0	0	0	0	0	10	10	15	5	11	2
Great Crested Grebe	0	0	0	0	0	0	1	2	10	2	9	2
Slavonian Grebe	0	0	0	0	0	0	0	0	3	0	0	0
Cormorant	0	0	0	0	0	0	7	13	3	1	3	4
Bittern	0	0	0	0	0	0	0	1	1	1	1	0
Little Egret	0	0	0	0	0	40	91	98	15	56	38	0
Grey Heron	0	0	0	0	0	5	27	41	13	20	5	3
Mute Swan	0	0	0	0	0	0	5	3	9	6	3	0
Dark-bellied Brent	0	0	0	0	0	0	0	1	79	9	73	8
Light-b. Brent (Canada)	0	0	0	0	0	0	0	0	29	4	5	0
Feral/hybrid Goose	0	0	0	0	0	0	0	12	12	0	12	12
Wigeon	0	0	0	0	0	0	0	3	0	0	0	0
Teal	0	0	0	0	0	1	23	54	65	45	41	11
Mallard	0	0	0	0	0	6	212	254	162	216	273	57
Shoveler	0	0	0	0	0	0	12	19	21	23	20	28
Pochard	0	0	0	0	0	0	10	14	19	30	16	14
Tufted Duck	0	0	0	0	0	0	110	50	90	48	80	21
Red-b. Merganser	0	0	0	0	0	0	0	0	14	0	4	0
Water Rail	0	0	0	0	0	4	25	22	19	25	15	0
Moorhen	0	0	0	0	0	30	105	100	152	173	156	40
Coot	0	0	0	0	0	0	91	36	69	58	226	65
Oystercatcher	0	0	0	0	0	0	0	0	1,842	1,472	1,623	0
Ringed Plover	0	0	0	0	0	0	0	0	195	259	264	0
Golden Plover	0	0	0	0	0	0	21	33	83	294	59	0
Grey Plover	0	0	0	0	0	0	0	0	464	326	393	0
Lapwing	0	0	0	0	0	0	95	207	438	1064	244	28
Sanderling	0	0	0	0	0	0	0	0	247	189	304	0
Purple Sandpiper	0	0	0	0	0	0	0	0	0	0	5	0
Dunlin	0	0	0	0	0	0	0	0	2,319	1,015	1,990	0
Snipe	0	0	0	0	0	3	36	65	77	75	84	5
Woodcock	0	0	0	0	0	0	0	4	13	18	12	0
Bar-tailed Godwit	0	0	0	0	0	0	0	0	106	139	133	0
Curlew	0	0	0	0	0	0	0	0	214	329	168	0
Redshank	0	0	0	0	0	0	0	0	280	213	186	1
Greenshank	0	0	0	0	0	0	0	1	11	17	23	0
Common Sandpiper	0	0	0	0	0	0	0	1	0	0	0	0
Turnstone	0	0	0	0	0	0	0	0	316	266	415	0
Black-h. Gull	0	0	0	0	0	0	11	176	127	80	39	0
Common Gull	0	0	0	0	0	0	1	0	0	1	0	0
Lesser B-b Gull	0	0	0	0	0	0	0	0	2	1	0	0
Herring Gull	0	0	0	0	0	0	127	230	150	195	113	63
Great B-b Gull	0	0	0	0	0	0	8	50	12	37	44	16
TOTALS												
WATERFOWL	0	0	0	0	0	89	881	1,044	7,405	6,398	6.894	301
GULLS	0	0	0	0	0	0	147	456	291	314	196	79

APPENDIX 9. TOTAL NUMBERS OF WATERFOWL RECORDED BY WeBS AT INLAND AND COASTAL SITES, 1997-98

GREAT BRITAIN: COASTAL SITES

	Apr	May	Jun	Jul	Aug	Sep	Oct	Nov	Dec	Jan	Feb	Mar
Sectors	532	404	347	363	452	623	712	667	719	722	726	688
Sites	127	97	85	96	102	160	181	170	180	182	174	173
Red-t. Diver	320	58	3	135	24	239	230	391	564	396	496	345
Black-t. Diver	11	12	0	0	0	1	27	39	34	12	19	20
Great N. Diver	34	20	2	2	1	16	15	19	58	34	27	37
Little Grebe	173	106	76	106	277	560	775	1,046	1,128	890	760	621
Great C. Grebe	622	167	137	343	819	1,155	1,627	1,642	2,097	1,776	1,720	1,353
Red-n. Grebe	10	2	0	19	68	31	44	26	34	29	37	26
Slavonian Grebe	43	3	0	0	3	27	105	113	296	211	175	132
Black-n. Grebe	1	1	1	1	2	1	5	19	45	43	9	16
Cormorant	2,705	1,973	2,026	2,491	4,828	6,742	7,104	5,985	5,531	5,315	5,057	4,271
Bittern	0	0	0	0	0	0	0	0	1	0	0	0
Little Egret	143	21	47	185	410	381	424	270	272	268	242	311
Great White Egret	0	0	0	0	0	0	0	0	0	1	0	1
Grey Heron	522	370	553	854	1,188	1,778	1,704	1,211	1,306	1,030	984	934
Spoonbill	0	2	2	3	2	2	5	5	11	8	12	6
Lesser Flamingo	0	0	0	0	1	1	1	1	1	1	0	1
Ful. Whist. Duck	0	0	0	0	0	0	0	0	0	0	0	1
Mute Swan	2,052	1,356	1,485	2,152	2,098	2,744	3,486	3,144	3,863	3,807	3,761	3,385
Black Swan	0	1	1	1	2	2	3	2	5	4	1	6
Bewick's Swan	2	0	2	0	0	0	40	197	890	665	892	23
Whooper Swan	21	2	4	3	1	2	240	137	208	243	252	305
Bean Goose	0	0	0	0	0	0	0	0	2	2	7	0
Pink-f. Goose	3,508	214	7	5	3	4,486	26,454	60,920	104,789	41,322	18,984	22,406
Euro. Whitefront	230	0	0	1	1	1	50	356	1,892	2,892	3,984	113
Greenl. Whitefront	110	0	0	0	0	8	48	62	178	1	111	95
Lssr White-f. Goose	0	0	0	0	0	0	0	0	0	0	1	0
Greylag Goose	1,455	826	386	942	1,914	3,507	9,287	3,619	9,083	5,283	5,399	2,840
Bar-headed Goose	0	0	1	3	1	1	0	1	1	0	0	0
Snow Goose	0	0	0	0	0	1	0	2	0	1	10	3
Ross's Goose	0	0	0	0	4	0	0	1	1	1	0	0
Emperor Goose	0	0	0	0	1	1	0	1	0	0	0	0
Canada Goose	1,708	929	1,546	1,774	4,682	5,759	7,308	7,453	6,461	6,122	4,791	3,312
Barnacle Goose	3,563	7,513	3	1	6	369	6,623	1,641	2,948	10,534	5,518	20,242
Brent Goose	0	0	0	0	0	1	2	1	3	11	0	1
Dark-b. Brent	16,651	2,012	40	26	41	261	49,598	76,058	95,627	88,497	77,350	51,634
Black Brant	0	0	0	0	0	0	0	1	0	1	1	0
Light-b. Brent	9	0	0	0	0	1,660	2,397	2,598	2,527	773	466	56
Egyptian Goose	0	1	0	0	6	18	8	3	1	2	2	5
Feral/hybrid Goose	1	1	2	0	0	5	6	3	2	10	8	6
Unidentified Goose	79	0	0	0	0	0	0	9	0	0	0	0
Ruddy Shelduck	0	0	0	0	0	0	3	4	7	6	4	4
Shelduck	27,125	13,462	14,838	15,570	30,758	33,551	57,675	62,482	73,318	64,534	62,665	48,478
Muscovy Duck	0	1	1	1	0	1	42	46	52	33	33	0
Wood Duck	0	0	0	0	0	0	1	1	0	0	0	1
Mandarin	1	0	0	0	0	1	11	40	30	38	11	0
Wigeon	2,824	321	123	59	449	24,799	154,796	167,388	186,719	163,208	129,777	71,273
American Wigeon	0	0	0	0	0	0	0	0	0	0	2	2
Chiloe Wigeon	0	0	0	0	0	0	1	1	0	0	0	0
Falcated Duck	0	0	0	0	0	0	0	0	0	0	0	1
Gadwall	214	88	52	41	129	315	375	807	870	854	1,079	774
Teal	5,345	150	130	314	6,300	22,641	38,057	54,652	64,491	47,441	30,650	15,826
Mallard	5,451	3,562	6,117	5,673	16,953	25,188	31,569	31,320	35,022	32,193	22,497	10,479
Pintail	393	12	11	6	125	6,303	12,978	15,790	21,233	10,622	10,410	4,215
Garganey	2	3	2	0	8	1	0	0	0	0	0	3
Blue-winged Teal	0	0	0	0	0	0	0	0	0	0	0	0
Shoveler	538	70	59	100	427	841	1,376	1,574	1,968	2,073	1,666	1,688
Red-c. Pochard	0	0	0	0	0	0	0	0	1	0	0	1
Pochard	281	104	67	103	153	220	414	1,257	2,217	2,953	2,192	718
Tufted Duck	1,060	432	263	359	535	914	1,452	1,432	2,436	2,277	2,269	1,754
Scaup	1,718	17	4	9	34	185	717	1,131	7,242	4,268	3,692	1,276
Eider	19,997	14,190	15,460	16,199	21,566	22,913	24,546	19,094	19,323	15,377	14,566	15,910
King Eider	0	0	0	0	0	1	1	0	0	0	0	0

	Apr	May	Jun	Jul	Aug	Sep	Oct	Nov	Dec	Jan	Feb	Mar
Long-tailed Duck	1,186	22	0	1	0	1	345	610	1,736	1,453	709	779
Common Scoter	4,053	1,268	530	501	889	3,433	3,869	5,239	8,560	5,434	3,413	2,267
Surf Scoter	3	0	0	0	0	0	1	3	6	3	2	3
Velvet Scoter	576	172	101	18	84	61	308	742	790	390	454	528
Unid. scoter sp.	0	0	0	0	0	0	0	1	0	0	0	0
Goldeneye	1,029	82	5	6	11	90	1,203	3,053	9,422	6,826	6,195	4,853
Smew	0	0	0	0	0	0	0	2	10	13	12	5
Red-b. Merganser	1,995	476	604	583	871	1,523	2,342	2,712	3,780	3,691	3,324	3,769
Goosander	67	88	173	491	564	232	130	136	151	202	222	99
Ruddy Duck	43	34	22	39	70	77	58	63	33	26	49	64
Unidentified duck	0	0	0	0	0	0	3	0	0	0	0	0
Water Rail	54	2	0	2	10	25	34	39	56	38	34	35
Spotted Crake	0	0	0	0	1	2	0	0	0	0	0	0
Moorhen	739	368	290	439	572	933	1,155	1,420	1,451	1,319	1,384	1,438
Coot	1,563	711	911	1,311	2,362	2,895	3,015	3,494	3,736	4,027	3,849	3,275
Oystercatcher	78,088	40,103	33,560	61,758	171,417	225,839	235,759	250,161	238,519	211,557	219,651	136,744
Black-winged Stilt	0	0	0	0	1	0	0	1	1	1	1	1
Avocet	607	96	147	452	787	1,534	2,472	2,345	3,852	3,458	3,190	2,125
Little R. Plover	13	22	14	22	17	5	3	0	0	0	0	0
Ringed Plover	4,147	9,115	1,023	1,482	22,468	12,158	8,993	8,816	8,210	7,044	7,072	4,550
Kentish Plover	1	0	0	0	0	0	0	0	0	0	0	0
Dotterel	0	4	0	0	0	0	0	0	1	0	0	0
Golden Plover	6,391	378	129	2,293	27,101	33,132	53,504	121,006	114,403	93,628	101,805	30,529
Grey Plover	33,948	16,790	1,034	4,856	27,104	35,517	33,144	44,902	34,383	45,690	46,758	42,653
Lapwing	5,543	2,177	6,805	33,616	41,569	58,703	107,212	255,602	295,250	259,649	208,542	16,445
Knot	71,324	8,291	3,106	12,751	70,539	136,086	184,523	294,024	257,329	179,211	225,010	131,116
Sanderling	7,749	11,322	136	9,218	7,529	8,427	8,982	6,481	6,713	5,884	5,251	7,252
Western Sandpiper	0	0	0	0	1	0	0	0	0	0	0	0
Little Stint	1	10	5	7	25	36	6	4	0	1	1	2
Pectoral Sandpiper	0	0	0	0	0	2	0	0	0	0	0	0
Curlew Sandpiper	0	5	2	11	38	89	20	0	0	0	1	0
Purple Sandpiper	706	22	0	12	40	55	225	473	1,000	1,056	653	622
Dunlin	97,237	91,877	1,691	44,498	79,006	87,401	230,212	371,322	458,657	446,250	402,016	166,983
Ruff	96	90	2	38	268	357	104	71	72	125	128	110
Jack Snipe	1	0	0	0	0	2	19	26	45	30	52	43
Snipe	257	29	9	33	376	478	1,129	1,839	2,603	1,647	1,298	905
Great Snipe	0	0	0	0	1	0	0	0	0	0	0	0
Long-b. Dowitcher	0	0	0	1	0	0	1	1	0	0	0	0
Woodcock	0	0	0	0	0	0	3	1	3	11	11	0
Black-t. Godwit	8,556	1,603	1,549	3,154	16,447	14,771	13,729	13,378	13,154	12,062	11,785	13,617
Bar-t. Godwit	4,265	1,302	1,312	8,981	17,218	28,996	21,044	40,681	36,399	48,312	42,721	16,419
Whimbrel	109	1,615	222	697	584	290	43	12	3	3	5	9
Curlew	31,582	4,804	11,405	51,671	72,720	84,389	83,006	57,400	75,088	77,848	83,490	60,873
Spot. Redshank	59	19	9	163	212	143	131	179	98	53	14	59
Redshank	37,385	2,997	3,542	18,055	49,854	69,975	82,421	71,353	73,336	70,887	78,820	69,202
Greenshank	97	82	17	658	1,543	1,538	619	236	176	518	138	153
Lesser Yellowlegs	0	0	0	0	0	0	0	0	0	0	2	0
Green Sandpiper	8	3	5	47	171	87	53	42	21	17	25	22
Wood Sandpiper	1	1	0	5	24	9	1	0	0	0	0	0
Common Sand.	10	104	57	480	729	134	25	20	20	16	37	28
Turnstone	7,918	2,381	353	1,263	5,438	9,659	13,821	12,687	12,588	10,684	10,203	11,253
Grey Phalarope	0	0	0	1	1	0	1	0	0	0	0	0
Mediterranean Gull	8	9	10	23	30	14	21	29	9	65	54	40
Little Gull	46	29	5	37	19	7	18	14	0	7	0	0
Sabine's Gull	0	0	0	0	1	1	0	0	0	0	0	0
Black-h. Gull	25,333	8,751	15,196	65,824	103,005	92,622	71,067	71,717	96,299	77,480	77,005	54,964
Ring-b. Gull	1	0	1	0	0	1	2	1	2	1	3	0
Common Gull	5,013	3,023	2,943	4,038	12,503	11,113	8,757	7,881	11,139	16,485	15,587	11,100
Lesser B-b Gull	33,251	37,428	47,299	56,081	15,939	3,703	3,518	3,776	2,633	3,214	17,235	33,423
Herring Gull	31,448	29,831	34,421	35,637	29,853	40,235	37,303	54,733	39,355	50,728	53,172	44,536
Iceland Gull	1	2	0	0	0	0	0	0	2	0	1	2
Glaucous Gull	2	0	0	0	0	1	0	1	0	2	1	1
Great B-b Gull	1,397	1,432	1,637	2,524	4,254	5,727	5,008	10,010	5,470	4,977	2,975	1,668
Kittiwake	1,296	1,077	401	856	3,151	888	894	235	197	175	15	336
Unid. gull	12	120	0	0	3,505	1,266	986	500	4,165	2,490	280	480
Sandwich Tern	223	697	428	3,402	4,939	2,633	80	2	0	1	0	6
Roseate Tern	0	0	0	0	2	0	0	0	0	0	0	0
Common Tern	172	1,249	1,299	2,800	3,102	863	40	1	0	0	0	0
Arctic Tern	5	244	302	983	1,321	70	5	1	0	0	0	0

	Apr	May	Jun	Jul	Aug	Sep	Oct	Nov	Dec	Jan	Feb	Mar
Little Tern	3	357	364	232	137	23	1	0	0	0	0	0
Black Tern	0	11	0	0	19	5	1	0	0	0	0	0
Unid. tern	0	2	0	416	2	20	0	0	0	0	0	0
Kingfisher	1	0	2	5	17	42	37	30	36	24	22	10

GREAT BRITAIN: INLAND SITES

	Apr	May	Jun	Jul	Aug	Sep	Oct	Nov	Dec	Jan	Feb	Mar
Sectors	933	837	771	804	804	1352	1760	1841	1823	1889	1886	1839
Sites	764	725	664	687	692	1128	1410	1465	1451	1491	1484	1449
Red-t. Diver	3	5	1	5	3	1	8	4	7	35	5	17
Black-t. Diver	5	2	0	0	3	1	1	0	1	3	2	4
Great N. Diver	0	0	0	0	0	0	5	5	15	9	8	7
Pied-billed Grebe	0	0	1	0	0	0	0	0	0	0	0	0
Little Grebe	777	463	540	768	1,434	2,761	2,498	1,977	1,875	1,557	1,593	1,748
Great C. Grebe	3,315	2,772	2,881	3,193	4,418	6,099	6,364	6,336	5,604	5,123	5,580	6,066
Red-n. Grebe	2	1	2	2	2	4	3	5	7	7	1	3
Slavonian Grebe	12	1	1	1	0	9	24	33	19	31	17	24
Black-n. Grebe	7	12	9	7	16	24	19	13	2	5	5	12
Cormorant	3,271	2,275	1,402	1,842	2,468	4,617	6,554	6,845	6,401	6,862	6,083	5,740
Bittern	1	0	0	1	0	0	8	7	13	12	10	7
Little Bittern	0	0	1	0	0	0	0	0	0	0	0	0
Little Egret	6	2	3	9	4	8	6	7	17	10	30	13
Great White Egret	1	0	0	0	0	0	0	0	0	1	0	0
Grey Heron	1,084	818	967	1,184	1,493	1,781	1,980	1,912	1,988	1,926	1,945	1,893
White Stork	1	1	1	1	1	1	3	3	3	3	2	2
Spoonbill	0	0	0	1	0	0	0	0	0	0	0	0
Chilean Flamingo	1	0	1	0	0	0	0	0	0	0	0	0
Greater Flamingo	1	1	1	0	1	1	1	2	1	1	0	0
Fulv. Whist. Duck	0	0	0	0	0	2	0	0	0	0	0	0
Mute Swan	6,264	5,087	6,620	7,510	9,120	11,098	13,197	15,026	13,501	12,956	11,652	10,983
Black Swan	11	5	12	9	12	18	18	24	26	12	14	20
Trumpeter Swan	0	0	0	0	0	3	3	3	3	2	3	2
Bewick's Swan	18	0	0	3	0	0	87	176	4,551	3,445	4,008	81
Whooper Swan	384	11	6	6	4	45	1,379	2,218	2,963	2,207	3,529	2,760
Swan Goose	0	0	11	13	14	16	33	34	26	38	30	25
Bean Goose	4	1	1	1	0	1	36	162	47	38	51	3
Pink-f. Goose	15,810	430	8	14	257	59,503	95,844	31,615	15,586	26,657	53,417	31,765
White-f. Goose	0	0	1	0	0	1	0	0	0	0	0	0
Euro. Whitefront	3	0	0	0	1	2	2	118	263	1,203	1,361	551
Green. Whitefront	79	0	0	0	0	0	497	615	422	591	92	365
Lser White-f. Goose	0	0	1	1	1	0	1	1	1	0	1	0
Greylag Goose	8,730	2,836	6,954	7,094	11,264	15,851	44,082	30,034	35,014	24,221	21,597	20,533
Bar-headed Goose	5	4	4	4	9	14	9	13	13	6	11	16
Snow Goose	27	10	24	25	4	15	70	66	35	48	54	66
Ross's Goose	2	1	1	0	2	0	0	1	1	0	1	1
Emperor Goose	0	0	1	1	1	1	3	3	6	5	3	3
Canada Goose	9,277	8,199	17,837	17,744	23,263	34,636	34,478	35,772	33,181	30,899	25,673	19,668
Barnacle Goose	190	44	29	106	147	667	10,298	7,838	1,159	5,133	8,557	3,441
Brent Goose	0	0	0	0	0	1	0	0	0	0	0	0
Dark-bellied Brent	22	0	1	0	0	1	2,595	8,345	3,418	4,072	1,678	448
Light-bellied Brent	1	0	0	0	0	0	1	0	3	2	0	0
Red-b. Goose	0	0	0	0	0	0	0	0	0	0	0	1
Egyptian Goose	27	16	29	93	367	201	224	164	162	150	161	96
Feral/hybrid Goose	32	41	52	63	44	162	273	260	337	429	302	254
Ruddy Shelduck	3	0	2	7	3	10	5	3	2	3	1	3
Cape Shelduck	0	2	2	4	0	0	1	0	0	0	0	0
Paradise Shelduck	0	0	1	0	0	0	1	0	0	1	0	0
Shelduck	2,933	1,612	904	304	254	215	285	657	1,034	2,000	3,154	3,238
Muscovy Duck	25	28	27	26	13	31	59	70	75	66	60	36
Wood Duck	3	0	0	1	1	0	3	7	8	2	5	8
Mandarin	76	54	122	72	49	155	169	275	236	197	162	181
Wigeon	8,462	113	88	1,046	569	12,397	53,303	90,642	128,102	163,891	138,889	111,827
Am. Wigeon	0	0	0	0	0	0	0	1	1	0	1	2
Chiloe Wigeon	0	0	0	1	0	0	2	3	1	3	4	4
Falcated Duck	0	0	0	0	0	0	0	0	0	0	0	1

	Apr	May	Jun	Jul	Aug	Sep	Oct	Nov	Dec	Jan	Feb	Mar
Gadwall	2,039	1,089	1,193	1,015	3,705	6,698	8,149	12,266	11,869	10,102	9,777	6,122
Teal	9,632	224	336	1,426	7,165	28,943	48,605	68,305	73,263	69,508	49,515	33,245
Speckled Teal	0	1	0	1	1	1	0	0	0	0	1	2
Mallard	20,117	16,441	24,304	32,147	52,972	75,848	91,501	108,893	99,653	90,762	65,671	44,204
Chestnut Teal	0	0	0	0	0	0	0	1	0	0	0	0
Pintail	340	4	2	61	37	918	1,519	1,682	3,284	4,212	5,352	3,781
Bahama Pintail	0	0	0	0	0	1	0	0	0	0	0	0
Cape Teal	0	0	0	0	0	0	1	1	1	0	0	0
Garganey	10	23	4	19	29	16	4	4	0	0	1	2
Shoveler	2,256	274	232	514	2,815	6,929	7,892	7,056	5,890	5,337	6,074	5,569
Ringed Teal	0	0	0	0	1	0	0	0	0	0	0	0
Red-c. Pochard	0	2	3	1	8	16	46	94	70	85	41	45
Pochard	930	410	818	1,932	6,483	7,882	15,101	22,819	27,515	39,138	27,686	10,597
Ring-necked Duck	0	0	0	0	1	2	0	1	1	1	2	3
Ferruginous Duck	1	1	0	0	1	1	2	1	2	1	2	1
Tufted Duck	18,178	6,372	7,492	16,798	28,610	36,843	37,337	46,522	49,568	48,260	43,289	38,361
Scaup	19	8	1	1	3	30	339	278	287	129	481	38
Lesser Scaup	1	0	0	0	0	0	0	0	0	0	0	0
Eider	60	31	24	28	11	52	33	43	26	4	3	71
Long-tailed Duck	0	0	0	0	0	0	31	60	57	61	23	47
Common Scoter	3	0	6	0	0	0	2	1	5	54	4	4
Velvet Scoter	0	0	0	0	0	0	3	2	2	0	0	0
Goldeneye	3,087	84	77	126	90	151	2,252	5,182	6,933	7,692	8,867	8,288
Hooded Merganser	0	0	0	0	1	1	0	0	0	0	0	0
Smew	6	2	0	0	0	0	3	13	139	287	224	67
Red-b. Merganser	345	228	267	118	161	441	421	322	388	325	573	501
Goosander	452	367	258	495	467	585	825	1,943	3,094	3,311	3,406	2,067
Ruddy Duck	1,283	594	530	606	1,262	2,110	2,445	2,724	3,552	2,966	3,056	2,592
Feral/hybrid duck	91	135	138	116	144	168	170	165	158	181	132	131
Hybrid Aythya	1	0	0	0	0	1	0	1	1	0	1	2
Unidentified duck	0	0	0	0	0	2	0	0	0	0	0	0
Water Rail	43	45	44	46	62	99	166	231	289	218	233	204
Spotted Crake	0	0	2	0	0	0	0	0	0	0	0	0
Moorhen	4,014	2,751	2,340	3,701	5,352	7,937	9,555	10,381	9,663	9,500	10,459	9,450
Coot	18,625	11,109	16,916	24,421	43,714	71,214	84,177	99,013	91,656	74,865	62,156	42,113
Crane	0	0	0	0	0	0	0	6	4	6	6	0
Oystercatcher	2,463	1,096	1,159	1,022	485	260	636	1,249	1,061	1,721	7,545	12,160
Avocet	428	131	90	1	32	16	82	64	7	6	42	344
Black-w. Pratincole	0	0	0	0	1	0	0	0	0	0	0	0
Little R. Plover	181	266	235	118	35	13	2	1	0	0	0	9
Ringed Plover	189	234	168	120	695	422	374	842	339	327	268	231
Dotterel	1	0	0	0	0	0	0	0	0	0	0	0
Golden Plover	2,102	17	37	67	701	3,481	17,146	43,671	61,042	44,533	45,123	13,706
Grey Plover	7	18	0	0	271	2	68	182	160	6	18	6
Unid. wader	0	0	0	0	0	0	0	0	0	0	1	0
Lapwing	4,145	1,753	6,307	21,013	27,200	41,512	65,582	143,134	169,216	175,839	138,736	20,846
Knot	0	1	0	2	16	15	24	1	68	6	0	15
Sanderling	0	7	1	0	80	4	41	1	0	32	22	36
Little Stint	0	2	0	1	3	11	9	1	2	0	0	0
Pectoral Sandpiper	0	0	0	0	0	1	0	0	0	0	0	0
Curlew Sandpiper	0	2	0	5	12	27	22	1	0	0	0	0
Purple Sandpiper	1	0	0	0	0	0	3	72	3	5	5	0
Dunlin	339	374	23	301	1,104	436	2,252	5,266	3,925	2,570	3,074	2,214
Ruff	186	4	2	101	212	230	111	112	191	259	296	289
Jack Snipe	11	1	0	0	0	1	46	29	33	30	51	31
Snipe	607	109	39	107	572	1,051	3,698	4,187	4,801	2,974	3,740	2,636
Long-b. Dowitcher	0	0	0	0	0	0	1	0	0	0	0	0
Woodcock	2	2	1	0	1	1	7	18	37	18	14	9
Black-t. Godwit	1,960	105	53	325	497	1,941	1,319	570	45	185	737	1,844
Bar-t. Godwit	6	4	0	3	9	2	2	0	7	1	40	6
Whimbrel	8	120	1	15	16	6	1	0	0	0	0	0
Curlew	933	320	218	474	701	1,969	2,615	4,560	6,479	7,844	8,147	7,318
Spot. Redshank	12	4	5	3	9	62	16	0	0	2	2	1
Redshank	2,979	588	539	775	1,902	1,475	2,238	2,258	2,415	1,913	1,803	3,220
Greenshank	8	17	9	54	287	247	94	26	16	9	20	27
Green Sandpiper	47	3	32	169	281	158	104	119	111	61	75	66
Wood Sandpiper	2	2	1	7	22	8	0	0	0	0	0	0
Common Sand.	37	357	267	432	607	168	18	9	6	4	7	13
Turnstone	126	62	0	4	45	53	278	239	152	70	60	97

	Apr	May	Jun	Jul	Aug	Sep	Oct	Nov	Dec	Jan	Feb	Mar
Grey Phalarope	0	0	0	0	0	1	2	0	0	2	0	0
Mediterranean Gull	9	2	3	7	0	5	29	6	6	8	11	8
Little Gull	1	1	0	2	9	2	0	2	0	2	0	1
Sabine's Gull	0	0	0	0	0	2	0	0	0	0	0	0
Black-h. Gull	28,141	23,501	19,937	16,422	17,767	48,300	66,962	179,857	157,622	154,987	113,028	99,358
Ring-b. Gull	0	0	0	0	0	0	0	0	1	1	2	4
Common Gull	3,686	472	264	1,076	1,077	12,631	24,354	63,796	61,811	53,605	70,941	29,107
Lesser B-b Gull	2,608	1,393	1,812	3,004	7,472	11,198	8,952	18,152	5,868	3,017	3,334	3,543
Herring Gull	3,717	2,456	1,980	4,296	2,917	8,929	8,863	12,697	12,039	12,517	8,618	5,298
Iceland Gull	1	0	0	0	0	0	0	0	0	2	4	1
Glaucous Gull	2	0	0	0	0	1	0	0	1	6	7	3
Great B-b Gull	314	223	169	280	981	772	1,334	3,840	3,578	4,529	1,481	1,137
Kittiwake	1	0	2	250	1,000	6	0	2	1	8	1	1
Unid. gull	7	190	53	305	237	350	401	430	2,330	752	5,480	120
Sandwich Tern	163	1,524	448	455	779	42	7	1	0	0	0	0
Common Tern	52	843	1,195	1,223	594	41	0	0	0	0	0	1
Arctic Tern	2	5	83	24	16	8	3	1	0	0	0	0
Little Tern	0	100	68	4	1	0	0	0	0	0	0	0
Black Tern	0	12	2	0	5	1	0	0	0	0	0	0
White-w. Black Tern	0	0	0	0	2	0	0	0	0	0	0	0
Unid. tern	0	16	16	29	62	0	0	0	0	0	0	0
Kingfisher	62	66	79	118	136	218	243	180	174	115	156	178

NORTHERN IRELAND: COASTAL SITES

	Apr	May	Jun	Jul	Aug	Sep	Oct	Nov	Dec	Jan	Feb	Mar
Sectors	*15*	*15*	*15*	*15*	*15*	*23*	*23*	*51*	*64*	*66*	*56*	*24*
Sites	*3*	*3*	*3*	*3*	*3*	*6*	*6*	*7*	*8*	*8*	*7*	*7*
Red-t. Diver	0	0	0	0	0	2	4	16	19	21	0	42
Great N. Diver	0	1	0	0	0	0	0	2	2	6	1	4
Little Grebe	1	2	0	0	0	105	78	101	129	112	74	35
Great C. Grebe	2	5	0	3	39	1,700	2,021	2,466	2,350	1,468	154	2,382
Slavonian Grebe	0	0	0	0	0	2	5	2	2	7	1	0
Black-n. Grebe	0	0	0	0	0	0	1	0	0	0	0	0
Cormorant	80	104	59	123	126	732	751	813	852	500	367	445
Grey Heron	16	35	24	48	43	168	182	102	172	150	66	58
Mute Swan	166	116	74	56	52	231	258	217	321	218	201	199
Bewick's Swan	3	0	0	0	0	0	0	0	5	16	0	2
Whooper Swan	67	3	0	0	0	0	362	445	76	337	151	566
Bean Goose	0	0	0	0	0	0	0	0	0	0	0	0
Pink-footed Goose	0	0	0	0	0	30	2	0	0	0	0	0
Greenl. Whitefront	36	0	0	0	0	0	111	8	0	19	3	37
Greylag Goose	88	0	0	0	0	115	100	86	379	34	357	581
Canada Goose	0	0	0	0	0	38	70	4	57	23	183	105
Barnacle Goose	2	0	0	0	0	148	134	123	130	131	126	117
Dark-bellied Brent	0	0	0	0	0	0	0	0	0	6	66	0
Light-bellied Brent	198	0	0	0	0	12,805	14,910	9,303	5,675	3,323	3,727	2,580
Shelduck	264	148	80	69	22	84	777	2,463	3,310	4,336	3,391	2,153
Mandarin	3	2	5	0	0	0	0	2	0	0	0	0
Wigeon	82	3	0	1	0	1,958	10,367	6,843	7,049	2,370	1,795	1,642
Gadwall	0	0	0	0	0	63	44	47	53	37	55	43
Teal	44	11	0	0	29	360	1,445	1,133	2,314	2,677	1,516	905
Mallard	75	96	184	126	810	3,599	2,913	2,227	3,185	2,120	1,273	599
Pintail	0	0	0	0	0	8	24	26	304	119	350	79
Shoveler	0	0	0	0	0	18	83	70	101	89	54	34
Pochard	0	0	0	0	0	17	35	101	69	161	116	48
Tufted Duck	0	2	0	0	1	81	113	240	162	232	236	245
Scaup	0	0	0	0	0	0	9	0	389	656	77	6
Eider	19	58	23	125	162	421	797	981	1,088	521	413	293
Long-tailed Duck	0	0	0	0	0	0	0	10	20	8	11	18
Common Scoter	0	0	0	0	0	0	0	0	0	1	0	1
Velvet Scoter	0	0	0	0	2	0	0	4	1	0	0	1
Goldeneye	25	0	0	0	0	0	59	498	940	683	578	776
Smew	0	0	0	0	0	0	0	0	0	1	1	1
Red-b. Merganser	12	6	15	11	296	381	559	449	585	402	258	550
Moorhen	1	3	2	1	1	7	6	0	4	1	1	2

	Apr	May	Jun	Jul	Aug	Sep	Oct	Nov	Dec	Jan	Feb	Mar
Coot	0	0	0	0	0	197	263	415	340	334	202	142
Oystercatcher	1,192	925	583	2,220	3,881	11,454	13,979	13,714	16,308	17,249	11,256	7,982
Ringed Plover	63	7	3	5	8	233	200	396	601	525	290	29
Golden Plover	1,614	0	0	0	1	815	1,874	8,809	10,772	9,756	8,915	6,514
Grey Plover	2	0	0	0	2	82	67	153	278	352	285	179
Lapwing	40	95	63	275	782	1,091	4,298	9,121	20,561	18,165	13,315	302
Knot	0	0	0	0	2	144	92	3,602	8,184	9,655	4,426	493
Sanderling	43	37	0	0	0	55	0	15	41	1	46	0
Curlew Sandpiper	0	0	0	0	0	6	3	0	0	0	0	0
Purple Sandpiper	0	0	0	0	0	0	10	44	76	70	20	29
Dunlin	45	55	16	50	113	842	1,018	11,708	16,748	13,198	14,236	1,324
Ruff	0	0	0	1	0	0	0	0	0	0	0	0
Jack Snipe	0	0	0	0	0	0	0	2	4	0	0	0
Snipe	5	0	0	0	0	0	22	135	119	56	104	51
Black-tailed Godwit	30	0	2	6	21	83	76	54	163	243	236	263
Bar-tailed Godwit	3	4	11	66	25	282	262	482	2,983	3,343	857	320
Whimbrel	0	331	1	14	2	5	0	0	0	0	0	0
Curlew	611	159	649	2,044	1,922	4,448	3,291	3,175	5,070	6,768	5,786	2,982
Spotted Redshank	2	0	0	0	0	4	1	2	1	1	0	1
Redshank	940	76	50	386	1,076	6,355	6,896	7,020	7,098	6,072	5,780	5,910
Greenshank	4	0	5	27	35	89	92	56	92	65	57	62
Green Sandpiper	0	0	0	0	0	9	0	0	0	0	0	0
Common Sandpiper	0	2	1	3	1	1	0	0	0	0	0	0
Turnstone	85	1	0	0	74	754	1,070	1,515	1,384	1,573	932	825
Little Gull	0	0	0	0	0	0	0	0	0	1	0	0
Black-h. Gull	149	110	375	1,518	1,960	7,610	6,238	6,204	7,276	9,848	6,245	5,947
Ring-b. Gull	0	0	0	0	0	0	0	0	0	1	0	0
Common Gull	81	119	109	469	1,515	2,911	2,226	1,337	1,149	1,825	3,104	233
Lesser B-b Gull	44	4	1	30	11	26	13	9	18	15	17	59
Herring Gull	96	230	210	287	460	3,419	2,746	3,575	2,472	3,712	1,665	3,131
Iceland Gull	0	0	0	0	0	0	0	0	0	4	0	1
Glaucous Gull	0	0	0	0	0	0	0	0	0	6	0	0
Great B-b Gull	110	482	114	126	173	471	335	285	221	449	119	296
Kittiwake	0	0	0	0	76	0	0	1	0	3	0	0
Sandwich Tern	26	69	130	296	606	479	6	0	0	0	0	0
Common Tern	0	10	0	0	0	0	0	0	0	0	0	0
Black Tern	0	0	0	0	1	2	1	0	0	0	0	0
Unid. tern	0	0	0	43	7	0	0	0	0	0	0	0
Kingfisher	0	0	0	0	0	0	1	1	0	1	0	0

NORTHERN IRELAND: INLAND SITES

	Apr	May	Jun	Jul	Aug	Sep	Oct	Nov	Dec	Jan	Feb	Mar
Sectors	*1*	*0*	*0*	*0*	*87*	*110*	*120*	*123*	*127*	*129*	*131*	*127*
Sites	*1*	*0*	*0*	*0*	*2*	*7*	*18*	*18*	*22*	*24*	*27*	*21*
Little Grebe	0	0	0	0	212	390	299	428	383	203	199	134
Great Crested Grebe	0	0	0	0	864	728	169	229	107	111	517	612
Slavonian Grebe	0	0	0	0	0	0	0	0	1	1	0	0
Cormorant	0	0	0	0	991	994	1,226	903	870	633	834	498
Grey Heron	0	0	0	0	217	166	101	77	113	122	122	89
Mute Swan	0	0	0	0	1,495	1,642	1,708	1,916	1,729	1,531	1,747	1,642
Bewick's Swan	0	0	0	0	0	0	0	47	41	117	75	44
Whooper Swan	87	0	0	0	1	34	502	697	600	1,233	1,976	1,251
Greenl. Whitefront	0	0	0	0	0	0	0	0	0	0	88	0
Greylag Goose	0	0	0	0	0	76	17	105	218	283	42	578
Canada Goose	0	0	0	0	0	3	3	2	0	0	273	13
Shelduck	0	0	0	0	12	61	32	101	135	349	177	348
Wigeon	0	0	0	0	2	201	911	3,130	3,011	1,195	2,944	2,096
Gadwall	0	0	0	0	61	66	36	69	89	70	71	111
Teal	0	0	0	0	47	405	625	1,284	2,509	1,120	2,202	1,438
Mallard	0	0	0	0	5,463	5,024	4,258	4,554	3,671	2,183	2,610	1,573
Pintail	0	0	0	0	0	2	5	0	14	0	8	27
Shoveler	0	0	0	0	14	18	29	28	106	18	27	46
Pochard	0	0	0	0	359	277	1,667	9,155	19,240	18,760	8,180	1,966
Tufted Duck	0	0	0	0	2,910	5,757	7,082	18,781	16,639	16,163	13,632	8,661
Scaup	0	0	0	0	0	1,427	3	1,330	493	3,160	3,671	2,244

	Apr	May	Jun	Jul	Aug	Sep	Oct	Nov	Dec	Jan	Feb	Mar
Eider	0	0	0	0	0	0	0	0	3	0	0	0
Goldeneye	0	0	0	0	21	69	497	5,609	3,839	4,205	4,214	4,918
Red-b. Merganser	0	0	0	0	23	103	47	52	24	23	1	27
Goosander	0	0	0	0	0	0	0	0	1	1	0	1
Ruddy Duck	0	0	0	0	23	24	8	7	23	0	28	14
Water Rail	0	0	0	0	0	0	0	1	1	0	2	0
Moorhen	0	0	0	0	132	226	206	266	185	168	257	262
Coot	0	0	0	0	3,023	5,303	5,548	6,230	5,005	2,806	3,093	2,793
Oystercatcher	0	0	0	0	6	1,521	1,340	1,903	1,320	550	0	494
Ringed Plover	0	0	0	0	0	0	32	0	58	0	0	12
Golden Plover	0	0	0	0	0	5	2,251	2,433	3,608	4,337	3,080	1,170
Grey Plover	0	0	0	0	0	0	0	0	2	0	0	2
Lapwing	0	0	0	0	628	1,478	1,580	4,355	8,375	10,098	1,876	299
Knot	0	0	0	0	0	22	45	170	0	0	0	19
Dunlin	0	0	0	0	2	3	43	121	55	498	77	10
Ruff	0	0	0	0	0	2	1	0	0	0	0	0
Jack Snipe	0	0	0	0	0	0	0	0	1	0	0	0
Snipe	0	0	0	0	8	12	31	75	70	79	69	128
Black-tailed Godwit	0	0	0	0	0	303	298	165	130	161	0	133
Bar-tailed Godwit	0	0	0	0	0	18	22	32	28	10	0	133
Curlew	0	0	0	0	82	714	670	576	902	861	859	1,114
Spotted Redshank	0	0	0	0	0	1	1	0	1	1	0	0
Redshank	0	0	0	0	0	75	213	139	31	22	11	64
Greenshank	0	0	0	0	0	2	1	1	1	0	0	0
Black-h. Gull	0	0	0	0	2,963	2,719	2,203	1,017	1,380	1,989	2,287	1,833
Ring-b. Gull	0	0	0	0	0	0	1	0	0	0	0	0
Common Gull	0	0	0	0	55	181	232	380	341	395	363	946
Lesser B-b Gull	0	0	0	0	706	998	586	266	19	160	63	143
Herring Gull	0	0	0	0	3	442	413	71	10	38	8	10
Iceland Gull	0	0	0	0	0	0	0	0	0	0	0	1
Glaucous Gull	0	0	0	0	0	0	0	0	0	0	0	1
Great B-b Gull	0	0	0	0	38	125	75	85	31	88	31	12
Sandwich Tern	0	0	0	0	0	58	7	0	0	0	0	0
Kingfisher	0	0	0	0	0	0	0	0	1	0	0	0

APPENDIX 10. LOCATIONS OF WEBS COUNT SITES MENTIONED IN THIS REPORT

The location of all counts sites or areas mentioned in this report are given here. Sites are listed alphabetically, with the 1km square OS grid reference for the centre of the site, the habitat (H) and the county or district. Note that this is not an exhaustive list of WeBS sites counted in 1997-98, simply those mentioned by name in this report. Figure A1 shows the location of many of the more important sites for waterfowl.

Habitat codes (the predominant habitat type is given for complex sites containing many different habitats)

L	Lake	M	Marsh
R	Reservoir	S	Sewage treatment works
P	Gravel or sand pit	E	Estuary
V	River	O	Open coast
C	Canal	N	Non-wetland

Site	1 km sq	H	County	Site	1 km sq	H	County
Abberton Reservoir	TL9818	R	Essex	Bough Beech Reservoir	TQ4947	R	Kent
Aberlady Bay	NT4581	E	Lothian	Brading Harbour	SZ6388	E	Isle of Wight
Alaw Reservoir	SH3968	R	Gwynedd	Bramshill Park	SK7560	L	Hampshire
Alde Complex	TM4257	E	Suffolk	Brent Reservoir	TQ2287	R	Gtr London
Aldford Brook & Eaton Park	SJ4059	V	Cheshire	Breydon Water & Berney Marshes	TG4907	E	Norfolk
Alloa Inch	NS8792	N	Central	Bridge of Earn	NO1417	N	Tayside
Alt Estuary	SD2903	E	Merseyside	Broad Water Canal	J1462	C	Antrim
Altofts Ings	SE3624	L	W Yorkshire	Buckden/Stirtloe Gravel Pits	TL2066	P	Cambs
Alton Water	TM1356	R	Suffolk	Buckenham Marshes	TG3505	M	Norfolk
Alvecote Pools	SK2504	L	Warwickshire	Burghfield Gravel Pits	SU6870	P	Berkshire
Ampton Water	TL8770	L	Suffolk	Burry Inlet	SS5096	E	W Glamorgan, Dyfed
Appin/Erriska/Benderloch	NM9043	O	Strathclyde				
Aqualate Mere	SJ7720	L	Staffordshire	Busbridge Lakes	SU9742	L	Surrey
Ardleigh Reservoir	TM0328	R	Essex	Bush River: Deepstown	C9434	V	Antrim
Arlington Reservoir	TQ5307	R	Sussex	Bute	NS0761	L	Strathclyde
Arran	NR9535	O	Strathclyde	Caerlaverock WWT	NY0565	E	Dumfries & Galloway
Arun Valley	TQ0314	V	West Sussex				
Ashford Common Waterworks	TQ0869	S	Surrey	Caistron Quarry	NU0001	P	N'th'mberland
Ash Levels	TR3162	M	Kent	Caithness Lochs	ND1859	L	Highland
Attenborough Gravel Pits	SK5234	P	Notts	Calf Hey Reservoir	SD7522	R	Lancashire
Avon Estuary	SX6745	E	Devon	Cambois to Newbiggin	NZ3084	O	N'th'mberland
Avon Valley (Lower)	SZ1499	M	Hampshire	Camel Estuary	SW9474	E	Cornwall
Avon Valley (Mid)	SU1510	M	Hampshire	Cameron Reservoir	NO4711	R	Fife
Ayr to Troon	NS3425	O	Strathclyde	Canary Road	H8755	M	Armagh
Ballyroney Lake	J229382	L	Down	Cardigan Bay	SH5020	O	Gwynedd, Dyfed
Ballysaggart Lough	H7961	L	Tyrone				
Balranald RSPB Reserve	NF7169	L	Western Isles	Carlingford Lough	J2013	E	Down
Bann Estuary	C7935	E	Londonderry	Carmarthen Bay	SN2501	E	Dyfed
Bardney Pits	TF1168	P	Lincolnshire	Carsebreck/Rhynd Lochs	NN8609	L	Tayside
Barleycroft Gravel Pits	TL3672	P	Cambs	Castlecaldwell Refuge Area	H0060	L	Fermanagh
Barn Elms Reservoir	TQ2277	R	Gtr London	Castle Howard Lake	SE7170	L	N Yorkshire
Barnstone Pool	SK7334	P	Notts	Castle Loch, Lochmaben	NY0881	L	Dumfries & Galloway
Baron's Haugh	NS7555	L	Strathclyde				
Barton Pits	SK2017	P	Staffordshire	Cefni Reservoir	SH4475	R	Anglesey
Baston/Langtoft Gravel Pits	TF1212	P	Lincolnshire	Cemlyn Bay	SH3393	O	Gwynedd
Bayfield Loch	NH8271	L	Highland	Chasewater	SK0307	R	W Midlands
Beadnell to Seahouses	NU2231	O	N'th'mberland	Chatsworth Park Lake	SK2670	L	Derbyshire
Beaulieu Estuary	SZ4298	E	Hampshire	Cheddar Reservoir	ST4454	R	Somerset
Beauly Firth	NH5848	E	Highland	Chew Valley Lake	ST5659	R	Avon
Beddington Sewage Farm	TQ2966	S	Gtr London	Chichester Gravel Pits	SU8703	P	West Sussex
Bedfont & Ashford Gravel Pits	TQ0872	P	Gtr London	Chichester Harbour	SU7700	E	West Sussex
Beesands Ley	SX8141	L	Devon	Chilham & Chartham Gravel Pits	TR0954	P	Kent
Belfast Lough	J4083	E	Down	Chillington Hall Pool	SJ8550	L	Staffordshire
Belvide Reservoir	SJ8610	R	Staffordshire	Chorlton Water Park	SJ8291	P	Greater Manchester
Benacre Broad	TM5383	L	Suffolk				
Benbecula	NF8150	N	Western Isles	Christchurch Harbour	SZ1792	E	Dorset
Besthorpe & Girton Gravel Pits	SK8165	P	Notts	Church Wilne Reservoir	SK4632	R	Derbyshire
Bewl Water	TQ6733	R	Sussex	Clachan	NR7656	N	Strathclyde
Bicton Reservoir	SM8407	R	Dyfed	Clarydale Water	SN0417	L	Dyfed
Black Cart Water	NS4767	M	Borders	Clea Lake 1	J506557	L	Down
Blackmoorfoot Reservoir	SE0912	R	W Yorkshire	Cleddau Estuary	SN0005	E	Dyfed
Blackwater Estuary	TL9307	E	Essex	Clifford Hill Gravel Pits	SP8061	P	Northants
Blagdon Lake	ST5150	R	Avon	Clumber Park Lake	SK6374	L	Notts
Blenheim Park Lake	SP4316	L	Oxfordshire	Clwyd Estuary	SJ0079	E	Clwyd
Blickling Lake	TG1729	L	Norfolk	Clyde Est.	NS3576	E	Strathclyde
Blithfield Reservoir	SK0524	R	Staffordshire	Coll	NM2055	N	Strathclyde
Blunham Gravel Pits	TL1551	P	Bedfordshire	Colliford Reservoir	SX1871	R	Cornwall
Blyth Estuary (Suffolk)	TM4675	E	Suffolk	Colne Estuary	TM0614	E	Essex
Blyth to Newbiggin	NZ3084	O	N'th'mberland	Colne Valley Gravel Pits	TQ0489	P	Gtr London
Bolton-on-Swale Gravel Pits	SE2498	P	N Yorkshire	Colonsay/Oronsay	NR3896	N	Strathclyde

Site	1 km sq	H	County	Site	1 km sq	H	County
Colwick Country Park	SK6039	L	Notts	Fala Flow	NT4258	L	Lothian
Colwyn Bay	SH9079	O	Clwyd	Fal Complex	SW8541	E	Cornwall
Combermere	SJ5884	L	Cheshire	Farmoor Reservoirs	SP4406	R	Oxfordshire
Connaught Water	TQ4095	L	Essex	Farmwood Pool	SJ8173	L	Cheshire
Conwy Estuary	SH7877	E	Caernarvon	Fen Drayton Gravel Pits	TL3470	P	Cambs
Coombe Pool	SP3979	L	Warwickshire	Ferry Meadows	TL1497	P	Cambs
Coquet Estuary	NU2706	E	Gwynedd	Fiddlers Ferry Power Station	SJ5585	P	Cheshire
Corby Loch	NJ9214	L	Grampian	Lagoons			
Cotswold Water Park (East)	SU1999	P	Glos, Oxon	Filey Bay	TA1279	O	N Yorkshire
Cotswold Water Park (West)	SU0595	P	Glos, Wilts	Fillingham Lake	SK9485	L	Lincolnshire
Cowgill Reservoirs	NT0327	R	Strathclyde	Fincastle Loch	NN8762	L	Tayside
Craigalea to Newcastle	J704337	O	Down	Findhorn Bay	NJ0462	E	Grampian
Cresswell to Chevington Burn	NZ2895	O	N'th'mberland	Fisherwick & Elford Gravel Pits	SK1710	P	Staffordshire
Crichel Lake	ST9907	L	Dorset	Fleet/Wey	SY6976	E	Dorset
Criddling Stubbs Quarry Pool	SE5120	P	W Yorkshire	Fleet Pond	SU8255	L	Hampshire
Cromarty Firth	NH7771	E	Highland	Fonthill Lake	ST9331	L	Wiltshire
Crombie Reservoir	NO5240	R	Tayside	Fordwich & Westbere Gravel Pits	TR1860	P	Kent
Cropston Reservoir	SK5410	R	Leicestershire	Foreland	SZ6584	O	Isle of Wight
Crouch/Roach Estuary	TQ8496	E	Essex	Foremark Reservoir	SK3224	R	Derbyshire
Crowdy Reservoir	SX1483	R	Cornwall	Fort Henry Ponds & Exton Park	SK9412	L	Leicestershire
Croxall Pits	SK1814	P	Staffordshire	Lake			
Cults Reservoir	NJ9002	R	Grampian	Forth Estuary	NT2080	E	Lothians,
Cuttmil Ponds	SU9145	L	Surrey				Central, Fife
Danna/Keills Peninsula	NR7383	O	Strathclyde	Forth/Teith Valley	NS7595	N	Central
Daventry Reservoir	SP5763	R	Northants	Foryd Bay	SH4559	E	Gwynedd
Deben Estuary	TM2942	E	Suffolk	Fowey Estuary	SX1254	E	Cornwall
Dee Estuary (England/Wales)	SJ2675	E	Merseyside,	Frainslake to Freshwater West	SR8898	O	Dyfed
			Cheshire,	Frenchess Road Pond	TO2851	L	Surrey
			Clwyd	Girvan to Turnberry	NS2002	O	Strathclyde
Dee Estuary (Scotland)	NJ9505	E	Grampian	Gladhouse Reservoir	NT2953	R	Lothian
Deene Lake	SP9492	L	Northants	Glenfarg Reservoir	NO1011	R	Tayside
Deeping St James Gravel Pits	TF1808	P	Lincolnshire	Grafham Water	TL1568	R	Cambs
Dengie Flats	TM0300	E	Essex	Great Cumbrae	NS1656	O	Strathclyde
Derwent Reservoir	NZ0251	R	Durham	Great Pool Westwood Park	SO8763	L	Hereford &
Derwent Water	NY2621	L	Cumbria				Worcester
Deveron Estuary	NJ6964	E	Grampian	Grimsthorpe Lake	TF0222	L	Lincolnshire
Didlington	TL7796	P	Norfolk	Grouville Marsh	WV6949	M	Channel Isles
Dinnet Lochs	NJ4800	L	Grampian	Guernsey Shore	WV27	O	Channel Isles
Dinton Pastures	SU7872	M	Berkshire	Gun Knowe Loch	NT5135	L	Borders
Ditchford Gravel Pits	SP9468	P	Northants	Gunton Park Lake	TG2234	L	Norfolk
Doddington Pool	SJ7146	L	Cheshire	Haddo House Lakes	NJ8734	L	Grampian
Don Mouth to Ythan Mouth	NJ9815	O	Grampian	Hamford Water	TM2225	E	Essex
Doon Estuary	NS3219	O	Strathclyde	Hamilton Low Parks	NS7257	L	Strathclyde
Dorchester Gravel Pits	SU5795	P	Oxfordshire	Hammer Wood Pond	SU8423	L	West Sussex
Dornoch Firth	NH7384	E	Highland	Hampton & Kempton Reservoirs	TQ1269	R	Gtr London
Dowlaw Dam	NT8569	R	Borders	Hanningfield Reservoir	TQ7398	R	Essex
Doxey Marshes	SJ9024	M	Staffordshire	Hardley Flood	TM3899	M	Norfolk
Draycote Water	SP4469	R	Warwickshire	Harewood Lake	SE3144	L	W Yorkshire
Drumgay Lough	H2448	L	Fermanagh	Hay-a-Park Gravel Pits	SE3658	P	N Yorkshire
Drummond Pond	NN8518	L	Tayside	Hayle Estuary	SW5537	E	Cornwall
Druridge Pool	NZ2796	L	N'th'mberland	Headley Mill Pond	SU8138	L	Hampshire
Duddon Estuary	SD2081	E	Cumbria	Heaton Park Reservoir	SD8205	R	Greater
Dundrum Bay	J4235	E	Down				Manchester
Dungeness Gravel Pits	TR0619	P	Kent	Heigham Holmes	TG4420	M	Norfolk
Dupplin Loch	NO0320	L	Tayside	Herne Bay	TR1768	O	Kent
Durham Coast	NZ4349	O	Durham	Hilfield Park Reservoir	TQ1596	R	Hertfordshire
Dyfi Estuary	SN6394	E	Dyfed	Hill Ridware Lake	SK0717	L	Staffordshire
Dysynni Estuary	SH5702	E	Gwynedd	Hillsborough Main Lake	J2458	L	Down
Earls Barton Gravel Pits	SP8966	P	Northants	Hirsel Lake	NT8240	L	Borders
Earlsferry to Anstruther	NO5302	O	Fife	Hogganfield Loch	NS6467	L	Strathclyde
Easterloch/Uyeasound	HP5901	O	Shetland	Holburn Moss	NU0536	L	N'th'mberland
East Fortune Ponds	NT5580	L	Lothian	Holden Wood Reservoir	SD7722	R	Lancashire
East Sanday Coast	HY7241	O	Orkney	Holkham	TF8845	E	Norfolk
Eccup Reservoir	SE2941	R	W Yorkshire	Holland Haven	TM2117	M	Essex
Eden Estuary	NO4719	E	Fife	Hollowell Reservoir	SP6872	R	Northants
Eglwys Nunydd Reservoir	SS7984	R	W Glamorgan	Holme Pierrepoint Gravel Pits	SK6239	P	Notts
Ellesmere Lakes	SJ4035	L	Shropshire	Hornsea Mere	TA1947	L	Humberside
Emberton Gravel Pits	SP8850	P	Bucks	Horsey Mere	TG4422	L	Norfolk
Erme Estuary	SX6249	E	Devon	Houghton Green Pool	SJ6292	L	Cheshire
Esthwaite Water	SD3596	L	Cumbria	Hule Moss	NT7149	L	Borders
Etherow Country Park	SJ9791	L	Greater	Humber Estuary	TA2020	E	Humberside,
			Manchester				Lincolnshire
Eversley Cross & Yateley GPs	SU8601	P	Hampshire	Hurleston Reservoir	SJ6255	R	Cheshire
Exe Estuary	SX9883	E	Devon	Inland Sea	SH2779	E	Gwynedd
Eyebrook Reservoir	SP8595	R	Leicestershire	Inner Clyde Estuary	NS3576	E	Strathclyde
Fairburn Ings	SE4627	P	N Yorkshire	Inner Moray Firth	NH6752	E	Highland

Site	1 km sq	H	County	Site	1 km sq	H	County
Irvine/Garnock Estuary	NS3038	E	Strathclyde	Loch of Stenness	NY2812	L	Orkney
Irvine to Saltcoats	NS2839	E	Strathclyde	Loch of Strathbeg	NK0758	L	Grampian
Islay	NR3560	N	Strathclyde	Loch of Swannay	HY3127	L	Orkney
Islesteps	NX9772	V	Dumfries & Galloway	Loch of the Lowes	NO0443	L	Tayside
Jersey Shore	WV6249	O	Channel Isles	Loch of Wester	ND3259	L	Highland
Jura	NR5672	N	Strathclyde	Loch Ryan	NX0565	E	Dumfries & Galloway
Kedleston Park Lake	SK3141	L	Derbyshire				
Kenfig Pool	SS7981	L	Glamorgan	Lochs Beg & Scridain	NM5027	L	Strathclyde
Kentra Moss/Lower Loch Shiel	NM6668	L	Highland	Loch Spynie	NJ2366	L	Grampian
Kessingland Levels	TM5185	L	Suffolk	Loch Tullybelton	NO0034	L	Tayside
Kilconquhar Loch	NO4801	L	Fife	Loch Watten	ND2256	L	Highland
Kilkeel to Lee Stone Point	J3214	O	Down	Loe Pool	SW6424	L	Cornwall
Killimster Loch	ND3056	L	Caithness	Longnewton Reservoir	NZ3616	R	Cleveland
Killough Harbour	J5437	O	Down	Longside Lake	TQ0168	P	Surrey
King George VI Reservoir	TQ0473	R	Surrey	Longueville Marsh	WV6748	M	Channel Isles
King George V Reservoir	TQ3796	R	Gtr London	Lossie Estuary	NJ2470	E	Grampian
Kingsbridge Estuary	SX7411	E	Devon	Lothing Lake & Oulton Broad	TM5292	E	Suffolk
Kings Bromley Gravel Pits	SK1116	P	Staffordshire	Lough Aghery	J2853	L	Down
King's Dyke Pits	TL2397	P	Cambs	Lough Foyle	C6025	E	Londonderry
Kings Mill Reservoir	SK5159	R	Notts	Lough Money	J5345	L	Down
Kirkby-on-Bain Gravel Pits	TF2360	P	Lincolnshire	Loughs Neagh & Beg	J0575	L	Down, Antrim, Londonderry, Tyrone, Armagh
Kislingbury Gravel Pits	SP7158	P	Northants				
Knight & Bessborough Reservoirs	TQ1268	R	Surrey				
Knockshinnock Lagoons	NS6013	L	Strathclyde				
Lackford Gravel Pits	TL7971	P	Suffolk	Lowbank Gravel Pit	NN9417	P	Tayside
Lade Sands	TR0921	O	Kent	Lower Bogrotten	NJ4861	N	Grampian
Lancaster Canal	SD4766	C	Lancashire	Lower Derwent Valley	SE6938	M	Humberside
Landbeach Gravel Pits	TL4865	L	Suffolk	Lower Windrush Valley GPs	SP4004	P	Oxfordshire
Langstone Harbour	SU6902	E	Hampshire	Lurgashall Mill Pond	SU9326	L	West Sussex
Langtoft West End Gravel Pits	TF1111	P	Lincolnshire	Lynford Gravel Pit	TL8194	P	Norfolk
Larne Lough	D4200	E	Antrim	Machrihanish	NR6522	N	Strathclyde
Lavan Sands	SH6474	E	Gwynedd	Maer Lake	SK2070	M	Cornwall
Lee Valley Gravel Pits	TL3702	P	Hertfordshire. Essex	Marsh Lane Gravel Pits	TL3069	P	Cambs
				Martin Mere	SD4105	L	Lancashire
Leighton Moss	SD4875	L	Lancashire	Marton Mere	SD3435	L	Lancashire
Leighton & Roundhill Reservoirs	SE1577	R	N Yorkshire	Meadow Lane Gravel Pits	TL3270	P	Cambs
Leventhorpe Flood Meadows	SE3629	M	W Yorkshire	Medway Estuary	TQ8471	E	Kent
Lindisfarne	NU1041	E	N'th'mberland	Mere Sands Wood	SD4415	L	Lancashire
Linford Gravel Pits	SP8442	P	Bucks	Merryton Ponds	NS7654	L	Strathclyde
Linne Mhuirich & Loch Na Cille	NR7080	O	Strathclyde	Mersey Estuary	SJ4578	E	Cheshire
Little Paxton Gravel Pits	TL1963	P	Cambs	Middle Tame Valley Gravel Pits	SP2096	P	Staffordshire, Warwickshire
Little Stour Valley	TR2056	M	Kent				
Livermere	TL8771	L	Suffolk	Middle Yare Valley	TG3504	M	Norfolk
Llangorse Lake	SO1326	L	Powys	Milldam & Balfour Mains Pools	HY4817	L	Orkney
Llyn Penrhyn	SH3077	L	Gwynedd	Minsmere	TM4666	L	Suffolk
Llyn Traffwll	SH3276	L	Gwynedd	Moorgreen Reservoir	SK4849	R	Notts
Llysyfran Reservoir	SN0324	R	Dyfed	Monach Isles	NF6262	O	Western Isles
Loch Branahuie & Aignish	NB4732	L	Western Isles	Monikie Reservoir	NO5038	R	Tayside
Loch Calder	ND0760	L	Highland	Montrose Basin	NO6958	E	Tayside
Loch Clunie	NO1144	L	Tayside	Moray Coast	NJ3067	O	Grampian
Loch Etive	NM9434	L	Strathclyde	Moray Firth	NH8060	E	Highland
Loch Ewe: Aultbea	NG8788	L	Highland	Morecambe Bay	SD4070	E	Lancashire, Cumbria
Loch Eye	NH8379	L	Highland				
Loch Fleet Complex	NH7896	E	Highland	Nafferton Mere	TA0558	L	Humberside
Loch Garten & Mallachie	NH9718	L	Highland	N-E Glamorgan Moorland Pools	SO0808	L	Glamorgan
Loch Gelly	NT2092	L	Fife	Nene Washes	TF3300	M	Cambs
Lochs Heilen & Mey	ND2568	L	Highland	Netherfield Gravel Pits	SK6339	P	Notts
Loch Indaal	NR3261	E	Strathclyde	Newark Bay	ND4689	O	Orkney
Loch Insh & Spey Marshes	NH8304	L	Highland	Newgale Beach	SM8421	O	Dyfed
Loch Ken	NX7168	R	Dumfries & Galloway	New Road Pits	TI1549	P	Bedfordshire
				Newtown Estuary	SZ4291	E	Isle of Wight
Loch Leven	NO1401	L	Tayside	North Killingholme Haven Pits	TA1619	P	Humberside
Loch Lomond	NS4388	L	Strathclyde	North Mainland Orkney	HY2915	O	Orkney
Loch Mahaick	NN7006	L	Central	North Norfolk Marshes	TF8546	E	Norfolk
Loch Mullion	NN9833	L	Tayside	North Ronaldsay	HY7655	N	Orkney
Loch of Boardhouse	HY2725	L	Orkney	North Uist	NF8370	N	Western Isles
Loch of Harray	HY2915	L	Orkney	North Warren & Thorpeness Mere	TM4658	L	Suffolk
Loch of Hempriggs	ND3447	L	Caithness	North West Solent	SZ3395	E	Hampshire
Loch of Hundland	HY2926	L	Orkney	Nosterfield Gravel Pits	SE2880	P	N Yorkshire
Loch of Isbister	HY2523	L	Orkney	Nunnery Lakes	TL8781	L	Norfolk
Loch of Kinnordy	NO3655	L	Tayside	Ogden Reservoir	SD7622	R	Lancashire
Loch of Lintrathen	NO2754	L	Tayside	Orkney	HY4010	N	Orkney
Loch of Skaill	HY2418	L	Orkney	Orwell Estuary	TM2238	E	Suffolk
Loch of Skene	NJ7807	L	Grampian	Osterley Park Lakes	TL1478	L	Gtr London
Loch of Spiggie	HU3716	L	Shetland	Ouse/Lairo Water	HY5019	L	Orkney

Site	1 km sq	H	County	Site	1 km sq	H	County
Ouse Washes	TL5394	M	Cambs	S-E Deerness	HY5606	N	Orkney
Outer Ards	J6663	O	Down	Seahouses to Budle Point	NU2231	O	N'th'mberland
Overstone Park Lakes	SP8065	L	Northants	Seaton Gravel Pits	TR2258	P	Kent
Pagham Harbour	SZ8796	E	West Sussex	Sennowe Park Lakes	TF9825	L	Norfolk
Pannel Valley	TQ8815	M	East Sussex	S-E Stronsay	HY6822	N	Orkney
Panshanger Estate	TL2812	L	Hertfordshire	Severn Estuary	ST5058	E	Glos, Avon,
Par Sands Pools	SX0853	L	Cornwall				Somerset,
Passfield Lake	SU8234	L	Hampshire				Gwent, Mid
Paultons Bird Park	SU3116	L	Hampshire				Glam, South
Pegwell Bay	TR3563	E	Kent				Glam
Pennington Flash	SJ6499	L	Greater	Shell Pond (Carrington)	SJ7591	L	Greater
			Manchester				Manchester
Pen Ponds	TQ1972	L	Gtr London	Shipton-on-Cherwell Quarry	SP4717	P	Oxfordshire
Pentney Gravel Pits	TF7013	P	Norfolk	Shrigley Lake	J518544	L	Down
Pirton Pool	SO8847	L	Hereford &	Shustoke Reservoir	SP2391	R	Warwickshire
			Worcester	Skelton Lake	SE3430	L	W Yorkshire
Pitsford Reservoir	SP7669	R	Northants	Skinflats	NS9284	E	Central
Poole Harbour	SY9988	E	Dorset	Slains Lochs	NK0230	L	Grampian
Portavo Lake	J5582	L	Down	Slamannan Plateau	NS8474	N	Central
Port Meadow	SP4908	M	Oxfordshire	Slapton Ley	SX8243	L	Devon
Portsmouth Harbour	SU6204	E	Hampshire	S Muskham & N Newark GPs	SK7956	P	Notts
Portworthy Mica Dam	SX5660	P	Devon	Snettisham	TF6535	E	Norfolk
Possil Loch	NS5870	L	Strathclyde	Solway Estuary	NY1060	E	Cumbria
Pugney Water	SE3218	P	W Yorkshire	Somerset Levels	ST4040	M	Somerset
Queen Elizabeth II Reservoir	TQ1167	R	Surrey	Sonning Gravel Pits	SU7475	P	Oxfordshire
Queen Mary Reservoir	TQ0769	R	Surrey	Sound of Harris	NF9788	O	Western Isles
Queen Mother Reservoir	TQ0076	R	Berkshire	Sound of Tarransay	NG0498	O	Western Isles
Ramsbury Lake	SU2671	L	Wiltshire	Southampton Water	SU4507	E	Hampshire
Ranworth & Cockshoot Broads	TG2515	L	Norfolk	South Down	J5036	O	Down
Revesby Reservoir	TF3067	R	Lincolnshire	South Ford	NF7747	O	Western Isles
Rhunahaorine	NR7049	N	Argyll	South Iver Gravel Pits	TQ0377	P	Bucks
Ribble Estuary	SD3825	E	Lancashire	South Milton Ley	SX6842	M	Devon
Ringstead Gravel Pits	SP9775	P	Northants	South Stoke	TQ0210	V	West Sussex
R Arrow/R Lugg Floodplain	SO5057	R	Hereford &	South Uist	NF8032	N	Western Isles
			Worcester	South Walls	ND3089	N	Orkney
R Avon: Britford Water Meadows	SU1628	M	Wiltshire	South West Lancashire	SD4015	N	Lancashire
River Avon: West Amesbury	SU1541	V	Wiltshire	South Westray	HY4646	N	Orkney
River Clyde: Carstairs Junction	NS9744	V	Strathclyde	Spade Oak Gravel Pit	SU8887	P	Bucks
River Clyde: Lamington	NS9833	V	Strathclyde	Spey Mouth	NJ3465	E	Grampian
River Derwent: Chatsworth	SK2569	V	Derbyshire	Sprotbrough Flash	SE5300	L	S Yorkshire
R Eamont:Watersmeet to Pooley Bridge	NY5329	V	Cumbria	Staines Reservoir	TQ0575	R	Surrey
				Staines Moor Gravel Pits	TQ0373	R	Surrey
Rivers Eamont & Eden: Honeypot to Edenhall	NY5631	V	Cumbria	Stainhill Reservoir	TQ1269	R	Gtr London
				St Andrews Bay	NO5121	O	Fife
R. Forth: W Carse Farm - R. Teith	NS7693	V	Central	Stanford Reservoir	SP6080	R	Leicestershire
River Foyle: Grange	C3606	V	Tyrone	Stanford Training Area	TL8695	L	Norfolk
River Frome: Wareham to Wool	SY8487	V	Dorset	Stanwick Gravel Pits	SP9772	P	Northants
River Idle: Bawtry to Miserton	SK7195	V	Notts	St Benets Levels	TG3815	M	Norfolk
River Lagan: Flatfield	J1961	V	Down	St Mary's Island	NZ3475	O	N'th'mberland
River Nith: Keltonbank to Nutholm	NX9774	V	Dumfries &	Stodmarsh	TR2061	L	Kent
			Galloway	Stoke Newington Reservoirs	TQ3287	R	Gtr London
River Soar: Leicester	SK5805	V	Leicestershire	Stour Estuary	TM1732	E	Essex, Suffolk
River Spey: Boat of Balliefirth	NH9922	V	Highland	Strangford Lough	J5560	E	Down
River Tay: Dunkeld	NO0042	V	Tayside	Stranraer Lochs	NX1161	L	Dumfries &
River Tay: Scone	NO1026	V	Tayside				Galloway
R Test: Fullerton to Stockbridge	SU3535	V	Hampshire	Stratfield Saye	SU6759	R	Hampshire
R Teviot:Kalemouth to Roxborough	NT7030	V	Borders	Strathearn	NN8819	N	Tayside
R Teviot: Nisbet to Kalemouth	NT6925	V	Borders	Strinesdale	SD9506	R	Greater
R Tweed: Kelso to Coldstream	NT7737	V	Borders				Manchester
River Tweed: Magdalenehall	NT6331	V	Borders	Studland Bay	SZ0383	O	Dorset
River Usk: Pencelli	SO0925	R	Powys	Summerleaze Gravel Pits	SU8982	P	Berkshire
R Wensum: F'knh'm to G't Ryburgh	TF9428	V	Norfolk	Summerston	NS5771	V	Strathclyde
River Wye: Bakewell to Haddon	SK2366	V	Derbyshire	Sutton/Lound Gravel Pits	SK6985	P	Notts
River Wye: Putson	SO5138	V	Hereford	Swale Estuary	TQ9765	E	Kent
Rostherne Mere	SJ7484	L	Cheshire	Swanbourne Lake	TQ0108	L	West Sussex
Rough Firth	NX8453	E	Dumfries &	Swanholme Lake	SK9468	L	Lincolnshire
			Galloway	Swanpool (Falmouth)	SW8031	L	Cornwall
Rufford Lake	SK6465	L	Notts	Swansea Bay	SS6391	O	Glamorgan
Rutherford	NT6431	V	Borders	Swillington Ings	SE3828	P	W Yorkshire
Rutland Water	SK9207	R	Leicestershire	Swithland Reservoir	SK5513	R	Leicestershire
Ryde Pier to Puckpool Point	SZ6092	O	Isles of Wight	Tabley Mere	SJ7276	L	Cheshire
Rye Harbour/Pett Level	TQ9418	E	East Sussex	Tamar Complex	SX4363	E	Devon,
Salford Docks	SJ8097	C	Greater				Cornwall
			Manchester	Tattershall Pits	TF2057	P	Linconshire
Sandbach Flashes	SJ7259	L	Cheshire	Taw/Torridge Estuary	SS4733	E	Devon
Scolt Head	TF8046	E	Norfolk	Tay/Isla Valley	NO1438	L	Tayside

Site	I km sq	H	County
Tay Estuary	NO3225	E	Fife, Tayside
Tees Estuary	NZ5528	E	Cleveland
Teign Estuary	SX8772	E	Devon
Temple Water	J5750	L	Down
Thames Estuary	TQ7880	E	Kent, Essex, Gtr London
Thanet Coast	TR2669	O	Kent
Theale Gravel Pits	SU6570	P	Berkshire
Thoresby Lake	SK6370	L	Notts
Thorpe Water Park	TQ0268	P	Surrey
Thrapston Gravel Pit	SP9979	P	Northants
Threave Estate	NX7362	V	Dumfries & Galloway
Thursley Lake	SU9239	L	Surrey
Timsbury Gravel Pits	SU3624	P	Hampshire
Tiree	NL9741	N	Strathclyde
Tophill Low Reservoirs	TA0748	R	Humberside
Tottenhill Gravel Pits	TF6311	P	Norfolk
Traeth Coch	SH5480	E	Anglesey
Traighear	NF8276	N	Western Isles
Traigh Luskentyre	NG0798	E	Western Isles
Tring Reservoirs	SP9113	R	Hertfordshire
Trinity Broads	TG4614	L	Norfolk
Tullynagee Lough	J4763	L	Down
Tundry Pond	SU7752	L	Hampshire
Tweed Estuary	NT9853	E	N'th'mberland
Twyford Gravel Pits	SU7875	P	Berkshire
Tyninghame Estuary	NT6379	E	Lothian
Tyrella	J4735	O	Down
Tyttenhanger Gravel Pits	TL1804	P	Hertfordshire
Upper Lough Erne	H3231	L	Fermanagh
Upper Quoile	J4745	V	Down
Upper Tay	NN9557	N	Tayside
Upton Warren LNR	SO9367	L	Hereford
Virginia Water	SU9769	L	Berkshire
Walland Marsh	TQ9824	M	Kent
Walmore Common	SO7425	M	Glos
Walthamstow Reservoir	TQ3589	R	Gtr London
Walton Lock	SJ6086	C	Cheshire
Wanstead Park Ponds	TQ4187	L	Gtr London
Wantsum Marshes	TR2366	M	Kent
Wash	TF5540	E	Lincolnshire, Norfolk
Water Sound	ND4394	O	Ornkey
Wath Main Ings	SE4302	P	S Yorkshire
Weirwood Reservoir	TQ3934	R	Sussex
Wellington Country Park	SU7362	L	Hampshire
Wemyss Bay to Fairlie	NS2059	O	Strathclyde
Westfield Marshes	ND0664	M	Highland
Westport Lake	SJ8550	L	Staffordshire
West Water Reservoir	NT1252	R	Borders
Whisby Gravel Pits	SK9167	P	Lincolnshire
Whittlesford Gravel Pits	TL4649	P	Cambs
Whitton Loch	NT7419	L	Borders
Widewall Bay	ND4292	O	Orkney
Wigtown Bay	NX4456	E	Dumfries & Galloway
Wilderness Pond	SS8277	L	Glamorgan
Willen Lake	SP8741	R	Bucks
William Girling Reservoir	TQ3694	R	Gtr London
Windermere	SD3995	L	Cumbria
Wintersett Country Park Lake	SE3716	P	W Yorkshire
Woburn Park Lakes	SP9632	L	Bedfordshire
Woodford River		V	Fermanagh
Woolston Eyes	SJ6588	P	Cheshire
Worsborough Reservoir	SE3403	R	Greater Manchester
Wraysbury Gravel Pits	TQ0073	P	Berkshire
Wraysbury Reservoir	TQ0274	R	Surrey
Wynyard Lake	NZ4224	L	Cleveland
Yarnton Gravel Pits	SP4710	P	Oxfordshire
Yarwell Gravel Pits	TL0797	P	Northants
Ythan Estuary	NK0026	E	Grampian
Ythan to Collieston	NK0226	O	Grampian

Key to Figure A1

Abberton Reservoir 169
Adur Estuary 243
Alaw Reservoir 125
Alde Complex 163
Almnouth 73
Alt Estuary 115
Artro Estuary 131
Attenborough GPs 140
Auchencairn Bay 81
Avon Estuary 219
Axe Estuary 225
Ballo Reservoir 33
Bann Estuary 89
Baston/Langtoft GPs 145
Beaulieu Estuary 231
Belfast Lough 98
Belvide Reservoir 136
Berney Marshes 159
Black & White Lochs (Loch Inch) 87
Blackwater Estuary 170
Blagdon Lake 195
Blithfield Reservoir 137
Blyth (Northumberland) Estuary 75
Blyth (Suffolk) Estuary 161
Brading Harbour 237
Braint Estuary 128
Breydon Water 160
Burry Inlet 200
Camel Estuary 206
Cameron Reservoir 32
Carlingford Lough 93
Carmarthen Bay 201
Carron Valley Reservoir 41
Carsebreck & Rhynd Lochs 40
Castle Lo. (Lochmaben) 79
Castle Semple & Barr Lochs 47
Cefni Estuary 127
Cheshunt Gravel Pits 181
Chew Valley Lake 194
Chichester Harbour 240
Christchurch Harbour 229
Clandeboye Lake 97
Cleddau Estuary 204
Clwyd Estuary 119
Coll 57
Colne Estuary 168
Colwyn Bay 120
Conwy Estuary 121
Coquet Estuary 74
Corby Loch 20
Cotswold Water Park East 190
Cotswold Water Park West 191
Cowgill Reservoirs 60
Cromarty Firth 9
Crombie Reservoir 27
Crouch/Roach Estuary 177
Cuckmere Estuary 245
Dart Estuary 221

Deben Estuary 164
Dee (Eng/Wal) Estuary 118
Dee (Scotland) Estuary 22
Deeping St James GPs 146
Dengie Flats 171
Deveron Estuary 14
Dinnet Lochs 24
Don Estuary 21
Dorchester Gravel Pits 188
Dornoch Firth 7
Draycote Water 139
Drummond Pond 39
Duddon Estuary 101
Dulas Bay 124
Dundrum Bay 94
Dungeness Gravel Pits 248
Dupplin Lochs 35
Durham Coast 76
Dyfi Estuary 134
Dysynni Estuary 133
Eden Estuary 31
Ellesmere Group 135
Erme Estuary 218
Exe Estuary 223
Eyebrook Reservoir 143
Fairburn Ings 109
Fal Complex 212
Fala Flow 64
Fedderate Reservoir 15
Fen Drayton Gravel Pit 149
Fiddlers Ferry Lagoons 113
Fleet Bay 84
Fleet/Wey 226
Forth Estuary 62
Foryd Bay 129
Fowey Estuary 213
Gadloch 42
Gannel Estuary 207
Gladhouse Reservoir 63
Grafham Water 147
Guernsey Shore 209
Gunton Park 155
Haddo House Lakes 17
Hamford Water 167
Hanningfield Reservoir 178
Hay-a-Park Gravel Pits 104
Hayle Estuary 208
Helford Estuary 211
Hickling Broad 156
Hirsel Lake 69
Holborn Moss 72
Hornsea Mere 107
Hoselaw Loch 68
Hule Moss 66
Humber Estuary 108
Hunterston Estuary 49
Inland Sea 126
Inner Clyde Estuary 46
Inner Moray Firth 10
Irt/Mite/Esk Estuary 100
Irvine/Garnock Estuary 48
Jersey Shore 210
King George V Res 180
Kingsbridge Estuary 220
Kinnordy Loch 28
Kirkcudbright Bay 83

Lackford Gravel Pits 152
Lake of Menteith 44
Langstone Harbour 239
Larne Lough 99
Lavan Sands 122
Lindisfarne 71
Little Paxton GPs 148
Llandegfedd Reservoir 197
Loch Clunie 36
Loch Druidibeg 59
Loch Eye 8
Loch Fleet 6
Loch Garten 25
Loch Gilp 51
Loch Gruinart 56
Loch Indaal 55
Loch Ken 82
Loch Leven 34
Loch Lomond: Endrick Mouth 45
Loch Mahaick Doune 43
Loch na Cille 52
Loch of Boardhouse 2
Loch of Harray 3
Loch of Lintrathen 29
Loch of Skene 23
Loch of Spiggie 1
Loch of Stenness 4
Loch of Strathbeg 16
Loch of the Lowes 37
Loch Quien 50
Loch Ryan 88
Loch Spynie 11
Loch Tullybelton 38
Loch Watten 5
Looe Estuary 214
Lossie Estuary 12
Lough Foyle 90
Loughs Neagh & Beg 92
Lower Derwent Valley 105
Lower Windrush Valley GPs 189
Luce Bay 86
Machrihanish 54
Martin Mere 116
Mawddach Estuary 132
Medina Estuary 235
Medway Estuary 175
Meikle Loch Slains 19
Mersey Estuary 114
Mid Avon Valley 228
Middle Tame Valley GPs 138
Middle Yare Marshes 158
Minsmere 162
Montrose Basin 26
Morecambe Bay 103
Nene Washes 150
Nevern Estuary 203
Newhaven Estuary 244
Newtown Estuary 234
North Norfolk Coast 154
North-West Solent 230
Ogmore Estuary 198
Orwell Estuary 165
Otter Estuary 224

Ouse Washes 151
Outer Ards Shoreline 96
Pagham Harbour 241
Pegwell Bay 173
Pitsford Reservoir 142
Plym Estuary 216
Poole Harbour 227
Portsmouth Harbour 238
Pulborough/Amberley Brooks 242
Queen Mary Reservoir 182
Red Wharf Bay 123
Rhunahaorine 53
Ribble Estuary 117
R Tweed: Kelso to Coldstream 67
Rostherne Mere 111
Rough Firth 80
Rutland Water 144
Rye Harbour/Pett Levels 246
Sevenoaks Wildfowl Reserve 179
Severn Estuary 193
Solway Estuary 78
Somerset Levels 196
Southampton Water 232
Spey Estuary 13
St Benets Levels 157
Staines Reservoir 183
Stour Estuary 166
Strangford Lough 95
Stratfield Saye 186
Swale Estuary 174
Swansea Bay 199
Swithland Reservoir 141
Tamar Complex 215
Taw/Torridge Estuary 205
Tay Estuary 30
Tees Estuary 77
Teifi Estuary 202
Teign Estuary 222
Thames Estuary 176
Thanet Coast 172
The Wash 153
Theale Gravel Pits 187
Thorpe Water Park 184
Tiree 58
Tophill Low Reservoirs 106
Traeth Bach 130
Tweed Estuary 70
Tyninghame Estuary 65
Upper Lough Erne 91
Walland Marsh 247
Walmore Common 192
Wath & Broomhill Ings 110
West Water Reservoir 61
Wigtown Bay 85
Windermere 102
Woolston Eyes 112
Wootton Estuary 236
Wraysbury Gravel Pits 185
Yar Estuary 233
Yealm Estuary 217
Ythan Estuary 18

Figure A1. Location of important WeBS sites. Circles show the central position of 248 key WeBS sites, including all estuaries, in the UK and the Channel Islands. Sites chosen include most internationally important sites, but also sites of regional importance in areas with few wetlands or few sites counted by WeBS. Thus, inclusion of a site does not imply any measure of relative conservation importance.